Heidi Keller   Klaus Schneider
Bruce Henderson (Eds.)

Curiosity and Exploration

With 122 Figures

Springer-Verlag
Berlin Heidelberg New York
London Paris Tokyo
Hong Kong Barcelona
Budapest

Prof. Dr. Heidi Keller
Fachbereich Psychologie - Entwicklungspsychologie
Universität Osnabrück
Seminarstraße 20
D-49069 Osnabrück, Germany

† Prof. Dr. Klaus Schneider
Fakultät für Psychologie
Allgemeine Psychologie und Entwicklungspsychologie
Universitätsstraße 150
D-44801 Bochum, Germany

Prof. Dr. Bruce Henderson
Western Carolina University
Department of Psychology
Cullowhee, NC 28 123, USA

Computergrafics and Layout by Jürgen Lemke

ISBN-13: 978-3-540-54867-6   e-ISBN-13: 978-3-642-77132-3
DOI: 10.1007/978-3-642-77132-3

Library of Congress Cataloging-in-Publication Data
Curiosity and exploration/edited by Heidi Keller, Klaus Schneider, Bruce Henderson. p. cm. Includes bibliographical references and index. ISBN 3-540-54867-X 1. Curiosity. 2. Curiosity in children. I. Keller, Heidi, 1945- . II. Schneider, Klaus, 1941- . III. Henderson, Bruce. BF323.C8C88 1994 155.4'18—dc20 94-103-5 CIP

This work is subject to copyright. All rights are reserved, whether the whole or part of the material is concerned, specifically the rights of translation, reprinting, reuse of illustrations, recitation, broadcasting, reproduction on microfilm or in any other way, and storage in data banks. Duplication of this publication or parts thereof is permitted only under the provisions of the German Copyright Law of September 9, 1965, in its current version, and permission for use must always be obtained from Springer-Verlag. Violations are liable for prosecution under the German Copyright Law.

© Springer-Verlag Berlin Heidelberg 1994

The use of general descriptive names, registered names, trademarks, etc. in this publication does not imply, even in the absence of a specific statement, that such names are exempt from the relevant protective laws and regulations and therefore free for general use.

Typesetting: Camera ready by editors
SPIN: 10025141   21/3130 - 5 4 3 2 1 0 - Printed on acid-free paper

Table of Contents

CHAPTER I: Introduction

CHAPTER 1.0

1 **Preface: The Study of Exploration**

Heidi Keller, Klaus Schneider, Bruce Henderson

2 Definitions and Distinctions
3 The Functions of Exploration: A Psychobiological Perspective "Ultimate Causes"
4 "Proximate Causes"
7 The Contents of This Book
8   1. Functional and Motivational Mechanisms of Exploration
10   2. Development and Individual Differences
11   3. The Interconnections of Exploratory Behavior and other Behavioral Systems
13   4. Applied Perspectives
13 Additional References

CHAPTER 1.1

15 **Fairy Tales and Curiosity. Exploratory Behavior in Literature for Children or the Futile Attempt to Keep Girls from the Spindle**

Rosemarie M. Rigol

17 Evidence of Curiosity and Exploratory Behavior in Fairy Tales
19 Types of Exploratory Behavior in Fairy Tales

19	Movement
21	Handling and Sensual Experience
24	Evaluation of Exploratory Behavior in Fairy Tales
25	Curiosity as Subject of the Story
26	Curiosity as a Part of Development
27	The Right to be Curious
28	Literary Culture and Children's Development

CHAPTER II: Function and Motivational Mechanics of Exploration

CHAPTER II.2

31 An Ethological Conception of Exploratory Behavior

Christiane Buchholtz, Andrea Persch

33	The Sensoric Part
34	The Motoric Part
34	The Action Readiness System
37	The Releasing of Exploration in this Model
38	Appetitive Behavior in the Behavior System of Exploration
40	References

CHAPTER II.3

43 The Neurobiological Foundation of Exploration

Finn K. Jellestad, Gry S. Follesø, Holger Ursin

44	Components of the Exploratory Behavioral System

44	Exploration
44	Orienting Responses
45	Habituation
46	Novelty and Fear
46	Situation-Specific Behavior
47	Other Topics Related to Exploratory Behavior
47	A Neurological Model of Exploration
47	General Anatomy of Cognitive Information
49	The Hippocampal Formation
49	Orienting Response and Habituation
50	Exploration and Novelty
52	Spontaneous Alternation
52	The Amygdaloid Complex
53	Orienting Responses and Habituation
54	Exploration and Amygdala Lesions: Sensory Changes or Emotional ?
54	Sensory Changes
55	Emotional Changes
56	Conclusion
56	Specific Transmitter Systems
58	Other Structures
58	Conclusion
59	References

CHAPTER II.4

65 Two Characteristics of Surprise: Action Delay and Attentional Focus

Michael Niepel, Udo Rudolph, Achim Schützwohl, Wulf-Uwe Meyer

67 The Experimental Paradigm

68	Effects of Schema Discrepancy
70	Duration of the Surprise Reaction
71	Generality of Effects
72	Number of Trials
75	General Discussion
76	References

CHAPTER II.5

79 Interest and Curiosity. The Role of Interest in a Theory of Exploratory Action

Andreas Krapp

80	The Neglect of Content in Curiosity and Exploration Research
80	The Research Perspective of General Psychology
81	The Research Perspective of Differential Psychology
82	The Theory of Action Research Perspective
82	Unsolved Problems in Curiosity and Exploration Research
84	A Concept of Object-specific Interests
85	Interest Object
86	Structural Components of the Interest-Oriented Person-Object-Relationship
87	Special Characteristics of an Interest-Oriented Person-Object Relationship
90	Interest and Exploration
90	Interests Determine the Content and Direction of Diversive Exploration
91	Interests Influence the Goal Orientation of Specific Curiosity and Exploration
92	Interests Influence the Design of an Exploratory Action
92	Interests Serve as the Basis of Content Continuity in a Sequence of Exploratory Actions

94	Interests Determine the Nature and Orientation of Cognitive Construction
96	Summary and Conclusion
97	References

CHAPTER II.6

101	**Interest and Exploration: Exploratory Action in the Context of Interest Genesis**

Benedykt Fink

102	The Interests of Individuals
104	The Characteristics of Beginning Interests
105	Structural Aspects of Early Interests
107	The Interest Genesis Project
107	Empirical Findings Relating to the Interest Genesis Model
108	Typical Structures of Interest-Oriented Person-Object-Relations
109	Global Developmental Principles of Interest Development
110	Specific Components of Development
112	Hypothetical Models of Structural Change
114	Discussion - Interest Genesis as both the Result and a Cond-itional Factor of Exploratory Behavior
115	Exploratory Behavior as a Factor in Interest Genesis
116	The Importance of Interests in the Development of Exploratory Behavior
117	Conclusions
118	References

CHAPTER III: Development and Interindividual Differences

CHAPTER III.7

123 The Relationship Between Attachment, Temperament, and Exploration

Dymphna C. van den Boom

- 124 The Balance between Attachment and Exploration
- 125 The Influence of Infant Irritability on Exploration
- 126 Exploratory Behavior in the First Year
- 130 Infant Exploration: Relations to Home Environment and Security of Attachment
- 133 Method
- 133 Subjects
- 134 Design
- 135 Variables and Data Collection Procedures
- 138 Results
- 138 Effect of the Intervention on Infant Exploratory Behavior
- 141 Effect of Intervention on the Quality of Attachment
- 143 Discussion
- 144 Explanatory Models
- 145 References

CHAPTER III.8

151 Interindividual Differences in the Development of Exploratory Behavior: Methodological Considerations

Clemens Trudewind, Klaus Schneider

154 The Behavior Assessment Approach
156 The Development of a Parents' Questionnaire
158 Analysis of the Dimensionality of the Parents' Questionnaire
165 The Incentive-Reactivity Assessment Approach
166 Development of a Puppet-Show Instrument Assessment Procedure
170 First Steps in Validating the Curiosity Motive-Scores
173 Conclusions
174 References

CHAPTER III.9

177 Preschoolers' Exploratory Behavior: The Influence of the Social and Physical Context

Klaus Schneider, Lothar Unzner

177 Children's Exploration in Natural Contexts
179 An Empirical Study
180 Method
181 Subjects
181 Procedure
181 Data Collection and Coding

184	Results and Discussion
184	Exploratory Behavior of the Children
186	Control Behaviors of the Mothers
188	Children's Exploration and the Distance between Mother and Children
195	References

CHAPTER III.10

199 A Developmental Analysis of Exploration Styles

Heidi Keller

203	An Empirical Study
207	Results
209	Discussion
210	References

CHAPTER III.11

213 Individual Differences in Experience-Producing Tendencies

Bruce Henderson

213	Individual Differences in the Tendency to Explore
215	The Category of Experience-Producing Tendencies
216	EPTs and Intelligence
219	A Recent Study of Curiosity/Exploration - Intelligence Relations
221	The Need for Developmental Designs
223	Moderating Influences on EPTs
223	Implications and Conclusion
224	References

CHAPTER IV: The Interconnections of Exploratory Behavior and other Behavioral Systems

CHAPTER IV.12

227 The Active Exploratory Nature of Perceiving: Some Developmental Implications

Ad W. Smitsman.

228 The Active Exploratory Nature of Perceiving
231 Perceptual Search and Information for Action
232 Developmental Implications
233 The Task Specificity of Early Exploratory Activities
235 Exploration and the Development of Action Skills
238 References

CHAPTER IV.13

241 The Process and Consequences of Manipulative Exploration

Axel Schölmerich

243 The Process of Exploration
243 Different Models in the Exploration Literature
244 A Comparison with Habituation
245 Problems of Obversation: from Quantities of Behavior to Quantities of Information
246 The Sequential Organization of Exploratory Behavior
247 An Empirical Study
247 Data Collection
247 The Object

248	The Sample
248	The Coding Procedure
249	Results
249	Amount of Exploratory Behavior
253	Amount of Information
254	Interrelationships between Exploration and Information Acquisition
256	The Regulation of Exploratory Action and Interindividuel Differences in Strategies of Exploration
257	References

CHAPTER IV.14

259 Motivational and Cognitive Determants of Exploration

Shulamith Kreitler, Hans Kreitler

260	The Nature of Exploration: The Five Factors of Exploration
265	The Motivational Determinants of Exploration: The Cognitive Orientation of Curiosity
270	Cognitive Determinants of Exploration: Patterns of Meaning Variables
279	Cognitive Motivation and Cognitive Dynamics of Exploratory Modes
281	Main Findings, Major Conclusions, and Some Afterthoughts
282	References

CHAPTER V: Applied Perspectives

CHAPTER V.15

287 Computer Systems as Exploratory Environments

Siegfried Greif

288	Simplicity in Human-Computer Interaction
289	Exploratory Environments and Minimalist Design
290	The "Simple is always Best" Hypothesis
290	Complexity and Exploration Theory
293	Definition of the Complexity of a Task Situation
295	Design of Systems with adaptable Complexity
296	The "Individual System"
297	Empirical Research
302	Research Perspectives: Exploratory Activities in Error Situations
304	References

CHAPTER V.16

307 Urban Development for Children Reexploring a New Research Area *

Dietmar Görlitz, Richard Schröder

308	Psychologists' Curiosity about the City and its Children
308	Special Features and Innovations
309	On the Tradition of the Subject
309	An Articulated Interest
311	A City Discovers its Children. Herten as a Model
311	Kinderfreunde "Children's Friends"
311	History

312	Formal Organization
312	The "Children's Friends" Office
313	Activities of "Children's Friends"
314	City Map for Children
318	Child-friendly Living
318	Basics of Planning and Herten's Philosophy
320	"Child-friendly Living": The Project
324	A University and a Town as Partners - Child-centered Planning as a Coooerative Task
324	On the Cooperation between the Town of Herten and the Technical University of Berlin
325	The Four Stages of the Herten Design
328	Outlook on Researchers Curiosity and Forms of Action Appropriate to Children
330	Postscript
330	References

CHAPTER VI: Epilogue

CHAPTER VI.17

333 Applause for Aurora: Sociobiological Considerations on Exploration and Play

Robert Fagen

339	References
341	Author Index
349	Subject Index

CHAPTER 1

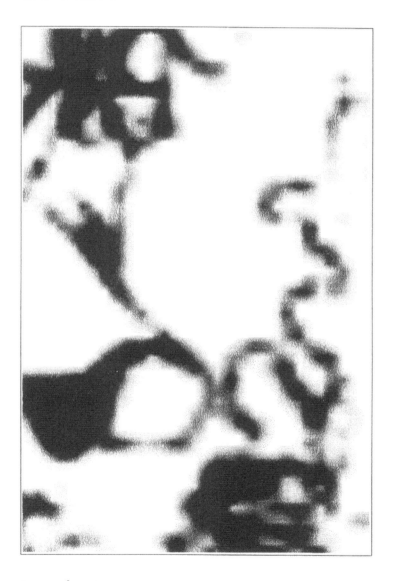

Introduction

CHAPTER 1.0

Preface:
The Study of Exploration

Heidi Keller, Klaus Schneider, Bruce Henderson

Exploration is a ubiquitous aspect of behavior in a wide range of settings. Curiosity and exploratory behaviors have been important in the study of perception and attention (c.f., Gibson, 1966, 1979; McCall, 1979), learning and motivation (Harlow, 1950), development (Ainsworth, 1973) and personality (Mayer, Caruso, Zigler, & Dryden, 1989). Some theorists have gone so far as to suggest that exploration is an essential characteristic of all human activity (Smitsman, 1991; Smitsman, van Leeuwen, & Peters, 1988). Like other common, yet important, constructs, such as learning or play, exploration has sometimes been difficult to succinctly define and distinguish from related behaviors. This may be one reason why the scientific interest in the nature of exploratory behavior has not matched its importance as a component of much animal and human behavior.

The study of a number of behavioral systems has included a conceptualization of exploration, but this conceptualization frequently has been incidental relative to the major focus of the research program. For example, Case (1985) introduced exploration as one of three postulated mechanisms in problem solving (the others being observation and imitation/interaction). However, the nature of exploration in regard to problem solving is not explicated in any detail and exploration is implicated only in the area of object-oriented problem solving.

The cursory or incidental treatment of exploration is apparent in other areas that are even more explicitly motivational in nature. In both the behavioristic and psychoanalytic traditions, exploratory behavior was thought of as resulting from a secondary, derived motivation of curiosity (see Brown, 1961; Freud, 1910). More recently, exploration has received attention from attachment theorists as a behavioral system that exists in tension with other systems, including fear or wariness and affiliation as well as attachment itself. Central to the theory is that the attachment to a caregiver provides a "secure base for exploration" that is essential to the development of subsequent competence. Yet in the most commonly employed operationalization of this tension, the Ainsworth "strange situation", exploratory behavior is defined negatively as the absence of attachment behavior (Ainsworth, 1973). Little attention has been given in this paradigm to the nature, function or context of the exploratory behaviors (though see van den Boom, Chapter 7 in this volume).

This book is offered in the belief that Lorenz (1943, 1949) was correct in arguing that exploration is an autonomous behavioral system with undeniable adaptive value for many species (also see Cowan, 1983). Before describing the contents of the chapters that build this argument, we will provide a brief introduction to some of the definitional and background issues relevant to the study of curiosity and exploratory behaviors.

Definitions and Distinctions

Many attempts have been made to draw distinctions between or define different types of curiosity and exploration. Berlyne (1960) differentiated "intrinsically-motivated exploration" and "extrinsically-motivated exploration". When an individual surveys an environment searching for food, water or a mate, extrinsic exploration constitutes the appetitive phase of many goal-directed behaviors (Barnett, 1963). Exploration of the environment for its own sake, in contrast, is intrinsically motivated as a behavioral system with its own end. However, when an animal is patrolling its home range, seemingly extrinsically-motivated search behavior may be observed at a time when vital goals are not being pursued. Yet the information gathered in the process might be used at a later time in pursuit of a clear goal or in the instigation of a latent disposition for an instinctive behavior not based on a tissue deficit (Cowan, 1983). Thus, the distinction between extrinsic and intrinsic exploration may not always be made easily.

Another common distinction made is between exploration and play. Extending the distinction made by Berlyne (1960), Hutt (1970) and other investigators (e.g., Hughes, 1978) have categorized play as a "diversive" activity in which an individual implicitly asks the question "what can I do with this object?" In contrast, exploration is an activity in which the individual's question is "what does this object do?" Although the exploration-play distinction has been a useful one for generating interesting empirical research, like the extrinsic-intrinsic exploration distinction, play and exploration are not always easily distinguished in the natural environment.

In the hope of being inclusive rather than exclusive, we have not attempted to arrive at a definition of curiosity and exploratory behavior upon which every contributor to this volume would agree. Given the state of the art in research and theory on curiosity and exploratory behavior, we thought to attempt to do so would be counterproductive. Thus, in chapters that follow, curiosity and exploratory behavior are conceptually and operationally defined in many different ways. There may come a time when researchers will have to find a common definition in the study of these important behaviors. At this time, however, only a broad framework that provides perspective on the multiple approaches to thinking about the area seems to be the wisest tack.

The Functions of Exploration: A Psychobiological Perspective "Ultimate causes"

A modern psychobiological approach to exploration represents a broad perspective on curiosity and exploratory behavior that addresses contextual influences on exploration, individual differences in exploratory tendencies, and the developmental course of exploratory behavior. The basic assumption in this approach is that the competence to explore the physical and social environment represents innate behavioral capacities in human beings and other animal species. However, the behavioral expressions and quality of exploratory behavior are presumed to change phylogenetically, across species, as well as ontogenetically as a function of particular developmental tasks encountered by the individual, its developmental competencies, and the particular environmental requirements encountered (see Fragen, Chapter 17).

Exploration (and play, which is not distinguished from exploration in much of the ethological literature, see Keller & Boigs, 1991) also is regarded as a basic means for acquiring social skills, on a part with cooperative behavior (Fragen, 1987; Trivers, 1985). Exploration thus serves to provide anticipatory preparation for adulthood because of the acquisition of capacities and strategies that increase the likelihood of reproductive success in adult life. Moreover, this exploratory learning is developmentally constrained. Not just any stimulus-response relationship, regardless of the relevance of the information acquired, is learned (Draper & Harpending, 1988). Exploratory learning occurs in an individual in regard to its developmental status in a particular context. Environmental information related to a specific developmental task is processed more easily during an appropriate developmental phase. More generally, because the ecological niche of human beings is social in nature (Chasiotis, 1990), social learning should be easier than learning outside a social context.

A psychobiological approach will also suggest that exploration, like play, can be expensive. Fragen (1981, 1987) has argued that the exploratory and playful activities of higher animals increase the risk of casualties, such as those who fall prey to predators and/or become separated from the group. However, regardless of the risks, the adaptive value of exploration and play apparently has prevailed over the expenses given the evidence of successful evolutionary dispersion of exploration and the fact that it constitutes a major activity among higher species.

"Proximate Causes"

Adaptational value, then, provides the basis for the "ultimate cause" of curiosity and exploratory behavior. What are the more immediate or "proximate" causes? Most theorists respond to this question by first looking at the environment. Those features of the environment that have been identified most frequently as causes of exploration fall into a category that Berlyne (1960) called "collative variables". These features of the physical and mental objects and events encountered by individuals include those that are new, complex, or ambiguous. They are "collative" in the sense that mental schemas of the perceiver are collated or compared to the stimulus and the comparison results in a relative assessment of novelty, complexity or ambiguity.

The comparison of mental schemas and objects/events results in some degree of similarity or discrepancy that is also relative to an individual's developmental status (Hunt, 1963). When there is a discrepancy between mental schemas and the experienced object or event, a state of subjective uncertainty is created (Berlyne, 1960, 1966, 1978). This state of subjective uncertainty is aversive. A person attempts to escape the aversiveness by reducing the subjective uncertainty. The major mechanism for gathering information that will serve to reduce the subjective uncertainty is exploratory behavior.

Does this mean that the curiosity or exploration motivation is aversive motivation? According to Berlyne (1960) the answer is "yes". However, there is precedent in conceptualizing other motivational systems for an intermingling of positive and negative qualities in acts. For example, in hunger motivation, food searching behaviors motivated by the aversive qualities of a hunger state lead to a positive consummatory act in the cases in which palatable food is found. In a similar manner, specific, object- or event-oriented exploration could be doubly motivated: first by the aversive state of subjective uncertainty and, second, by the anticipation of satisfaction made possible through successful acquisition of new knowledge and the release of tension as uncertainty is reduced. Depending on particular situational circumstances and/or individual differences, the balance of the aversive and positive motivational components would vary.

Especially in human exploratory behavior, both anticipated positive emotions and previously experienced ones are likely to be causal factors in exploration. Wohlwill (1981), for example, differentiated between "inspective" and "affective" exploration. Inspective exploration is aimed explicitly at the gathering of information (behaviorally) or the reduction of uncertainty (motivationally). Affective exploration, in contrast, is engaged in for its own sake, just for the fun of it. Berber (1935) made a similar distinction many years ago.

Unfortunately, we know little about the affective qualities of different levels of subjective uncertainty. There is clear evidence from sensory deprivation studies conducted in the 1950's that in the long run, very low sensory input and, presumably, a concomitant very low state of subjective uncertainty is experienced as very painful (Schultz, 1965). It also is reasonable to believe that very high states of uncertainty are aversive (Berlyne, 1966) and that during waking hours a moderate state of tension or activation caused by intermediate levels of collativity of the stimulus input is a preferred state for animals and human beings alike.

Berlyne's (1966) explanation for this preference for moderate levels of stimulation is that the assumed "arousal level" of objects, caused by their collative value, first instigates a "rewarding system", and then an "aversion system" as the arousal increases. The summation of these two opposing functions determines the hedonic value attached to an object with a given level of arousal potential. Liberating this general motivational model from the outdated notion of a general arousal mechanism in the central nervous system, the preference for intermediate levels of collative values can be explained as determined by the outcome of the two opposing evaluations, a positive and a negative one, of the very same object (Coombs & Avrunin, 1977a, 1977b).

Just as exploratory behavior has costs in the ultimate causation terms of survival, it has potential costs in terms of proximate causation. These costs, such as mental and physical effort or work, are likely to increase with positive acceleration at the high end of the collativity continuum. These costs may escalate as the cognitive systems are increasingly unable to assimilate novel objects encountered (Hunt, 1965). Furthermore, at higher collativity values, and, therefore, higher degrees of schema-percept discrepancy, anxiety might be instigated (Hebb, 1955; James, 1890; McDougall, 1908; Montgomery, 1955). As the costs of assimilation effort, anxiety, and percept-schema discrepancy all accelerate, inverted U-shaped exploration functions with a single peak on the dimension of collativity will be highly likely (Coombs & Avrunin, 1977a, 1977b).

This model of excitatory and inhibitory forces in curiosity and exploratory motivation renders previous conceptualizations based on highly speculative and simplistic psychophysiological concepts obsolete. The simple Coombs and Avruinins model, with very few assumptions, but equal predictive power relative to the more elaborate psychophysical models based on inadequate notions of the central and peripheral nervous system (see Schneider, in press).

Contexts and Individual Differences

The brief description of a psychobiological perspective on exploratory behavior provided above emphasizes the complexity of the functions of exploration and the importance of considering the relation of exploration to other behaviors. There are two other important implications of this approach. One is the need to pay attention to individual and developmental differences. Although exploratory behavior is presumed to have adaptive value, there is no assurance that the predisposition to respond to collativity in the environment will be equally present in all members of the species. Moreover, the discussion of ultimate causes above suggests that developmental differences will be a major source of variation in exploratory behavior. Both individuals and the tasks they are required to perform will change with age.

A second additional implication of the psychobiological model is the importance of social context on exploration. The social nature of the ecological niche of human beings and the possible mediating effects of conspecifics directly on the individual and indirectly through modifications of the physical environment (e.g., to reduce anxiety or mediate assimilation effort). A full understanding of the nature of curiosity and exploration will require a careful explication of the social and socially-mediated factors in the environment that influence the riskiness of the behavior and its relative collativity.

The Contents of This Book

The discussion above provides an orientation to the 17 chapters of this book. These chapters are based on papers presented at a special symposium on curiosity and exploration held in Osnabruck, Germany in March 1990. The purpose of the symposium was to bring together researchers to begin to formulate an overview of the state of the art in the study of curiosity and exploration. It was designed to continue the tradition established by Dietmar Görlitz and Jack Wohlwill at their "Curiosity, imagination, and play" symposium in Berlin in 1981 (Gorlitz & Wohlwill, 1987). We deeply regret that Jack Wohlwill, who contributed many creative ideas to the study of exploration, is no longer among us.

The individual chapters cover a wide range of research topics concerning curiosity and exploratory behavior. The authors represent backgrounds in psychology, psychobiology and other fields that can provide a fresh view of the topic. The book is framed by two chapters that provide particularly novel perspectives on curiosity and exploration. In the first, (Part I, Chapter 1: "Fairy tales and curiosity: Exploratory behavior in literature, or, The futile attempt to keep girls from the spindle".) Rosemarie M. Rigol, a German-language and literature scientist from the University of Osnabruck, Germany, analyzes notions of exploratory behavior in fairy tales. She classifies different forms of exploratory behavior and evaluates them with special attention to gender differences. Her contribution stimulates the consideration of contextual variables in evaluating curiosity and exploratory behavior in our general culture and in specific belief systems.

The other framing chapter (Part IV, Chapter 17, "Sociobiological considerations on exploration and play") by Robert Fragen of the University of Alaska-Fairbanks in Juneau, Alaska, USA, also addresses the everyday relevance of exploration and play. He argues that exploration and play interact in a dialogue in which they "interpret the individual to the environment, the environment to the individual (so that each is defined dialectically), and both to selection..." He argues further that these interpretations can lead to the emergence of individual and social traits that affect differential survival and to a radically different view of evolutionary dynamics. In a playful, curious species, phenotypic expression is contingent upon the individual's style and creativity.

These two novel, creative chapters both highlight our main concern for the future of research on curiosity and exploration: a stress on the contextual embeddedness of exploration and analyzing its functional importance in the life cycle.

The other 15 chapters are organized into four major categories: (1) Functional and motivational mechanisms of exploration; (2) Development and individual differences; (3) The interconnections of exploratory behavior and other behavioral systems; and (4) Applied perspectives.

1. Functional and Motivational Mechanisms of Exploration

Function has two aspects as we use it here. First, in terms of evolutionary, biological adaptation, and, second, in the systemic sense of the role of exploration for various psychic systems. In Chapter 2, Christiane Buchholtz and Andrea Persch of the Philipps University of Marburg, Germany, present "An ethological conception of exploratory behavior". They address the question of how ethology can help us better understand exploratory behavior through the use of a model, illustrated with a flow diagram, that integrates a sensory component, a motor component, and what they call an action readiness system. A level of action readiness is defined by different internal and external factors and then transformed into specific motor activity.

Specific motivational mechanisms in human exploration are discussed in Chapters 3-6. Chapter 3, "Neurobiological foundation of exploration", by Finn K. Jellestad, Gry S. Folleso, and Holger Ursin of the University of Bergen, Norway, presents, as described in the title, the physiological and neurological bases for exploratory behavior. The chapter focuses, in particular, on the role of a variety of brain structures, including the hypothalamus, the amygdaloid complex, and neurotransmitters, in the acquisition of information. The hypothalamic network is presented as largely involved in the recognition of novelty while the amygdaloid structure is seen as influential in the affective aspects of exploratory behavior. The authors conclude, on the basis of studies on the detection of novelty, that the brain functions as a homeostatic system in regard to information acquisition. Their homeostatic, neurobiological model fits well with the discussion of subjective uncertainty and optimal arousal presented earlier in this preface.

Chapter 4 is co-authored by Michael Niepel of the University of Bielefeld, Udo Rudolph of the University of the Bundeswehr in Hamburg, and Achim Schutzwohl and Uwe Meyer, also of the University of Bielefeld in Germany. Their chapter, entitled "Two characteristics of surprise: Action delay and involuntary atttentional focus", discusses the evidence for the motivational significance of the surprise reaction. They report on a program of research based on a two-stage paradigm. A specific schema expectation is first generated in subjects, and then is subsequently violated by the presentation of discrepant information. The results of a variety of experiments support the idea that events that conflict with a consolidated schema initiate processes leading to an impairment of other ongoing processes, such as delaying an action. Attention is focused on the surprising event, a possible first step in an exploration sequence. The surprise response is shown to be generalizable to responses in the auditory as well as the visual mode.

Andreas Krapp of the University of the Bundeswehr in Munich, Germany discusses interests in Chapter 5, entitled "Interest and curiosity: The role of interest in the theory of exploratory action". He critiques the extant research on exploration for its neglect of the content of exploration. He argues that exploration is always the exploration of something, but that the something has not been the focus of those who have studied exploration. Moreover, he argues that developmental changes in the content of exploration have also been ignored. He proposes a person-object theory of interest as an alternative to the content-general motivational conceptualizations of Berlyne and other theorists. He applies his interest theory to both diversive and specific exploration and their development.

Benedykt Fink of the University of the Bundeswehr in Munich, Germany, also addresses the exploration-interest relation in the final chapter in this section, one entitled "Interest and exploration: Exploratory action in the context of interest genesis". He addresses the development of interests in terms of children's duration and frequency of engagement with objects, their subjective value and children's fondness for particular objects and activities. He then describes a 5-year longitudinal study of a small group of children's interests and discusses recent empirical examples with respect to the application of a variety of models of interest development. The genesis of interests is then evaluated as a determinant of the direction of exploratory behavior.

2. Development and Individual Differences

Part III includes five chapters that address individual differences, developmental differences, or both. In these chapters, the development of exploratory behavior is conceptualized within the context of the development of the primary attachment relationship. The role of social support and the general ecology of development in determining the occurrence of exploratory behavior is stressed. Chapter 7, "The relationship between attachment, temperament, and exploration", by Dymphma van den Boom of the University of Leiden, The Netherlands, clearly shows this emphasis. She reports on an intervention study involving 100 irritable infants and their mothers in a treatment-control group design. Her results demonstrate that maternal sensitive responsiveness fosters a sense of security in their infants. This sense of security motivates subsequent processing of information from the environment. Infant irritability seems to constitute an important contributor to variations in experienced security. Van den Boom extends the classical view of the attachment-exploration balance by relating her results to the model of Gubler and Bischoff (1991) which postulates the interaction of the security system, the arousal system, and the autonomy system.

Clemens Trudewind and Klaus Schneider of Ruhr-University of Bochum, Germany (Chapter 8: Interindividual differences in the development of exploratory behavior: Methodological considerations) present data on a research project in progress that is aimed at developing instruments to assess curiosity. The goal of their program is to obtain indices of children's dispositional curiosity at three levels: (1) behavioral manifestations in structured situations; (2) parent and teacher questionnaires in which respondents are asked to report on children's exploratory responses in prototypical, everyday environments, and (3) children's attentional and emotional reactions to curiosity-inducing events in a puppet show. Preliminary results supporting the concurrent validity of these measures are presented.

In Chapter 9 ("Preschoolers' exploratory behavior: The influence of the social and physical context"), Klaus Schneider and Lothar Unzer report the results of a field study in which they observed toddlers together with their mothers at home and at a supermarket several times at two-month intervals. Curiosity incentives at home were toys, brought by an observer. Incentives in the supermarket were manifest in the abundance of goods exhibited on the shelves. The supermarket, but not the home, tended to elicit strong exploratory tendencies in the toddlers. It appeared that in the supermarket, the physical environment rather than social relationship with the mother controlled the child's exploratory behavior.

Heidi Keller of the University of Osnabruck, Germany presents data from her nine-year longitudinal study of the development of exploratory behavior in Chapter 10, "A developmental analysis of exploration styles". The manipulative and visual orientations in person-environment interactions are characterized as behavioral styles that can be found at different loci during childhood. A stable visual mode is identified as dysfunctional for purposes of long-term information intake and can be related to poor interactional quality in the behavioral exchange between children and their primary caregivers during early development.

In the last chapter in this section, Bruce Henderson of Western Carolina University in Cullowhee, North Carolina, USA, discusses the relation between exploration and intellectual competence in (Chapter 11:"Interindividual differences in experience-producing tendencies"). Individual differences in curiosity and exploration are interpreted as an important category of experience-producing tendencies (EPTs). Exploratory behaviors are considered to aid the child's interaction with the environment to produce new experiences and, thus, opportunities for learning. He relates the EPT notion to recent reconceptualizations of the nature-nurture issue.

3. The Interconnections of Exploratory Behavior and other Behavioral Systems

Three chapters are included in Part IV. The inclusion of topics on the interrelatedness of exploration with other systems necessarily has to be selective because as indicated earlier, exploratory behavior could conceivably be related to almost any other behavior category. The bias here is toward the relation between exploration and the domain of cognitive development. As described above, outside the study of exploration per se, investigators use common sense understandings of exploratory behaviors rather than specifying their unique features. The chapters presented here are aimed at clarifying the specific contributions of exploratory behavior within certain definitional constraints. However, the interrelatedness of exploratory behavior with emotional behavior, such as in the instance of attachment, is considered as part of the developmental process of exploration (see Part III).

In the first chapter in this section (Chapter 12: Exploration and perception: An ecological approach to exploratory behavior), Ad Smitsman of the University of Nijmwegen, The Netherlands, discusses the relation between exploration and the ecological approach to perceiving and acting originally developed by J. J. and E. J. Gibson. He points out the central concept of active exploration in perception and presents a framework for its empirical study. Included is a discussion of the ways children's exploratory activities can be conceived developmentally.

"The information intake function of exploration" is the title and scope of the presentation in Chapter 13 by Axel Scholmerich of the National Institute of Mental Health, Bethesda, Maryland, USA. He presents an information-intake analysis from nine-year-old children, observed exploring a novel box. He reports on the system for analysis he has developed to assess the amount of information extracted by the child. He discusses the relationship between manipulative exploration, usually assessed as a measure of the quality of exploration and the information intake parameters.

Shulamith and Hans Kreitler of the Tel Aviv University inIsrael present a review of "Motivational and cognitive determinants of exploration" in Chapter 14. Their approach to the study of curiosity and exploration is based on two theoretical frameworks they have developed in their work: the theory of cognitive orientation and the theory of meaning. They describe specific beliefs within cognitive orientations that provide the motivation for different types of exploration and what they call "patterns of meaning assignment" for each exploratory mode. They argue that their perspective provides both a basis for the diagnosis of reasons for low levels of exploration and a means for increasing curiosity and exploration.

4. Applied perspectives

In Part V, we have included chapters that provide examples of how exploration can be applied, specifically, to the nature of computer usage and to environments that encourage children's exploration in their free time. In Chapter 15, "Computer systems as exploratory environments", Siegfried Greif of the University of Osnabruck, Germany, discusses human-computer interaction in terms of an error-prone exploratory style of learning. He uses concepts from research on exploration, especially novelty and complexity, to describe the behavior of computer users and the design of "optimal exploratory environments" in this context.

A different applied perspective is introduced in Chapter 16, "Urban development for children: Reexploring a new research area". Dietmar Gorlitz and Richard Schroder of the Technical University of Berlin, Germany, and Model-Project Herten introduce a research program that represents a collaborative effort between the Technical University and the city of Herten. They describe their efforts to create an environment that allows children to explore and learn within a general theoretical context that emphasizes the importance of curiosity and exploration.

We hope that this book will stimulate readers to find new ways to study and explain curiosity and exploratory behavior. The overview provided here is necessarily limited. We encourage the planning of future symposia that will address areas of theory and research not represented here as well as opportunities for reports of developments in the programs included here. Our own biases are toward a need for further research and theory on the differential, developmental, and contextual aspects of exploration.

Additional References

Hughes, M. (1978). Sequential analysis of exploration and play. International Journal of Behavioral Development, 1, 83-97.

Mayer, J. D., Caruso, D. R., Zigler, E., & Dreyden, J. I. (1989). Intelligence and intelligence-related personality traits. Intelligence, 13, 119-133.

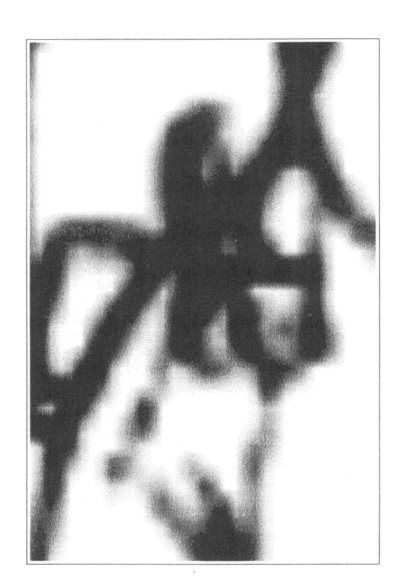

CHAPTER I.1

Fairy Tales and Curiosity. Exploratory Behavior in Literature for Children or the Futile Attempt to Keep Girls from the Spindle

Rosemarie M. Rigol

The everlasting attraction that fairy tales hold for children can be explained in several ways. For example, the habit of story-telling in some families or on radio and television or tape may create interest. The tradition of handing down only well-loved types of children's literature to the next generation may give rise to common preferences. Or the fact that fairy tales are simply structured may make them particularly attractive to children.

Despite these possibilities, we need to consider that the main reason children turn to stories is because they are looking for something related to their own lives in them. Obviously children find information of vital significance in the tales that keeps them returning to the same stories again and again and often makes them insist upon word-for-word repetition of the stories. None of the words should be lost; the unchangeable text seems to stimulate contemplation about one's own situation. We can only guess at the function the fictional representation of human traits may have for the mental development of children. May they serve as examples, warnings, or encouragement.

In any case, we may assume a specific interaction between child and text that only happens during the early stages of social development. The texts may have a certain anticipatory quality that helps the child focus on his own problems and conflicts with his environment. The child probably uses this literary disguise to develop his[1] capacity for a

---

[1] Using "his" does not mean that I solely consider the male part of young humanhood when speaking of children, certainly they are female, too. But space and the fact that I am an authoress make the lopsidedness bearable for the reading persons, I hope.

growing awareness of the world, the persons acting in it and the rules of behavior around him. Thus, he gains social experience without the risk of failure and punishment, using literature as a field of simulating exploration and experience.

Looking for these characteristics of texts we leave the field of literary criticism and even more so the limits of philological propriety. No doubt, formal and historical analysis have revealed interesting aesthetic dimensions in fairy tales[2]. Our approach cannot be compared with them in elegance and formality. But since stories which are appreciated by children, are not simply phenomena of fine arts, but more or less advisers and mirrors, we may as well accompany the child on his way to the discovery of everyday information in the tales. So we take a look at the stories with the question: What do they tell us about exploratory behavior and the function of curiosity?

We will expound this point of interest in three sections, evidence, types, evaluation of exploratory behavior and curiosity in fairy tales.

I shall therefore concentrate upon that handful of the so-called magic fairy tales of the Grimm brothers[3] that are the best and most widely known, in Germany at least. They belong to the small number of titles that form the basis of literary knowledge in the country. Everyone knows who *Hänsel and Gretel* were. The method is very simple. I look through the stories from the point of view "curiosity" and "exploration", document the findings, and try to classify them; nothing grandiose, just out of curiosity itself and with a view to placing myself in the child's situation while listening to stories.

---

[2]Lately the most impressive studies have been done by Heinz Röllecke, Wuppertal, but there has been a tradition of fairy tale philosophy, from Jakob Grimm up to Felix Karlinger, from Bolte-Polivka, Aarne-Thompson, Propp to Max Lüthi.
[3]They are: Hänsel and Gretel, Red Riding Hood, Little Snow White, Sleeping Beauty, The Frog Prince, Cinderella, Rapunzel (= Rampion), The Miller's Daughter, Little Brother and Little Sister, The Seven Ravens, together with the less magic stories which show the cleverness of the hero, as there are the thumbling tales and the Brave Little Tailor.

Evidence of Curiosity and Exploratory Behavior in Fairy Tales

The basis of this rather trivial question, as to whether curiosity as a motif is to be found in fairy tales at all, is the widely spread conviction we find in scholarly literature[4], that characters in fairy tales do not change or develop. They are supposed to represent elements of the action in the plot, thus being a kind of personification rather than persons in themselves. Change can be expressed by the introduction of a new character that takes over for the coming events in the plot. Since however curiosity as a human trait is closely bound up with an individual person that finds his or her representation in a literary figure, we first have to look for the existence of curiosity in the fairy tales' hero.

In the *Sleeping Beauty* we find a case of exploration. The girl had been put under a spell to sleep for hundred years at the age of fifteen:

It happened that on her fifteenth birthday the princess was alone in the castle (...). The girl amused herself by running about in the corridors and rooms and exploring all sorts of out of the way corners. At last she came to a small ancient tower, she climbed the winding staircase and found herself in front of a little door. There was a rusty key in the lock, and directly she turned it the door sprang open and there in the tiny room sat an aged dame before a spinning-wheel, spinning her flax industriously.

"Good day, old motherkin," said the princess, "what are you doing?"

"I am spinning," replied the old woman, and nodded her head.

"What is that thing called that goes around so merrily?" asked the princess, and she took hold of the spinning wheel to see if she could spin, too. Scarcely had she touched it when the spindle pricked her finger, and at the very same instant she sank on the couch behind her in a profound slumber. And this slumber spread over the whole castle. (203-204)[5]

---

[4] Compare Max Lüthi, ([7]1981); he speaks of "Flächenhaftigkeit".
[5] The numbers in brackets after each quotation are taken from: Grimm's Fairy Tales for children and the Household. Collected by The Brothers Grimm. Translated by Beatrice Marshall. London-New York and Melbourne: Ward, Lock and Co. Ltd. without year (1893) [= The Rainbow Serres].

The girl explores unknown areas of her environment, without the knowledge of her parents. Everything looks prearranged: the tower which had been hidden up until then, the door that springs open, the old woman and the forbidden spinning wheel. The result of the girl's curiosity is the coming of the sleep for all.

Almost as dramatically ends the exploration of *Hänsel and Gretel* when they arrive at the witch's home,

(...) which (..) was built of bread and thatched with cake, and the windows were made of barley-sugar (...). Hänsel climbed up and broke off a bit of the roof, to see what it tasted like, and Gretel stood by a window and nibbled it.
Then a voice called from inside -
Nibble ! Nibble ! Nibble !
Who's nibbling at my house ?
The children answered -
The wind , the wind ,
the child of heaven.
and ate on, quite unconcerned. (...)
All of a sudden the door opened and a very old woman, leaning on a cruch, hobbled out. Hänsel and Gretel shook in their shoes out of fright and let the good stuff fall out of their hands. (76)[6]

They obviously broke through a borderline between reality and magic, taking hold of parts of a house which looked rather strange to them.

We find a different form of exploration in *The Seven Ravens*, wherein the birth of a daughter makes the father banish the seven brothers by turning them into birds. When she is older, the girl leaves home to rescue her brothers. She walks to the end of the world and turns to the sun and the moon, asking for the whereabouts of the ravens.

She came to the sun who was so hot and terrible he ate little children. Quickly she ran away on to the moon, but she was horribly cold, and cruel and wicked, too, and when she saw the child, remarked " I smell human flesh ". So the child ran off as far as she could and came to the stars. They were friendly and kind (...).

The morning star rose and gave her a little wooden leg, and said, "Without that leg you can't get into the glass mountain, and in the glass mountain are your brothers."(112-113)

---

[6]Sometimes, Mrs. Moina Rück, who did proof-reading for me, found the cited English so unbearable that we changed a few propositions, for example "of" instead of "with" and "things" instead of "stuff" in the original Grimm's Fairy Tales as cited above.

Much could be said about the "hot he" and the "cold she" and the girl's turning to the stars, who are as small and childlike as herself, and as dependent on the sun as the girl upon the help of others. The girl's long search shows the purposeful variety of exploration, not as playful as that of Hänsel or the Sleeping Beauty.

These few examples which stand for many other stories, too, show that curiosity and exploration exist in fairy tales. They actually seem to belong to the central motifs that bring the new, strange and dangerous elements into the plot. The curiosity behavior is presented rather artistically, transformed into fictional actions, often combined with the miraculous and stressing the daring attitude of the acting figure towards the unknown. In this way the child is confronted with his own wishes for novelty and mystery in the stories in which other people's experiences and troubles are contained.

Types of Exploratory Behavior in Fairy Tales

Since exploratory behavior in fairy tales is mostly concealed in actions, we shall now look for the types of fictional events that can be considered as expressing curiosity.

Movement

In general the characters tend to move away from their familiar vicinity. One has to leave the commonplace to get into touch with the unknown, and only then is there a chance to experience one's own ability to cope with strange situations.

Movement as a metaphor includes several elements of exploration: the opening-up of new horizons that present different possible ways of life, the necessity of being active, otherwise there is no movement, the courage to give up security and meet the challenge, and the ability to bear fear and danger. Movement with the implicit meaning of exploration needs a certain framework of restriction, at least according to the norms of the time when the fairy tales originated. We have a few examples when the heroes move away voluntarily. There is for example the story of *The Brave Little Tailor* who left his home, over-estimating his abilities. After he had killed seven flies, he pondered

"What a fellow you are!" he said, admiring his own bravery," the whole town shall know of your exploit. And the little tailor with all speed cut himself out a belt for himself, hemmed it, and on it inscribed the words, "Seven at one stroke"! "Not only the town," he exclaimed, "but the whole world shall know it." The tailor (went) out into the world, because the workshop has become too small for his great bravery. (97)

Usually the heroes are driven away or are forced to move, as in *Hänsel and Gretel*. Their mother[7] plans

(...) tomorrow morning early we will lead the children into the wood, and where it is the densest, light them a fire, give each of them a bit of bread, and then go to our work and leave them alone. They will never find their way home, and we shall be rid of them. (72)

And *Little Snow White* who was threatened with death by her (step)mother.

Then she (the queen - RR) summoned a huntsman and said to him: "Here, take the child out into the forest. I can bear the sight of her no longer. Kill her and bring me her liver and lungs as proof that she is dead. (214)

Since the hunter spared her, she was bound to move away from her dangerous home against her will.

Meanwhile the poor child was wandering, desolate and alone, in the large forest, and was so fearful that she peeped behind every leaf to see who was there. At last she set off running, and ran over sharp stones and through thorns and brambles. (214)

In some fairy tales the children are not thrown out, but driven away by the unbearable family situation. For example in *Little Brother and Little Sister* who were neglected and ill-treated by their stepmother and decided

"Come, let us go into the wide world together." (54)

Or in *Dame Holle* (188 f) wherein the good girl desperately fled into the well because the stepmother was never satisfied with her work.

In other cases the hero is apparently given unrealizable tasks by his master that remove him from his usual environment. Often the orders barely conceal the intention to destroy the young man or woman. However, by handing him or her over to danger, this seems to provoke the activity and the intelligence of the person sent off. With the help of good people, he or she overcomes the dreadful situations and returns

---

[7]In the first edition of the fairy tales it was mainly the mother who planned the children's troubles. Since Biedermeier mothers protested against this image, Wilhelm Grimm changed it into "stepmothers".

home as a different person. The inner change is presented as a change of social status: *Snow White* becomes queen, *Hänsel and Gretel* take the place of their stepmother who had died in the meantime, *Red Riding Hood* declares her moral change by promising lasting obedience to her mother.

But the heroes do not always move away from home. Sometimes the outside world breaks into their familiar life and challenges it. We find this form of movement in *Rampion* (61 ff) where the family is replaced by the world of the witch and finally the prince comes to rescue the girl.

Snow White settles down with the seven dwarfs and finds almost the same situation she is used to there: a kind of family. Only when the stepmother again tries to kill her by intruding into her world and persuading her to eat the poisoned apple does she enter a new state of life and now is ready for the prince who rescues her because of her beauty.

There is a tendency in the fairy tales to reserve the active exploration for the male and enforcement or waiting for the female. But this culturally specific usage is not strictly maintained. The topoi of fairy tales are much older than their written form. Human characteristics and experiences which are the main subject of the stories behind the metaphorical scenes are not so much considered as being gender-dependent than one would expect from the time of the formalization of the stories.

So the movement of the main character, whether it is voluntary or enforced by a dangerous command or by the sudden intrusion of the strange element into familiar life, shows, as a metaphor, the necessity of exploratory behavior for the development of one's own capacity. To grow up and become an adult, one has to meet the unknown and grapple with it.

Handling and Sensual Experience

Meeting a new situation in fairy tales means, meeting a new situation means to approach it and to act. A form of taking up the challenge is to use one's own senses: the hero looks, listens, smells, tastes, touches with skin and body to acquaint himself with the matter in question. As an action he usually tries out former behavior patterns to get the new situation under control.

- *Hänsel and Gretel* taste the roofing tiles because they smell and look edible. As a result of their childlike approach, a friendly old woman appears who later turns out to be a rather disagreeable witch who wants to eat the children. Both of the children had to learn that you cannot integrate new phenomena into old patterns of interpretation or you will be deceived.

- *Little Snow White* tests the dwarfs' home by taking vegetables and bread from the waiting plates, by drinking from each mug and by lying down in each bed until she has found the right one. The familiar patterns, applied to the pseudo-home, prove unsafe. They do not protect her against temptation by the stepmother, and, above all, against her own growing curiosity and eagerness for change.

- *Red Riding Hood* is tempted by the wolf to look for the flowers and not for the right way to her grandmother's. Meanwhile the familiar situation changes into a hostile one which the girl has then to cope with.

In the examples above the hero and the heroine use their familiar perception of the world. This proves to be only helpful as the first step into a new field of experience, but must finally fail. The patterns have to be changed during exploration. Thus the purely sensory or motor experience takes on the less obvious structure of the friendly or hostile tendencies underlying the strange situation the actor meets.

Language - One prominent instrument to satisfy curiosity is the use of language: ask, enter into a dialogue, attack with words. We find this language game in quite a number of fairy tales, for example in *Red Riding Hood* (144), *Little Snow White* (213), *The Wolf* and *The Seven Kids* (29), *Rampion* (61), *Hänsel and Gretel* (70), *Dame Holle* (108). Some of the dialogues have a rather idiomatic quality. They are familiar to everyone are even used in advertising and politics as a means of creating a basis of common understanding. This is the case with the dialogue between *Red Riding Hood* and the wolf who pretends to be the grandmother

"Aye, grandmother, what big ears you have got!"
"So that I can hear better."
up until
"And grandmother, I never knew you had such a terribly big mouth before!"
"All the better to eat you up with." (115-116)

Red Riding Hood approaches the strangeness of the situation with the help of questions, slowly coming closer and closer to the dangerous centre.

The *Miller's Daughter* only succeeds with her questions after a third person has found out the real name of Rumpelstilzchen. The third day

she asks the little man whose name she well knew by then:
"Are you called Kunz?!
"No"
"Then Hinz?"
"No"
"Then perhaps you are called Rumpelstilzchen?
"The devil told you, the devil told you!" shrieked the little man, and struck his right foot so deep in the earth that he fell and catching hold of his left foot to save himself, split himself in two." (229)

For once we see the ritual procedure of questioning: you have to ask three times at least. The structure shows how close questions are to other magic practice: they may disclose what is usually concealed. The reaction of the little man makes clear how unbearable the inquisition is for him: it even destroys the person questioned. Here the fairy tale presents the two sides of verbal curiosity, the right to know and to explore, and the risk of destroying.

At other times the hero meets demands by those persons and objects that are involved in his exploration process. There is the case of the endangered child who receives a warning as in *Little Brother And Little Sister*. After they had wandered a long time, *Little Brother* grew thirsty and wanted to drink from the first brook, but *Little Sister* heard the water say

"He who drinks of me is changed into a tiger - he who drinks of me is changed into a tiger."

She asked her brother

"I beg of you, brother, don't drink, or you will be changed into a wild beast and tear me to piece" (55)

The third time, however, the boy could not resist and drank. Immediately he was turned into a deer, a wild but gentle animal. This spell opened up quite a new way of life for the animal brother and the sister. Only the change made it possible for the girl to give up her child's life. Though the commands of the brook had been in vain, they aroused the curiosity of the children and made them step out of their former state.

Another form of demand consists of restrictions that constitute the conditions under which the situation can be changed. In *The Twelve Brothers* the girl whose brothers had been turned into ravens went out to rescue them and met an old woman who demanded from her:

"(...) if you would release them you must keep dumb for seven years. You must neither speak nor laugh, and one hour in which the rule is broken will make all the rest in vain. If you speak a single syllable it will be the death of your brothers." (51)

The spell was so binding that the sister even resisted the threat of death and did not speak before the seven years were up.

Once again we are confronted with the magic power of language. The girl experienced it with the help of the old woman . Knowledge comes from listening, is handed down and need not always be acquired at first hand.

When the heroine herself demands something, as in *Cinderella*, she knows that she needs what she asked for to change her situation. But here there is nobody to take over the task for her. She herself has to realize the power of her word and use it.

"Little tree shake, little tree shake throw gold and silver upon me." (Grimm II, 157)

She got the necessary robes and won the prince. Language is used primarily as a means of opening up new fields of possible experience, and, as such, is an instrument of exploration in those fields which cannot be treated in other ways.

To sum up the types of exploratory activities we found in fairy tales, we come to the following register:

- **Movement**. The hero or the heroine move away from the familiar towards the strange world, or strange elements break into the familiar surroundings of the hero or the heroine. In either case habitual behavior has to be changed by finding out adequate ways of handling the new situation.

- In the beginning the **Senses** are used as the first instrument to explore the new situation.

- On their way to investigate the unknown, the hero or the heroine meet verbal requests, warnings and advice, and they use **Language** to question the new situation.

Thus not only the obvious, but the underlying structure of the strange situation can be revealed.

Evaluation of Exploratory Behavior in Fairy Tales

We find exploratory activities evaluated in different ways. Sometimes curiosity itself is the main subject of the stories. In many cases it is merely a characteristic of the hero and as such is part of his or her development. Finally, there is a socially-specific usage of the right to be curious in fairy tales.

Curiosity as Subject of the Story

In *The Mary Child*, a pious and harmless variety of the bluebeard-story, the Virgin Mary forbids the girl to open the thirteenth door of the heavenly house while she is absent. Nevertheless, she hands over all the keys to the child thus tempting the girl.

The child spends her time visiting one of the permitted rooms each day, until only the thirteenth is left.

The little girl felt a great desire to know what was hidden there, and she said to her angel playfellows, "I won't open it quite, but I will just unlock it and open it a crack that I can peep through."

"O no," said the little angels, "that would be wrong (...) if you do it, you will suffer for it."

The child said no more then, but the desire in her heart gnawed and pricked her, and gave her no peace. (16)

Some days later the child can no longer resist and opens the forbidden door. There she finds the Holy Trinity sitting in "fire and pomp" (16). Her finger turns gold on touching a heavenly cloud. In the presence of the Virgin Mary she denies her disobedience and has to leave heaven, stricken with muteness.

Punishable curiosity? An offence against the hierarchy of adult norms? Yes and no. Certainly, obedience ranks higher than independent exploration, but still worse is the denial of one's own trespasses.

In this case the time of the written form may have influenced the evaluation of the child's curiosity. The Biedermeier period in the first half of the nineteenth century fostered the image of the good child, the innocent child, a child not yet involved in desire and other rebellious inclinations. Curiosity, however, is founded in the urge to find out, to know more than is granted by the adults. So it breaks up the framework set by rules prescribing the ways for a child to acquire awareness of the world and society. Denial of the deed represents the child's claim to be curious despite the rules of the adults.

We find a similar attitude towards curiosity in *Red Riding Hood, Little Snow White, Sleeping Beauty* and especially in those stories in which there is a heroine. This, of course, is in keeping with the general restrictive expectations towards girls and women in that historical setting.

Curiosity as a Part of Development

There are about thirty fairy tales in Grimm that have curiosity and exploration of the world as the central motif of the plot, which means in this kind of story, it is the main behavioral trait of the successful hero.

In some stories the quality of the structurizing motif is strengthened by presenting the character only as big as a thumb. There the size forms an unfavorable prerequisite for the exploration of the (adult) world around. Usually the listener sympathizes with the thumbling, mainly because he accosts the world despite his handicap, thereby proving to be rather clever, as for example in the story of the *Thumbkin* who was the only child of a peasant couple.

One day the peasant was preparing to go into the forest to fell wood, and he said to himself, "I do wish there were someone who could bring the wagon after me."

"O father!" cried Thumbkin,

"I'll bring you the waggon. Depend on me, and you will see it shall be in the forest at the proper time."

The peasant burst out laughing. (...) answered the Thumbkin, "I will sit in the horse's ear and cry to him how he is to go."

"Well," said his father, "for once in a way let us try it." (159-160)

Here we have quite a positive attitude towards the child's attempt to master a new task: a courageous boy, humorous and helpful parents who trust in their little son's self-assessment.

Other stories show the hero apparently minimized in his intellectual capacity, at least according to the first impression he gives. He appears simple-minded and discouraged, having already accepted his role of the stupid one among the cleverer brothers (sisters never seem to be around in such cases).

These exploratory activities therefore soon come to a standstill and have to be prompted anew by another person or object.

There were once two king's sons who went out in quest of adventure. They fell in with bad company so that they did not return to their father's roof. Their youngest brother, whom they called the little blockhead went in search of them.

The Queen Of Bees (275). Since the "retarded" boy shows greater understanding when meeting ants, ducks and bees, he receives competent help from the animals when needed. Thus he not only rescues his brothers, but wins, too, the most beautiful princess and a kingdom.

In the *Three Feathers* (277) the youngest brother is described as being "silent and simple", nicknamed "the duffer" (277). He can depend upon the sympathy and help of magic instances because he is sufficiently courageous to get involved with the eerie and inexplicable. When he blows the feather that is to lead him to success and happiness and it falls down rather close to the starting point, he finally recovers from his gloom and does something decisive.

The poor duffer sat down, and was very sorrowful. But after a few minutes he noticed a trap door near where the feather lay. He raised it, and finding a staircase, went down it. Then he came to another door, knocked, and he heard a voice inside singing.

The door was opened, and he saw a great fat toad sitting surrounded by a crowd of little toads. (277-278)

The toad helps him to fulfil all tasks his father set him, and at the end the "duffer" gains the kingdom.

Here exploration is the key to the hero's good luck. The overt success is only one result, the change in the person counts even more. The outcast thumbling becomes an accepted (since useful) member of the community. The stupid boy, the blockhead, the duffer, changes into a wise kind because he shows sufficient courage, voluntarily or softly enforced, to enter upon unknown adventures.

In these fairy tales, childlike exploration is considered as a central element of development. However, it is the sensitive type of curiosity that makes the heroes deal with the new phenomena according to their character, thus creating an emphatic experience and an adequate integration into knowledge of the world.

The Right to be Curious

In fairy tales there is no common right to be curious. If the high and mighty show curiosity, and that means kings, parents, adults, older brothers and sisters - simply anyone who has already reached a firm social status, it is often negatively evaluated. The young and inexperienced may use exploration as a means of solving a problem or of self-developing. Their investigation of the new phenomenon is an accepted activity. If the established people, however, indulge in curiosity, they only do so to augment their control of the status quo, to secure their social predominance.

One of the drastic examples here is *Little Snow White*, wherein the queen's explorations finally fail while the girl's come to good end. The fairy tale is on the side of the weak, the small and the simple-minded, who need not necessarily be the morally good ones. It legitimizes their curiosity by granting support through other people or magic instances. As a result of this overview we can observe three aspects in the evaluation of curiosity and exploratory behavior in Grimm's fairy tales.

1. In stories with a heroine there is often a warning against curiosity. But even then, the main figure shows a certain development. So the punishment described may therefore more indicate the difficulties and dangers of exploration than the general disapproval of the behavior itself.

2. We mainly find affirmation of exploratory activities as a way of change to the better in thumbling tales and in the stories of the stupid third son, wherein the situation of an outcast can only be overcome by courageous involvement in the challenging tasks.

3. Curiosity is generally positively evaluated, if it is a characteristic of the young and inexperienced. With socially established persons it is considered as an attempt to control the situation, whereas with children and adolescents it is seen as creative behavior.

Literary Culture and Children's Development

As we have seen, fairy tales contain important elements of children's behavior. Because literature does not allow direct experience as for example in the manipulation of an object and its change or in the reaction of another person after you have spoken to, we have to question the role literature may possibly play in children's development.

For one thing, fairy tales, like any other text, represent part of the historical experience with human behavior. Thus the presence of the topic referred to shows that it is considered to be a basic characteristic, especially important in early stages of development. To read the stories, as a child, as a parent or adult person, means, too, that everybody is reminded of the existence of the trait in question. Thereby, adults learn to keep in mind that children have to be curious. Thus a tradition of developmental necessities and educational tolerance is handed down from one reading generation to another, as a canon of expectable human behavior, collected in fiction about and for children.

On the other hand, the child will experience behavior in a metaphorical way while listening to the stories. This again means a kind of learning without sanctions: to get acquainted with behavior, to test outcomes of behavior in imagination to weigh different possibilities. Children will find their own desires and activities in texts that are accepted by other children, too. Thus, texts form the basis of common understanding and, in the long run, the prerequisites of cooperation.

Because children are in general not able to assimilate experience cognitively, stories open the door to a more pictorial and emotional relationship to information when facing their own traits of character that sometimes cause trouble.

In addition, the tales are not prescriptive. This, and the fact that they are only loosely connected to historical dates, allow free access to the content and its interpretation. That this is exercised on a fictional level, introduces at the same time a method of anticipation and simulation without risk and opens up the field of symbolic exploration.

The conclusion we can reach in regard to curiosity in fairy tales is that the listener will meet exploratory behavior in the stories, he or she will experience the outcomes symbolically and will need curiosity as an attitude to follow the story and compare its contents with their own experiences.

CHAPTER II

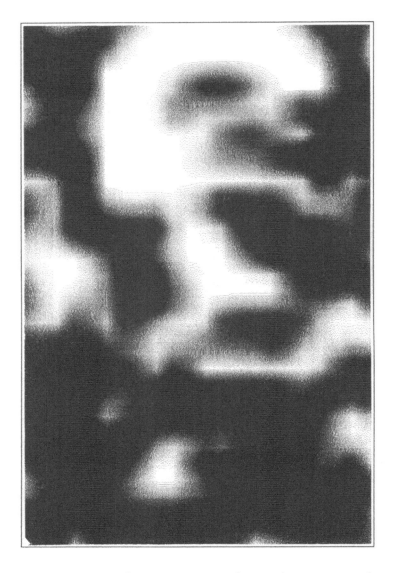

Function and Motivational Mechanisms of Exploration

CHAPTER II.2

An Ethological Conception of Exploratory Behavior

Christiane Buchholtz, Andrea Persch

If an animal is placed in a new environment or perceives unknown objects, it begins to explore. Wünschmann (1963) described exploratory behavior in many vertebrates from fish up to chimpanzees. Tembrock (1983) pointed out that this behavior is species specific. Based on the behavioral repertoire of the species, the animals can explore the unknown objects in the form of smelling, e.g., pigs (Stolba & Wood-Gush, 1980), monkeys (Joubert & Vauclair, 1986), and mice (Piechulla, 1990); biting, e.g., rats (Barnett & Cowan, 1976) and wolves (Fox, 1973); nibbling, e.g., mice (Eibl-Eibesfeldt, 1950) and cows (Murphey et al., 1981); picking, e.g., birds (Wünschmann, 1963); or touching and manipulating the unknown object with the extremities, e.g., mice (Persch, 1989), polecats (Weiss-Bürger, 1981), and monkeys (Welker, 1956). One special characteristic of exploratory behavior is that an animal tries many different behavioral elements from different behavioral systems in exploring one unknown object.

A raven that perceives an unknown object carries out a series of actions in a programmed sequence. It carefully starts with actions of mobbing at a bigger beast of prey as it approaches the unknown object. While ready for escape, it creeps sidewards and a bit backwards, gives the object an awful blow with its beak, and turns away as quickly as possible. Is the object alive and runs away, the raven runs after it and acts as it does when approaching large prey. If the object is already dead, which the raven tests with heavy blows of its beak, it grasps at the unknown object with its talons and tries to tear it into pieces. If it proves to be edible, the actions of eating and hiding prey occur. If the object is not useable, it becomes uninteresting to the raven and will possibly be used as a seat or torn into pieces for hiding more interesting objects (Lorenz, 1978).

Lorenz (1965) saw the importance and efficiency of exploratory behavior, especially in animals he characterized as "specialized nonspecialists." They make use of exploration to improve the chances of finding necessities such as food and water. They handle unknown objects as if they have biological relevance. If they recognize some aspect of the object as being relevant, these nonspecialists classify the object with regard to its biological significance. Thus, exploration enables animals to adapt to different areas having different sources of food. For example, the common raven can live in a desert as a carrion eater, in Central Europe as a hunter of small animals, or on an island as a predator of birds' nests (Lorenz, 1978).

Is exploration a separate behavior system? A behavior system integrates a group of behavioral elements with the same or similar task and effect (Immelmann, 1983). The exploration system integrates behavior elements to investigate novel objects (the task) and to classify the objects according to their biological significance (the effect). Exploratory behavior appears in situations where there are no other impinging needs such as hunger, or the need for social contact or comfort (Barnett & Cowan, 1976), but it also occurs in other situations in which other needs, such as hunger, are present (Zimbardo & Montgomery, 1957). Therefore, it seems clear that exploration has its own motivation, its own action readiness.

The fact that exploration has reward value, as has been shown in experiments on learning, provides additional evidence in support of classifying exploration as its own behavior system. Like food (Skinner, 1974) or the opportunity to play (Buchholtz, 1952; Schaffer, 1976), the opportunity for visual exploration has a reinforcing effect, as Butler (1953), for example, showed in experiments involving the operant conditioning of rhesus monkeys. Furthermore, v. Holst and v. Saint Paul (1960) have shown that electrophysiological stimulation releases an "emotion to explore" in hens. This "emotion" results in exploratory behavior when special parts of the hens' brains are stimulated. Based on these findings, it seems legitimate to classify exploration as a separate behavioral system with its own motivation or action readiness.

How can ethology help in the understanding of the behavior system of exploration? A model based on ethological, neuromorphological, and neurophysiological findings will be presented that should show how ethological concepts can inform an understanding of exploratory behavior. It is a model which shows the organism in communication with its environment. It is the authors' special task to describe the phenomena of exploration using this model.

The organism is an open system facing its environment (Buchholtz, 1982). It is obvious that there are environmental features (stimuli) which influence the behavior of an organism. To show the possible effects of these influences inside the organism, a first step is to separate the organism into a sensoric and a motoric part and the connection between them, the action readiness system (ARS). The separation is based on functional, not on morphological aspects. The model is based on theoretically-demanded functional units. Neuromorphologists and neurophysiologists can largely characterize these units (see below). Figure 1 presents the functional units in a diagram of the functional organisation of behavior. The organism is surrounded with its environment. In the following, we describe the organism's particular functional units.

Figure 1. Diagram for the functional organisation of behavior. Basical functional units and their relations within the organism.

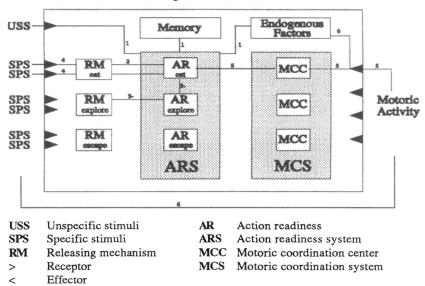

USS	Unspecific stimuli	AR	Action readiness
SPS	Specific stimuli	ARS	Action readiness system
RM	Releasing mechanism	MCC	Motoric coordination center
>	Receptor	MCS	Motoric coordination system
<	Effector		

Further explanations are in the text.

The Sensoric Part

The first unit of the sensoric part are the receptors which can receive environmental stimuli. These are specific stimuli (SPS) and unspecific stimuli (USS). Specific stimuli release a specific behavior within a behavior system. It is known that the afferent data processing system sums up the effectiveness of several SPS ('stimulus summation'; Bower, 1966; Leong, 1969). Prerequisite for the effectiveness of the specific stimuli are the unspecific stimuli like temperature or humidity. As a consequence of learning processes, USS can also gain significance for the organism and can change into specific stimuli (Buchholtz, 1973).

In order to make it possible to recognize the specific stimuli there must be filter mechanisms (Lorenz, 1935; Schleidt, 1962; Tinbergen, 1948) which do not release reactions until those specific stimuli occur (Ewert, 1976; Hubel & Wiesel, 1962; Lettvin et al., 1959). These filter mechanisms, the releasing mechanisms (RM), belong to three categories.
1. Innate releasing mechanism (IRM): Based on an innate filter effect, the organism reacts innately to specific stimuli, e.g., 'Child schema.'
2. Innate releasing mechanism modified by experience (IRME): The IRM enlarges as a result of learning processes. The animal learns additional stimuli (USS become SPS). The organism reacts by to a the combination of innately recognized and learned specific stimuli, e.g., 'Imprinting.'
3. Acquired releasing mechanism (ARM): The organism reacts only to learned specific stimuli, every participation of innate release is ruled out, e.g., 'Discrimination learning.'

The Motoric Part

The released behavior (motoric activity) is composed of single actions. Therefore, the organism activates different groups of muscles. To coordinate behavior in a sensible spatial and temporal orientation there must be motoric coordination centers (MCC) which activate the adequate effectors (v. Holst, 1939; v. Holst & v. Saint Paul, 1960).

The Action Readiness System

The action readiness system (parts of the limbic system, cf. Schmidt & Thews, 1983) is the connecting functional unit between the sensoric and the motoric part. With this system the level of simple input-output-correlation is left. It is the fundamental processing level where one can explain spontaneous behavior, such as appetitive behavior or vacuum activity. Within this system are soecific action readinesses.

In order to keep the presentation of the model comprehensible every single action readiness (AR) and releasing mechanism belongs to one behavior system, e.g., food acquisition, exploration, or escape. It is probable that the particular level of the described action readiness is based on, among other things, influences of subordinate units (Tinbergen, 1951).

The action readiness is an animal's motivation to do something special. A large number of factors, introduced in Figure 2, influence the ARs (Becker-Carus et al., 1972).

Figure 2. Diagram for the action readiness (adapted from Becker-Carus et al., 1972). Further explanations are in the text.

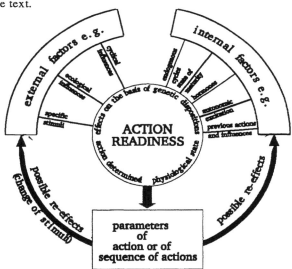

Based on the actual genetic disposition of the organism, internal and external factors influence the action readiness. Included in the internal factors are endogenous factors (endogenous cycles, state of maturity, hormones, autonomic excitation) and memory. External factors can be specific stimuli and unspecific stimuli (cyclical and ecological influences). The processing of these influences within the action readiness leads to an action-determined physiological state, the particular level of action readiness. A high level of action readiness can lead to an activation of the corresponding motoric coordination center and therefore to adequate motoric activity. Measuring the occuring parameters of an action or of a sequence of actions, it is possible to define the particular level of the organism's action readiness. Additionally, the released motoric activity can also re-affect the internal and external factors. Beside the effects on the particular action readinesses already mentioned, there are also influences among each other in the action readiness system. A high action readiness for escape can, for example, inhibit the action readiness for food acquisition. Figure 1 describes an occuring motoric activity as a consequence of perceived stimuli.

External (unspecific stimuli) and internal (endogenous factors, memory) factors and the influences within the action readiness system (1) define the processing result of an action readiness (e.g., food acquisition) which exists at a determined point of time. A high action readiness (food acquisition) activates the adequate releasing mechanism (2). Generally, it inhibits other ARs (e.g., exploration) and corresponding RMs (3). This is the basis of the 'reaction specificity' (Tinbergen, 1966).

The effectiveness of the specific stimuli belonging to the releasing mechanism (food acquisition) is again influencing the action readiness (food acquisition) (4). This leads to an activation of the adequate MCC and therefore to a motoric acticity (5). The motoric activity re-affects the endogenous factors and the spatial orienting attitude facing environmental stimuli (6).

It is obvious that there are more possible connections among the units in the diagram, but only the important ones for understanding the momentary problems are included here. Also, the model shows only a momentary state of the occuring connections. The organism is "frozen" at this point of time and circumstance.

One can also describe the occurence of exploratory behavior using the model for the functional organisation of behavior. Specific stimuli can release exploration. Characteristic features (SPS, invariants) of the objects that release exploration are novelty or the unusual arrangement of known stimuli. Therefore, one must assume that this is a special task of pattern recognition. The releasing mechanism for exploration must have another processing method than other RMs, like the ones for food acquisition or social behavior. By exploring, the releasing mechanism selects stimuli, but it leads the excitation to the corresponding AR only, if the stimuli imply novelty. If the organism learns those stimuli, it builds up and stores new or additional invariants. The releasing mechanism becomes an innate releasing mechanism modified by experience or the organism builds up a new acquired releasing mechanism. It cannot release exploration related to these specific stimuli any more, but one must take into account the limitations of "forgetting processes." Additionally, the organism can, as a consequence of learning processes, link those stimuli to other behavior systems or store them in the form of latent learning (Hinde, 1973; Hull, 1952).

The Releasing of Exploration in this Model

Analyses of exploration among a breed of albinotic mice (Persch, 1989) show that, when a mouse is confronted with an unknown object, one can observe the following behavior. The mouse approaches the object slowly with an extremely streched out body and a "stiff" tail. After a short approach, it turns back and cautiously approaches again. While sniffing and touching the object, the mouse can always show turn-back-behavior in between. Some time later the animal will thoroughly explore the object while sniffing and touching it from all sides, climb on it, gnaw on it, and eventually drag it away.

Figure 3. Diagram for the functional organisation of behavior. The release of exploration and escape after perceiving specific stimuli.

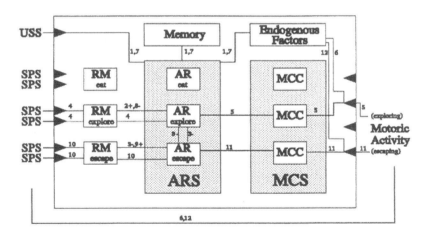

USS	Unspecific stimuli	AR	Action readiness
SPS	Specific stimuli	ARS	Action readiness system
RM	Releasing mechanism	MCC	Motoric coordination center
>	Receptor	MCS	Motoric coordination system
<	Effector		

Further explanations are in the text.

In the model, the unknown object has specific stimuli for exploration as well as for escape (see Figure 3). The influences of endogenous factors, memory, unspecific stimuli (1), and the influences inside the action readiness system lead to a higher AR for exploration than for escape. As a result, the AR activates the releasing mechanism for exploration (2) and inhibits the action readiness for escape and, therefore, the releasing mechanism for escape (3). The selection of the unknown stimuli in the RM for exploration causes an excitation which leads to the action readiness for exploration (4), this activates the corresponding MCC and the motoric activity, exploring, occurs (5). Due to the re-effects of the motoric activity (6) and the other factors (7) this can lead to displacements within the action readiness system. The re-effects can lead to a higher AR for escape which can inhibit the AR for exploration and also the corresponding RM (8). At the same time the releasing mechanism for escape is activated (9), the specific stimuli are worked up and the excitation leads to the action readiness (10). The activation of the corresponding MCC results in the motoric activity of escaping (11). The re-effects of this motoric activity can lead to new displacements within the action readiness system and exploration occurs again (see above). Gradually, if the object is recognized as harmless, the action readiness for escape lowers and the organism can thoroughly explore the object, storing the specific stimuli of the object. As a result of learning processes, the releasing mechanisms for escape and exploration change into IRMEs or the organism builds up new ARMs.

The part of exploration looked until now applies to object-related behavior. A prerequisite is that the organism can perceive specific stimuli, that is, these stimuli must be present. But one also can see spontaneously occuring exploratory behavior in the case the organism cannot perceive specific stimuli. This, then, is appetitive behavior.

Appetitive Behavior in the Behavior System of Exploration

Appetitive behavior is a goal-directed searching behavior carried out to find a stimulus situation belonging to an action readiness (Buchholtz, 1982). In exploration, appetitive behavior is the search for change, which is the same as searching for novelty and, therefore, as searching for specific stimuli belonging to exploration. In a familiar environment, animals can show a lasting appetitive behavior, which can eventually give them the opportunity to explore (Chapman & Levy, 1957; Hinde, 1973; Krechevsky, 1937). Rats even cross an electric grating to reach a labyrinth that contains different objects to explore (Nissen, 1930).

Figure 4. Diagram for the functional organisation of behavior. The release of appetitive and object-related exploratory behavior.

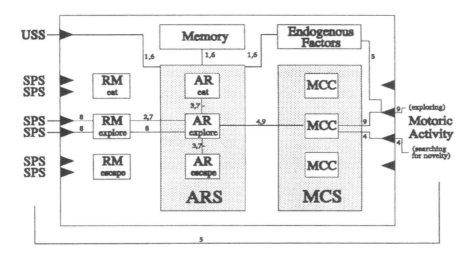

USS	Unspecific stimuli	AR	Action readiness
SPS	Specific stimuli	ARS	Action readiness system
RM	Releasing mechanism	MCC	Motoric coordination center
>	Receptor	MCS	Motoric coordination system
<	Effector		

Further explanations are in the text.

We can explain the releasing of appetitive behavior in this model as follows (see Figure 4). Because of the influences of endogenous factors, memory, and unspecific stimuli, there is a rise in the action readiness for exploration (1). It activates the releasing mechanism for exploration (2) and inhibits other action readinesses (3). While there is a lack of perceived SPS and an excitation from the RM, the AR for exploration activates a MCC which leads to a motoric activity, the search for novelty (4). The re-effects of the motoric activity on the endogenous factors and on the perceiving of stimuli (5) cause new influences leading to the ARS (6). The result is, together with influences of the USS and of memory, a continuing high level of AR for exploration. The AR inhibits the other action readinesses and activates the releasing mechanism for exploration (7). If the organism really perceives unknown specific stimuli while searching, the motoric activity exploring occurs (8, 9). One can now observe the object-related exploratory behavior.

Consequently, the spontaneously appearing appetitive behavior and the object-related behavior both belong to the one behavior system of exploration. The level of action readiness, as a result of the influencing factors, determines the adequate motoric activity within the entire repertoire of an organism's behavioral elements.

References

Barnett, S.A., & Cowan, P.E. (1976). Activity, exploration, curiosity, and fear: An ethological study. *Interdisciplinary Science Reviews, 1*, 43-46.
Becker-Carus, C., Buchholtz, Chr., Etienne, A., Franck, D., Medioni, J., Schöne, H., Sevenster, P., Stamm, R.A., & Tschanz, B. (1972). Motivation, Handlungsbereitschaft, Trieb. *Zeitschrift für Tierpsychologie, 30*, 321-326.
Bower, T.G.R. (1966). Heterogeneous summation in human infants. *Animal Behaviour, 14*, 395-398.
Buchholtz, Chr. (1952). Untersuchungen über das Farbensehen der Hauskatze (Fellis domestica L.). *Zeitschrift für Tierpsychologie, 9*, 462-470.
Buchholtz, Chr. (1973). *Das Lernen bei Tieren. Verhaltensänderungen durch Erfahrung*. Stuttgart: Fischer.
Buchholtz, Chr. (1982). *Grundlagen der Verhaltensphysiologie*. Braunschweig, Wiesbaden: Vieweg.
Butler, R.A. (1953). Discrimination learning by rhesus monkeys to visual exploration motivation. *Journal of Comparative and Physiological Psychology, 46*, 95-98.
Chapman, R.M., & Levy, N. (1957). Hunger drive and reinforcing effect of novel stimuli. *Journal of Comparative and Physiological Psychology, 50*, 233-238.
Eibl-Eibesfeld, I. (1950). Beiträge zur Biologie der Haus- und der Ährenmaus nebst einigen Beobachtungen an anderen Nagern. *Zeitschrift für Tierpsychologie, 7*, 558-587.
Ewert, J.-P. (1976). *Neuro-Ethologie. Einführung in die neurophysiologischen Grundlagen des Verhaltens*. Heidelberger Taschenbücher 181. Berlin, Heidelberg, New York: Springer.
Fox, A. (1973). Physiological and biochemical correlates of individual differences in behaviour of wolf cubs. *Behaviour, 46*, 129-140.
Hinde, R.A. (1973). *Das Verhalten der Tiere I, II*. Frankfurt: Suhrkamp.
Holst, E.v. (1939). Die relative Koordination als Phänomen und als Methode zentralnervöser Funktionsanalysen. *Ergebnisse der Physiologie, 42*, 228-306.
Holst, E.v., & Saint Paul, U.v. (1960). Vom Wirkungsgefüge der Triebe. *Naturwissenschaften, 18*, 409-422.
Hubel, D.H., & Wiesel, T.N. (1962). Receptive fields, binocular interactions, and functional architecture in the cat's visual cortex. *Journal of Physiology, 160*, 106-154.
Hull, C.L. (1952). *A behavior system*. New Haven: Yale University Press.
Immelmann, K. (1983). *Einführung in die Verhaltensforschung*. Berlin, Hamburg: Parey.
Joubert, A., & Vauclair, J. (1986). Reaction to novel objects in a troop of guinea

baboons: Approach and manipulation. *Behaviour, 96*, 92-104.
Krechevsky, I. (1937). Brain mechanisms and variability II. *Journal of Comparative Psychology, 23*, 139-164.
Leong, C.Y. (1969). The quantitative effect of releasers on the attack readiness of the fish Haplochromis burtoni (Cichlidae). *Zeitschrift für Vergleichende Physiologie, 65*, 29-50.
Lettvin, J.Y., Maturana, H.R., McCulloch, W.S., & Pitts, W.H. (1959). What the frog's eye tells the frog's brain. *Proceedings in Instruments of Radio Engeneers, 47*, 1940-1951.
Lorenz, K. (1935). Der Kumpan in der Umwelt des Vogels. *Journal für Ornithologie, 83*, 137-413.
Lorenz, K. (1965). *Über tierisches und menschliches Verhalten. Aus dem Werdegang der Verhaltenslehre. Gesammelte Abhandlungen II.* München: Piper.
Lorenz, K. (1978). *Vergleichende Verhaltensforschung. Grundlagen der Ethologie.* Wien, New York: Springer.
Murphey, R.M., Duarte, F., Noaves, W.C., & Torres Penedo, M.C. (1981). Age group differences in bovine investigatory behavior. *Developmental Psychobiology, 14*, 117-125.
Nissen, H.W. (1930). A study of exploratory behaviour in the white rat by means of the obstruction method. *Journal of Genetic Psychology, 37*, 361-376.
Persch, A. (1989). *Verhaltensanalysen bei der Exploration von Mäusen des Stammes Han:NMRI.* Unpublished Diplomarbeit. Marburg: University of Marburg.
Piechulla, D. (1990). *Die Auswirkungen verschiedener Vorerfahrungen in unterschiedlichen Phasen der frühen Ontogenese auf das Lernverhalten bei einem Mäusestamm (Han:NMRI).* Unpublished Doctoral Dissertation. Marburg: University of Marburg.
Schaffer, W. (1976). *Vergleichende Untersuchungen einer Spiel- und Futterdressur im Rahmen einer operanten Konditionierung bei Iltisfrettchen.* Unpublished Staatsexamensarbeit. Marburg: University of Marburg.
Schleidt, W.M. (1962). Die historische Entwicklung der Begriffe "Angeborenes auslösendes Schema" und "Angeborener Auslösemechanismus" in der Ethologie. *Zeitschrift für Tierpsychologie, 19*, 697-722.
Schmidt, R.F., & Thews, G. (1983). *Einführung in die Physiologie des Menschen.* Berlin, Heidelberg, New York: Springer.
Skinner, B.F. (1974). *Die Funktion der Verstärkung in der Verhaltenswissenschaft.* München: Kindler.
Stolba, A., & Wood-Gush, D.G.M. (1980). Arousal and exploration in growing pigs in different environments. *Applied Animal Ethology, 6*, 382-383.
Tembrock, G. (1983). *Spezielle Verhaltensbiologie der Tiere. Wirbeltiere.* Stuttgart: Fischer.
Tinbergen, N. (1948). Social releasers and the experimental method required for their study. *Wilson Bulletin, 60*, 6-52.
Tinbergen, N. (1951). *The study of instinct.* Oxford: University Press.
Tinbergen, N. (1966). *Tiere und ihr Verhalten. Life, Wunder der Natur.* Nederland: Time-Life International.
Weiss-Bürger, M. (1981). Untersuchungen zum Einfluß des Erkundungs- und Spielverhaltens auf das Lernen bei Iltisfrettchen. *Zeitschrift für Tierpsychologie, 55*, 33-62.
Welker, W.I. (1956). Some determinants of play and exploration in chimpanzees. *Journal of Comparative and Physiological Psychology, 49*, 84-89.
Wünschmann, A. (1963). Quantitative Untersuchungen zum Neugierverhalten von Wirbeltieren. *Zeitschrift für Tierpsychologie, 20*, 80-109.
Zimbardo, P.G., & Montgomery, K.C. (1957). The relative strength of consummatory responses in hunger, thirst, and exploratory drive. *Journal of Comparative and Physiological Psychology, 50*, 504-508.

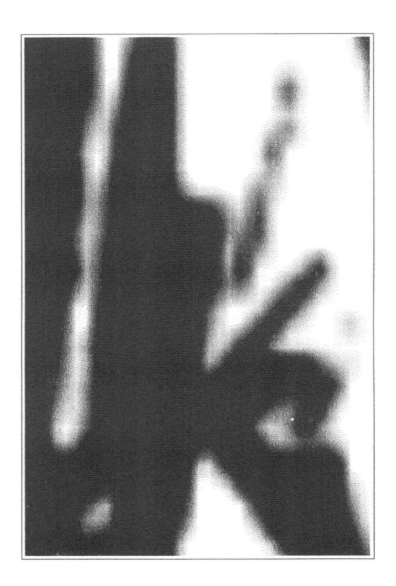

CHAPTER II.3

The Neurobiological Foundation of Exploration

Finn K. Jellestad, Gry S. Follesø, Holger Ursin

The search for the neurobiological foundations of exploratory behavior is interesting for several reasons. Insight in this area may contribute to the understanding of neurobiological mechanisms involved in other related phenomena such as attention, perception, learning, memory, emotion, and their interrelations. However, the search for substrates has clear definitions as a prerequisite. A better understanding of the neurobiological substrate may in itself contribute to an improved conceptual system. We are aware that we inevitably will have to deal with epistemological controversies. We will try to define our concepts as closely as we can to current usage in the neurobiological literature. We realize that a dialogue between neurobiological sciences and cognitive sciences is needed, and that our usage of terms may not make sense to cognitive scientists. Even so, we adhere to the terminology with which we are familiar, but realize that many of the research questions we deal with occur in the literature under different headings and under different research paradigms. We will briefly describe the terms we use, and the type of research situations in which these behaviors are observed. We will then deal specifically with brain areas that are assumed to be of special importance for these behaviors.

Components of the Exploratory Behavioral System

Exploration

A satiated or even a blind animal will, given the opportunity, explore a novel environment. It approaches novel items, and sniffs, bites, licks, looks, touches, and even manipulates discrete and new objects or other environmental features. We will refer to this as exploration.

This phenomenon has been explained by postulating the existence of motivational drives like a curiosity or exploratory drive. Montgomery and Monkman (1955) showed that exploratory behavior is neither explained by a hunger or thirst drive, nor by a general activity drive. Their results provided strong support for the hypothesis that the novelty of a situation arouses an exploratory drive which leads to exploratory behavior.

To define or explain exploratory behavior as "curiosity" or "novelty-seeking behavior" may not be useful for the neuroscientist who is trying to untangle the neural substrates for this particular type of behavior. This calls for an operational approach that builds on description and analysis of the actual behavior its various sequences and consequences. On the other hand, such an approach must be considered an intermediate step, because the neurobiological findings also warrant theories on what brain mechanisms and psychological mechanisms really are involved.

Orienting Responses

Sudden sensory stimulation may also instigate exploratory behavior. Pavlov first described the orienting reflex in 1910, and it is regarded as a part of the complex exploratory behavior of animal and man (Sokolov, 1963). Even though this reflex is a fairly stereotyped response, we will argue that it is not a passive reflex, but an active response. We, therefore, prefer the term orienting response (OR).

OR includes two phases, the first is suppression or arrest of ongoing behaviors. This reduces "noise" from other activities and puts the individual in an optimal state of alertness. The second stage involves orientation of the head and body toward the new stimulus and is usually called the information-seeking phase. Internally, the information-seeking phase is characterized by a slow pupillary dilatation, desynchronized electroencephalographic activity and other changes that reflect a general activation of the sympathetic nervous system.

Brain scientists initially got interested in this response when it was realized that brain stimulation might change the activation level, or level of wakefulness. The first reports on stimulation of the mesencephalic reticular formation dealt with the electroencephalographic (EEG) "activation" response. Ward (1949), and, 335 pages later in the same volume, Moruzzi and Magoun (1949), described this phenomenon to follow electrical stimulation of brain stem reticular structures. In the work that followed, brain stimulation was also done in freely moving animals. Many observers noted behavioral changes characterized by arrest of ongoing movements, and behavior patterns described as "searching" or "attention" (Kaada, 1951; Kaada, Jansen, & Andersen, 1953; H. Ursin & Kaada, 1960). Unfamiliar as the neurophysiologists were with the Pavlovian tradition at that time, it was not until 1960 that Fangel and Kaada (1960) realized that the response was close to, or even identical with, Pavlov's "orienting reflex".

Habituation

With repeated stimulus presentations, the OR will gradually diminish or habituate if the new stimulus has no consequences for the subject. This process serves as a filter for unnecessary sensory information or information that requires no action from the subject.

Rheinberger and Jasper first registered the habituation phenomenon for central nervous system (CNS) activity in 1937. They found that the pattern of electrical activity from all cortical areas showed clear changes in response to both external and internal (autonomic) stimulation. They also observed that this pattern habituated to repetition of the same stimulus and was abruptly brought back by a novel stimulus, a phenomenon that has been called dishabituation.

The orienting response elicited from CNS structures shows the habituation phenomenon unless it is elicited from structures believed to be directly involved in the arousal systems themselves (brain stem reticular formation, intrathalamic nuclei) or with strongly reinforcing areas (septum) or emotional areas (parts of the amygdala) (H. Ursin, Sundberg, & Menaker, 1969; H. Ursin, Wester, & R. Ursin, 1967; Wester, 1971). The electrical stimulation is believed to give rise to unexpected "information," which is treated the same way as when arising from regular sensory stimulation. If contiguous in time to significant events, a conditioned response will occur, if not contiguous to such events, habituation occurs (Sundberg, 1971).

Novelty and Fear

When an animal is introduced to a novel environment or if a change in a familiar environment has taken place, it will not immediately start to explore the new environment or the novel object. Berlyne (1960) did make an distinction between absolute novelty that has a quality which is not experienced before and relative novel or familiar items that are arranged in a nonfamiliar way. Except for strange and unfamiliar odours, adult animals seldom experience absolute novelty. Relative novelty is only possible to define operationally with reference to the animal's earlier experience (see Ennaceur & Delacour, 1988).

In a novel environment, animals will usually exhibit a gradual increase in exploratory activity with increasing time, thus the peak of exploratory activity is usually delayed. Several authors have suggested that this delayed onset reflects fear. In a review article, Russell (1973) discussed two central theories about the relationship between exploratory behavior and fear. The first one is the so-called two-factor theory: A novel stimulus arouses both curiosity and fear, and exploration is seen as the result of competing tendencies to approach or avoid. The other is the Halliday-Lester theory which suggests that novel stimuli arouse fear either in terms of approach (low fear) or in terms of avoidance (strong fear). Russell agreed with Halliday (1966) (and so do we) that exploration may serve to reduce fear, but he refrained from concluding that it is fear which motivates exploration. This issue is crucial, but remains unsolved, for many of the neurobiological findings for exploration.

Situation-Specific Behavior

Exploratory behavior differs in different situations. It depends on whether or not the animal is forced to explore (placed in a novel environment with no opportunity to escape), or is free to make its own choice (enter the novel environment or remain in the familiar) (Welker, 1957). It is also important whether the rat is allowed to enter from its own cage, or is placed in a totally new environment without any reference to previously experienced topographical or spatial marks. In this latter case the behavior may be characterized as looking or searching for the home cage, rather than exploring a new environment. The interpretation and the consequences of brain manipulation seem to vary dependent on these factors being considered (Eilam & Golani, 1989; Walsh & Cummins, 1976).

Other Topics Related to Exploratory Behavior

Exploration of the environment also involves learning about spatial relationships in the subject's surroundings. Blodgett demonstrated the existence of this type of learning in 1929 when he showed that rats that had previously been allowed to explore a maze without any motivational incentive were later superior to naive rats in navigating through the maze to obtain a food reward. This type of learning is usually called latent learning, which is a type of learning that requires memory for topographical and spatial relationships.

Although this chapter will discuss possible neural substrates for exploratory behavior as such, it will also be necessary to touch upon neural substrates for complex behavioral terms like fear, learning, and memory. Our presentation will draw mainly upon knowledge from animal research.

A Neurological Model of Exploration

General Anatomy of Cognitive Information

All primary sensory stimulus modalities (visual, auditive, somatosensory, olfactory, or gustatory) may evoke OR and exploratory behavior, but it is beyond the scope of this chapter to describe the anatomy of these primary modalities in any detail. However, we believe it useful to describe briefly the neuronal circuits that encode the internal representations of these different modalities. These are the so-called association cortices that are usually adjoining their respective primary and secondary sensory areas.

The association cortices are thought to process information at increasing levels of globality. The general rule is that a unimodal association cortex is assigned to a single sensory modality. A unimodal association cortex propagates information to polymodal areas, which in turn may send the information to supramodal areas where a more global level of information processing is assumed to take place (Mesulam, Van Hoesen, Pandya, & Geschwind, 1977).

Information from the primary sensory areas (visual, auditive, and somatosensory) is funnelled into either the frontal or the inferior temporal association cortex, or into both areas. These two association areas are interconnected by a massive fibre bundle called the uncinate fasciculus. The inferior temporal cortex projects to the entorhinal area which is considered to be the cortical gateway to the hippocampus. The inferior temporal cortex also projects to the amygdala. Olfactory information bypasses neocortical areas and is conveyed directly from the primary olfactory cortex to the amygdala and to the hippocampus via the entorhinal cortex (Nauta & Feirtag, 1986). There are, however, differences in the levels of integration in the sensory input that reaches the amygdala and the hippocampus. The hippocampus is thought to receive the most highly processed sensory information since it receives no input from unimodal association cortices, but is directly interconnected with both polymodal and supramodal association areas. The amygdala receives high-level unimodal and polysensory innervation (Amaral, 1987).

Figure 1. Diagram showing sensory information processing at increasing levels of globality. The hippocampal formation receives the most highly processed sensory information from both polymodal and supramodal association areas, whereas the amygdaloid complex receives unimodal and polymodal processed information.

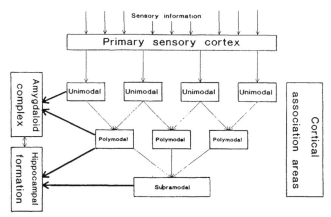

In the following we will therefore concentrate the discussion around the amygdala and the hippocampus, and when necessary, to other closely related anatomical areas.

The Hippocampal Formation

The hippocampus consists of two interlocking layers of cortical cells, the pyramidal cells (CA 1 to 4), and the dentate area. The dentate cells receive information from the entorhinal area and project to the CA cells, from which the information is relayed back to the retrohippocampal area. This loop is believed to be involved in comparing sensory input (Actual Value - AV) with the expected or preset values (Set Value - SV) for that particular event.

There is also a second loop. The pyramidal cells of CA3 project through the fornix to hypothalamus and midline structures. Some of these afferent fibres go to the septal area, which projects back to the hippocampus. This loop is probably involved in producing and regulating the particular electrical activity occurring in the hippocampus during arousal, the theta activity. This loop is, therefore, believed to be involved in the exploration or other behaviors elicited when hippocampus registers unexpected events.

Orienting Response and Habituation

Electrical stimulation of the hippocampus produces "arrest" of locomotion and orienting behavior (Kaada 1951; Kaada, Jansen, & Andersen, 1953). However, stimulation of this structure is very likely to produce electrical after-discharges, epileptic activity, and sensitization to the stimulation ("kindling", Goddard, 1964). Habituation studies of this phenomenon have, to our knowledge, not been carried out.

Following lesion studies, Köhler (1976a, 1976b) concluded that the hippocampus participates in the OR at least to auditory stimuli. However, he emphasized that the afferent control of the hippocampus upon the lateral septum plays a critical role in the habituation of the OR in that damage of both these two structures prolong habituation of OR. Raisman (1966) has shown that it is particulary the CA3 area of the hippocampus that sends fibres to the lateral septal nuclei. Thus, the effect on habituation observed after lateral septal lesions may be due in part to damage of hippocampal efferents.

More recent investigations by Myhrer (1988) are in line with Köhler's findings (Köhler, 1976a, 1976b) in that lateral septum lesioned rats exhibit increased exploration of objects whereas medial septal lesions reduce exploration. Both types of lesions attenuate preference for a novel object as compared to a control group.

Exploration and Novelty

The possible inhibitory role of the hippocampus is still controversial. The initial somato-motor arrest from hippocampus stimulation may be part of the orienting response. However, there are also findings suggesting a far more pronounced inhibitory role for the hippocampus. Gray (1982a), reviving the response inhibition theory of Kaada and McCleary (see Ursin, 1976, for a critical review), considers the septo-hippocampal circuit to be a substrate for the freezing type of fear behavior, which he believes is a reasonable animal model for anxiety. Loss of this inhibition should lead to a less fearful animal, and more rapid exploration.

Indeed, hippocampal lesions usually lead to increased locomotor activity in the open field test. However, lesions of the dorsal hippocampus in rats usually lead to abnormally long duration of exploratory activity - at least one hour, whereas normal rats habituate after 10-15 min (Leaton, 1965; Teitelbaum & Milner, 1963). This also suggests that other possible explanations should be considered, for instance, loss of place memory, loss of recognition memory, or hyperactivity.

The best known alternative hypothesis to that of Gray is the O'Keefe and Nadel (1978) theory of the hippocampus as the substrate for cognitive and spatial maps. This position is compatible with the prolonged exploration time; the lesioned rats have more difficulties in the establishment of cognitive or spatial maps, and require more time exploring the new environment.

There is strong evidence that the hippocampus is directly involved in detecting novelty, and in eliciting the related behavior. According to Sokolov (1963), perception is the matching of sensory input to a neuronal model of the environment. The orienting response occurs when the input is not matching the model. Ample evidence now exists to support the notion that neurons within the hippocampus (CA3) are involved in detection of both novel and familiar stimuli (Vinogradova,

1975), and in eliciting the response to the first exposure of a rat to a novel spatial location (Hill, 1978). O'Keefe and Dostrovsky (1971) first reported place- specific firing in hippocampal cells. These cells have been named "place cells", suggesting that their activity reflects the learning of spatial relations in a particular environment and reflects the ability to distinguish that environment from other environments.

More recently, Rolls and coworkers (1989) recorded the activity of 994 single hippocampal neurons in the monkey during performance in a serial multiple object-place memory task. They found that 2.4% of the neurons responded more the first time a particular object was seen in any position. They concluded that there are neurons in the hippocampus that respond to position in space and that also combine information about stimuli and their position in space, and that *only* respond to a stimulus the *first* time it is seen in a position in space.

There seems to be a reasonable consensus that the two hippocampal loops are involved in aspects of novelty detection and related behavior. There is also reasonable consensus that the two loops have different subroles. The entorhinal-hippocampus loop is involved in the input from the external world, and furnishes the system with the AV which should be compared with the SV. Discrepancies produce arousal or activation (H. Ursin, 1988), in which the second loop is involved. However, there is reason to believe that for the affective aspects of such situations, the amygdala, another limbic structure, is involved.

The main and dominant afferent connections to the hippocampus originate in the entorhinal area. The entorhinal cortex receives projections from several regions of the temporal, frontal, and parietal lobes as well as from the insula (Amaral, 1987). The entorhinal cortex integrates information from association cortices which again receive and integrate information from all sensory modalities and visceromotoric control systems (MacLean, 1975; Swanson, 1983). The information is then transmitted to the hippocampus.

The comparison between the AV and the SV is thought to take place in the subiculum (Gray, 1982b), the area between the hippocampal pyramidal cells and the entorhinal cortex. The AV information reaches the subiculum from a loop that comprises the entorhinal cortex, the dentate area, and CA3-CA1. In the subiculum the AV can be compared with the SV, the predicted information, which may be available from information in the hippocampus (O'Keefe & Nadel, 1978), or from the Papez circle (fornix-mammillary bodies-anteroventral thalamus-cingulate-subiculum; past experience).

Spontaneous Alternation

When an animal is running for two trials in a T-maze, without any specific reinforcement, there is a high tendency for the two responses to alternate if the rat chose the right arm on trial one it will turn left on trial two (Barnett & Cowan, 1976; Douglas, 1967). Such an alternation may be regarded as expressing exploratory behavior. According to O'Keefe and Nadel (1978), alternation behavior can be viewed most simply as reflecting a tendency on the part of the animal to acquire information about unknown parts of its environments. The longer the animal is allowed to explore one part of the environment, the less attraction this area will hold relative to other, unknown areas.

Rats with lesions in the hippocampus, and rats with lesions in the septal area, perseverate their choices in this situation. This finding seems to support notions about the importance of both structures for response inhibition. Dalland (1970) exploited the fact that normal animals alternate the place they go to, or the stimuli they approach (Walker, Dember, Earl, Fawl, & Karoly, 1955). By starting the animals either from North or from South, and letting them choose between East and West, it is possible to distinguish response perseverators (perseverate response, alternate stimulus) and stimulus perseverators (alternate response, perseverate stimulus). Surprisingly, the hippocampus-lesion rats showed the response perseveration, and the rats with septal lesions, the stimulus perseveration phenomenon. Again, there was a striking specificity in the behavioral effect, beyond what was predicted from current theories. One should be careful in ascribing molar functions to chains of neurons. Behavioral effects of lesions in a net of neurons may be different, even if there is only one synapse between them in the chain, as has been demonstrated for the dentate area and the pyramidal cells, within the hippocampus (Livesey & Bayliss, 1975; R. Ursin, H. Ursin, & Olds, 1966).

The Amygdaloid Complex

The amygdala complex consists of several nuclei which may be divided into a phylogenetically old, corticomedial group, and a younger, basolateral group. Fibres to and from overlying cortex course through the complex into the internal capsule and medial brain structures. There is a convergence of fibres in the region of the central nucleus, which consists of a medial, well recognizable part, and a lateral part which is difficult to discriminate from the putamen. The region of the central nucleus is particularly interesting for our discussion. In this area there is also a well defined fibre bundle from the corticomedial part to septal and hypothalamic regions, the stria terminalis, which crosses the previous fibres almost at a right angle. Finally, there is a third, less well known system, also perpendicular to the other two systems, running towards hippocampus and the entorhinal area. This "longitudinal association bundle" communicates with the angular bundle which, again, communicates with the cingulum bundle. This system connects the amygdala with several other important limbic areas, and may be involved in the affective parts of the explorative systems or memory systems.

Orienting Responses and Habituation

Almost all sites stimulated electrically in the amygdaloid complex, at least in the basolateral region, elicit orienting behavior (H. Ursin & Kaada, 1960). Repeated stimulation leads to habituation, unless the site also elicits emotional behavior (flight, defense) at higher intensities (H. Ursin, Wester, & R. Ursin, 1967).

Recently, Gallagher, Graham, and Holland (1990) have pointed out that neurotoxic lesions in the central amygdala nucleus produce a defect in a very specific class of conditioned behavior. The lesions did not alter the unconditioned elicitation of orienting responses, or the habituation of such responses. However, the lesions restricted to the neurons in the central nucleus abolished a conditioned response to the conditioned stimulus, a response which resembled the original orienting response. Holland has postulated earlier that there may be different behavioral mechanisms involved in conditioned responses depending on whether they are generated by the unconditioned or the conditioned stimulus. The results from this group support notions about the importance of the amygdala, particularly the central nucleus, for conditioned or affective aspects of situations evoking orienting or exploration.

Exploration and Amygdala Lesions: Sensory Changes or Emotional?

There are a number of reports of "hyperactivity" as a result of amygdala damage (Corman, Meyer, & Meyer, 1967; Glendenning, 1972; Schwartzbaum & Gay, 1966). Large lesions in the amygdala lead to a loss of aspects of sensory discrimination. Lesions in the amygdala complex also lead to taming effects, reduced flight or defense behavior (or both), and defective avoidance learning. These deficits affect both active and passive types of avoidance behavior. Lesions producing passive avoidance deficits also produce other types of perseveratory or "disinhibitory" responses. These behavioral changes are well established, and form important aspects of the "Klüver-Bucy syndrome."

Changes in exploratory behavior, therefore, may be explained in terms of emotional changes, particularly in fear, or in terms of response disinhibition, or as a consequence of loss of sensory abilities. We will discuss these options separately.

Sensory Changes

The "hypermetamorphosis" and "visual blindness" described by Klüver and Bucy support the latter explanation. Monkeys without the ability to discriminate visually put everything in their mouth and seem to try to analyze the objects that way. They still are unable to discriminate even simple food types (H. Ursin, Rosvold, & Vest, 1969). The most significant losses of visual discrimination are due to lesions of the inferotemporal cortex (Mishkin 1954; Mishkin & Delacour, 1975). It is, however, still an open issue whether the amygdala or parts of it contribute to even higher forms of sensory discrimination.

Amygdala neurons respond to visual, auditory, somesthetic, and gustatory stimuli. Ono, Fukuda, Nishijo, and Nakamura (1988) have suggested that amygdala neurons are involved in recognizing the affective significance of a stimulus. Some neurons respond to physical properties of objects, some to the affective significance (rewarding or nonrewarding), some to both aspects. The Ono et al. data suggest that visual information is sequentially processed from the inferotemporal cortex to the lateral hypothalamus. The inferotemporal cortex seems involved in the analysis of physical properties, the amygdala in the association of affective significance to the stimuli.

Visual orienting toward unfamiliar stimuli is greatly reduced by amygdala damage (Bagshaw, Mackworth, & Pribram, 1972). On the other hand, it is possible to teach the amygdala-lesioned animals to attend to visual stimuli like normal animals with reward (Schwartzbaum & Pribram, 1960). It may be that amygdala lesions alter the animal's "willingness" to respond to novel events unless there is a substantial reward for doing so (Isaacson, 1982, p.112). Again, there seems to be a consensus that whatever role the amygdala is playing for exploratory behavior and information processing, it is related to the affective significance of the stimulus.

Emotional Changes

Werka, Skår, and H. Ursin (1978) placed lesions in specific areas in the amygdala and overlying cortex in rats. Three different types of lesions were tested: central, basolateral, and cortex lateral to the amygdala. Lesions restricted to the central nucleus produced increased activity on all parameters studied in an open-field test, whereas the two other lesioned groups were not changed. In one-way active avoidance, all three groups with lesions showed deficits. The most pronounced change was observed in the central group. All groups showed the same degree of retention loss, but in forced extinction of one-way active avoidance after retraining, the cortical and basolateral groups were most defective. A fear-reduction hypothesis was proposed for the central lesion, in line with previous neuroendocrine work (Coover, H. Ursin, & Levine, 1973). The basolateral and cortical areas may be more specifically involved in passive avoidance behavior (Jellestad & Bakke, 1985; Jellestad, Markowska, Bakke, & Walther, 1986). Jellestad et al. (1986) found that ibotenic acid lesions (destroying only neurons and sparing passing fibres) restricted to the central did not produce any passive avoidance impairments and that additional destruction of fibres (electrolytic lesions) was required to produce an impairment that only affected the initial phase of the passive avoidance acquisition. Both ibotenic acid and electrolytical central amygdala lesions increased exploratory activity in an open field with a home cage. This increase in activity could not be attributed to inhibitory control (McCleary, 1966), because both groups learned the passive avoidance response, although at a slower rate in the electrolytic group.

The relative roles of the central and the basolateral nuclei for fear, passive avoidance, and exploration remain an open issue. Based on appetitive conditioning, Kesner, Walser, and Winzenried (1990) suggested that central, but not basolateral amygdala mediates memory for positive affective experiences. LeDoux and collaborators (LeDoux, Cicchetti, Xagoraris, & Romanski, 1990; LeDoux, Iwata, Cicchetti, & Reis, 1988) concluded from studies of conditioned emotional responses that the lateral nucleus is an essential link for endowing affective properties to auditory stimulation, whereas the central nucleus mediates the autonomic and the behavioral concomitants of conditioned fear differentially to the lateral hypothalamus and the midbrain central grey, respectively. This also suggests a reasonable explanation for the Coover, Werka, and Jellestad studies, with separate, but related, roles for the lateral and central nuclei for fear behavior, and, therefore, also for exploration involving fear or fear reduction.

Conclusion

The changes in exploratory behavior produced by amygdala lesions may still be explained in terms of emotional changes, particularly in fear, or in terms of response disinhibition, or as a consequence of loss of sensory abilities. No final conclusion is offered by the literature. However, we believe that there is a convergence in the literature, and the options may not necessarily be incompatible. The amygdala contains neurons that are involved in the affective aspects of stimulation (Nishijo, Ono & Nishino, 1988). The role of the nuclei seems to be to recognize the affective significance of stimuli, this is why amygdala affects memory (Aggleton & Mishkin, 1986). This will also affect exploratory behavior. The amygdala neurons are important both for the drive to explore, and for the association between neutral stimuli, and those stimuli carrying important signals for the individual.

Specific Transmitter Systems

There is a rapidly growing literature on specific transmitter systems in the brain, and, also, on CNS-mediated effects of a series of biologically active peptides. These signal systems are to a great extent common for the CNS, the immune system, and for the endocrine systems. We will not attempt to review this literature, even though exploration is used as a research tool in some of this research.

Some of the findings are supplementing our knowledge of specific structures discussed above. For the hippocampus, and related diencephalic areas, much of the attention has been focussed on the general memory effects from manipulation of the cholinergic system on memory, and, lately, also from manipulation of the serotonergic input to the hippocampus (Vanderwolf, 1987).

The amygdala complex is unusually rich in the number of transmitters and biologically active peptides found in the neuropile. This is particularly true for the central nucleus. There is no clear story on the relations between these biologically active substances and the functions of the complex. However, a few points seem reasonably well established.

Rodgers and File (1979) proposed that within the amygdala complex there are two distinct naloxone-sensitive opiate systems involved in modulating behavior responses to different kinds of stimulation. Morphine injected into the central nucleus produced naloxone-reversible reductions in exploration and activity, whereas similar injections in the medial nucleus produced a reduced exploration that was not reversed by naloxone. Greidanus, Croiset, Bakker, & Bouman (1979) reported that amygdaloid lesions blocked the effect of neuropeptides (vasopressin, adrenocorticotropic hormone) on the extinction of a conditioned avoidance response. Cholinergic components of the amygdaloid complex may be involved in the mediation of escape-avoidance behavior (Grossmann, 1972). Gallagher, Kapp, Frysinger, & Rapp (1980) also have suggested a role for an amygdala beta-adrenergic system in memory.

The most striking relationship, however, is the decisive influence of ascending catecholamine projections from the reticular formation of the brain stem. It has long been known that structures within or close to the brain stem reticular formation are involved in the regulation of arousal or activation, and, therefore, also in processes like exploration. In the last decade, particular interest has been focused upon the ascending catecholamine pathways. Both the noradrenergic neurons in the locus coerulues and the dopamine neurons in the ventral tegmentum of the midbrain are necessary for normal exploration and attention to external stimuli (see Clark, G.M. Geffen, & L.B. Geffen, 1987, for an extensive review). The relationship between these neuronal systems and serious psychopathology and psychosis is an obvious motivation for the intensive studies in this area. We have not reviewed this literature here because we believe these systems to be more basically related to the very function of staying awake and regulate the arousal or activation level, rather than being specifically involved in exploration as such.

Other Structures

The hypothalamus has an influence on other brain areas by altering the tendency to begin behavioral acts. In a sense, it influences the willingness of the animal to stop ongoing behaviors and to undertake different ones (Isaacson, 1982). These characteristics are shared by other structures.

Cholinergic stimulation of the midline of the thalamic nuclei affects fear responses in rats (D.C. Blanchard & R.J. Blanchard, 1972). Stokes and Best (1985) found that dorsomedial thalamic lesions produced postoperative impairment of radial maze performance, brightness-discrimination learning, and changes in open-field activity. The role of the dorsomedial thalamus in memory is not yet clearly understood. Through its connections to frontal cortical and hippocampal systems, a role in spatial information processing might be postulated.

The frontal lobes may be involved, particularly in situations that engage spatial memory. This may be related to, but is not identical to, similar functions of the dorsomedial thalamic nucleus, which projects to the frontal lobe (Kolb, Pittman, Sutherland, & Whishaw, 1982). There are also similarities between spatial exploration and problem solving in rats with lesions in the frontal lobes and lesions in the septal area (related to the hippocampus area) (Poucet & Herrmann, 1990).

Conclusion

Many brain structures are involved in the acquisition of information from the environment. We have concentrated on brain circuits believed to be involved in the exploration of the unknown. This requires identification of the fact that the situation is new, there has to be a template of what is known which is to be compared with the actual stimulus configuration.

In general arousal or stress theory, one of us (H. Ursin, 1978, 1988) has suggested that the brain functions as a homeostatic system for information. Whenever there is a discrepancy between the set values and the actual values for a given variable, activation or arousal is produced. This response serves to turn itself off. Behavior is elicited which should lead to the discrepancy being eliminated, and normal arousal levels re-established.

We have concentrated on the two structures we believe to be the most important for these functions. The hippocampus network seems involved in the recognition of novelty, and the initial elicitation of activity in diencephalic areas. The amygdala is more involved in the affective aspects of exploratory behavior.

References

Aggleton, J.P., & Mishkin, M. (1986). The amygdala: Sensory gateway to the emotions. In R. Plutchik and H. Kellerman (Eds.), *Emotion: Theory, research, and experience.* (pp. 281-299) New York: Academic Press.

Amaral, D.G. (1987). Memory: Anatomical organization of candidate brain regions. In F. Plum (Ed.), *Handbook of physiology: Section 1. The nervous system: Vol. 5. Higher functions of the brain, Part 1.* (pp. 211-294). Bethesda, MD: American Physiological Society.

Bagshaw, M.H., Mackworth, N.H., & Pribram, K.H. (1972). The effect of resections of the inferotemporal cortex and of the amygdala on visual orienting and habituation. *Neuropsychology, 10*, 153-158.

Barnett, S.A., & Cowan, P.E. (1976). Activity, exploration, curiosity, and fear: An ethological study. *Interdisciplinary Science Reviews, 1, 43-62.*

Berlyne, D.E. (1960). *Conflict, arousal, and curiosity.* New York: McGraw Hill.

Blanchard, D.C., & Blanchard, R.J. (1972). Innate and conditioned reactions to threat in rats with amygdaloid lesions. *Journal of Comparative and Physiological Psychology, 81*, 281-290.

Blodgett, H.C. (1929). The effect of the introduction of reward upon the maze performance of rats. *University of California Publication in Psychology, 4*, 113-134.

Clark, C.R., Geffen, G.M., & Geffen, L.B. (1987). Catecholamines and attention: 1. Animal and clinical studies. *Neuroscience and Biobehavioral Reviews, 11*, 341-352.

Coover, G., Ursin, H., & Levine, S. (1973). Corticosterone and avoidance in rats with basolateral amygdala lesions. *Journal of Comparative and Physiological Psychology, 23*, 716-726.

Corman, P.E., Meyer, P.M., & Meyer, D.R. (1967). Open-field activity and exploration in rats with septal and amygdaloid lesions. *Brain Research, 5*, 469-476.

Dalland, T. (1970). Response and stimulus perseveration in rats with septal and dorsal hippocampal lesions. *Journal of Comparative and Physiological Psychology, 1*, 114-118.

Douglas, R.J. (1967). The hippocampus and behavior. *Psychological Bulletin, 67*, 416-442.

Eilam, D., & Golani, I. (1989). Home base behavior of rats (Rattus norvegicus) exploring a novel environment. *Brain Research, 34*, 199-211.

Ennaceur, A., & Delacour, J. (1988). A new one-trial test for neurobiological studies of memory in rats: 1. Behavioral data. *Behavioral Brain Research, 31*, 47-59.
Fangel, C., & Kaada, B.R. (1960). Behavior "attention" and fear induced by cortical stimulation in the cat. *Electroencephalography and Clinical Neurophysiology Journal, 12*, 575-588.
Gallagher, M., Graham, P.W., & Holland, P.C. (1990). The amygdala central nucleus and appetitive pavlovian conditioning: Lesions impair one class of conditioned behavior. *The Journal of Neuroscience, 10*, 1906-1911.
Gallagher, M., Kapp, B.S., Frysinger, R.C., & Rapp, P.R. (1980). ß-Adrenergic manipulation in amygdala central nucleus alters rabbit heart rate conditioning. *Pharmachology, Biochemistry, and Behavior, 12*, 419-426.
Glendenning, K.K. (1972). Effects of septal and amygdaloid lesions on social behavior of the cat. *Journal of Comparative and Physiological Psychology, 80*, 199-207.
Goddard, G.V. (1964). Amygdaloid stimulation and learning in the rat. *Journal of Comparative and Physiological Psychology, 58*, 23-30.
Gray, J.A. (1982a). Precis of the neuropsychology of anxiety: An enquiry into the functions of the septo-hippocampal system. *Behavioral Brain Science, 5*, 469-484.
Gray, J.A. (1982b). *The neurophysiology of anxiety*. New York: Oxford University Press.
Greidanus, T.B.V.W., Croiset, G., Bakker, E., & Bouman, H. (1979). Amygdaloid lesions block the effect of neuropeptides (vasopressin, ACTH) on avoidance behavior. *Physiology & Behavior, 22*, 291-295.
Grossman, S.P. (1972). The role of the amygdala in escape-avoidance behaviors. In B.E. Eleftheriou (Ed.), *The neurobiology of the amygdala* (pp. 537-551). New York: Plenum.
Halliday, M.S. (1966). Exploration and fear in the rat. *Symposia- Zoological Society of London, 18*, 45-59.
Hill, A.J. (1978). First occurrence of hippocampal spatial firing in a new environment. *Experimental Neurology, 62*, 282-297.
Isaacson, R.L. (1982). *The limbic system* (2nd ed.). New York: Plenum Press.
Jellestad, F.K., & Bakke, H.K. (1985). Passive avoidance after ibotenic acid and radio frequency lesions in the rat amygdala. *Physiology & Behavior, 34*, 299-305.
Jellestad, F.K., Markowska, A., Bakke, H.K., & Walther, B. (1986). Behavioral effects after ibotenic acid, 6-OHDA, and electrolytic lesions in the central amygdala nucleus of the rat. *Physiology & Behavior, 37*, 855-862.
Kaada, B.R. (1951). Somato-motor, autonomic, and electrocorticographic responses to electrical stimulation of "rhinencephalic" and other structures in primates, cat and dog. *Acta Physiologica Scandinavica, 24 ,(Suppl. 83)*, 1-285.
Kaada, B.R., Jansen, J.,Jr., & Andersen, P. (1953). Stimulation of hippocampus and medial cortical areas in unanesthetized cats. *Neurology, 3*, 844-857.
Kesner, R.P., Walser, R.D., & Winzenried, G. (1990). Central but not basolateral amygdala mediates memory for positive affective experiences. *Behavioral Brain Research, 33*, 189-95.
Köhler, C. (1976a). Habituation after dorsal hippocampal lesions: A test dependent phenomenon, *Behavioral Biology, 18*, 89-110.
Köhler, C. (1976b). Habituation of the orienting response after medial and lateral septal lesions in the albino rat. *Behavioral Biology, 16*, 63-72.
Kolb, B., Pittman, K., Sutherland, R.J., & Whishaw, I.Q. (1982). Dissociation of the contributions of the prefrontal cortex and dorsomedial thalamic nucleus to spatially guided behavior in the rat. *Behavioral Brain Research, 6*, 365-378.

Leaton, R.N. (1965). Exploration behavior in rats with hippocampal lesions. *Journal of Comparative and Physiological Psychology, 59*, 325-330.
LeDoux, J.E., Cicchetti, P., Xagoraris, A., & Romanski, L.M. (1990). The lateral amygdaloid nucleus: Sensory interface of the amygdala in fear conditioning. *The Journal of Neurosciences, 10*, 1062-1069.
LeDoux, J.E., Iwata, J., Cicchetti, P., & Reis, D. (1988). Different projections of the central amygdaloid nucleus mediate autonomic and behavioral correlates of conditioned fear. *The Journal of Neurosciences, 8*, 2517-2529.
Livesey, P.J., & Bayliss, J. (1975). The effects of electrical (blocking) stimulation to the dentate of the rat on learning of a simultaneous brightness discrimination and reversal. *Neuropsychologia, 13*, 395-407.
MacLean, P.D. (1975). An ongoing analysis of hippocampal inputs and outputs: Microelectrode and neuroanatomical findings in squirrel monkeys. In R.L. Isaacson & K.H. Pribram (Eds.), *The hippocampus* (Vol. 1, pp. 177-211). New York: Plenum Press.
McCleary, R.A. (1966). Response-modulating functions of the limbic system: Initiation and suppression. In E. Stellar & J.M. Sprague (Eds.), *Progress in physiological psychology* (Vol. 1, pp. 209-272). New York: Academic Press.
Mesulam, M.M., Van Hoesen, G.W., Pandya, D.N., & Geschwind, N. (1977). Limbic and sensory connections of the inferior parietal lobule (area PG) in the rhesus monkey: A study with a new method for horseradish peroxidase histochemistry. *Brain Research, 136*, 393-414.
Mishkin, M. (1954). Visual discrimination performance following partial ablations of the temporal lobe: 2. Ventral surface vs. hippocampus. *Journal of Comparative and Physiological Psychology, 47*, 187-193.
Mishkin, M., & Delacour, J. (1975). An analysis of short-term visual memory in the monkey. *Journal of Experimental Psychology and Animal Behavioral Processes, 1*, 326-334.
Montgomery, K.C., & Monkman, J.A. (1955). The relation between fear and exploratory behavior. *Journal of Comparative and Physiological Psychology, 48*, 132-136.
Moruzzi, G., & Magoun, M. (1949). Brain stem reticular formation and activation of the EEG. *Electroencephalography and Clinical Neurophysiology Journal, 1*, 455-473.
Myhrer, T. (1988). Exploratory behavior and reaction to novelty in rats: Effects of medial lateral septal lesions. *Behavioral Neuroscience, 103*, 1226-1233.
Nauta, W.J.H., & Feirtag, M. (1986). *Fundamental neuroanatomy*. New York: W.H. Freeman and Company.
Nishijo, H., Ono, T., & Nishino, H. (1988). Single neuron responses in amygdala of alert monkey during complex sensory stimulation with affective significance. *The Journal of Neuroscience, 8*, 3570-3583.
O'Keefe, J., & Dostrovsky, J. (1971). The hippocampus as a spatial map: Preliminary evidence from unit activity in the freely-moving rat. *Brain Research, 34*, 171-175.
O'Keefe, J., & Nadel, L. (1978). *The hippocampus as a cognitive map*. Oxford: Clarendon Press.
Ono, T., Fukuda, M., Nishijo, H., & Nakamura, K. (1988). Plasticity in inferotemporal cortex-amygdala-lateral hypothalamus axis during operant behavior of the monkey. In C.C. Woody, D.L. Alkon, & J.L. McGaugh (Eds.), *Cellular mechanisms of conditioning and behavioral plasticity*. New York: Plenum Press.
Poucet, B., & Herrmann, T. (1990). Septum and medial frontal cortex contribution to spatial problem-solving. *Behavioral Brain Research, 37*, 269-280.

Raisman, G. (1966). The connections of the septum. *Brain, 89,* 317-348.

Rheinberger, M.B., & Jasper, H.H. (1937). The electrical activity of the cerebral cortex in the unanestethized cat. *American Journal of Physiology, 119,* 186-196.

Rodgers, J.D., & File, S.A. (1979). Exploratory behavior and aversive thresholds following intra-amygdaloid application of opiates in rats. *Pharmacology & Behavior, 11,* 505-511.

Rolls, E.T., Miyashita, Y., Cahusac, P.M.B., Kesner, R.P., Niki, H., Feigenbaum, J.D., & Bach, L. (1989). Hippocampal neurons in the monkey with activity related to the place in which a stimulus is shown. *The Journal of Neuroscience, 9,* 1835-1845.

Russell, P.A. (1973). Relationships between exploratory behavior and fear: A review. *British Journal of Psychology, 61,* 417-433.

Schwartzbaum, J.S., & Gay, P.E. (1966). Interacting behavioral effects of septal and amygdaloid lesions in the rat. *Journal of Comparative and Physiological Psychology, 61,* 59-65.

Schwartzbaum, J.S., & Pribram, K.H. (1960). The effects of amygdalectomy in monkeys on transposition along a brightness continuum. *Journal of Comparative and Physiological Psychology, 53,* 396-399.

Sokolov, E.N. (1963). *Perception and the conditioned reflex.* New York: Pergamon Press.

Stokes, K.A., & Best, P.J. (1985). Dorsomedial thalamic lesions: Post-operative impairment of radial maze performance, brightness discrimination learning, and changes in open-field activity. *Society for Neuroscience Abstracts, 11,* 833.

Sundberg, H. (1971). Electrical stimulation of the brain and learning in the cat. *Acta Physiologica Scandinavica,* (Abstract No 28-2A).

Swanson, L.W. (1983). The hippocampus and the concept of the limbic system. In W. Seifert (Ed.), *Neurobiology of the hippocampus.* London: Academic Press.

Teitelbaum, H., & Milner, P.M. (1963). Activity changes following partial hippocampal lesions in rats. *Journal of Comparative and Physiological Psychology, 56,* 284-289.

Ursin, H. (1976). Inhibition and the septal nuclei: Breakdown of the single concept model. *Acta Neurobiologiae Experimentalis, 36,* 91-115.

Ursin, H. (1978). Activation, coping, and psychosomatics. In H. Ursin, E. Baade, & S. Levine (Eds.), *Psychobiology of stress: A study of coping men* (pp. 201-228). New York: Academic Press.

Ursin, H. (1988). Expectancy and activation: An attempt to systematize stress theory. In D. Hellhammer, I. Florin, & H. Weiner (Eds.), *Neurobiological approaches to human disease* (pp. 313-334). Toronto, Huber.

Ursin, H., & Kaada, B.R. (1960). Subcortical structures mediating the attention response induced by amygdala stimulation. *Experimental Neurology, 12,* 1-20.

Ursin, H., Rosvold, H.E., & Vest, B. (1969). Food preference in brain lesioned monkeys. *Physiology & Behavior, 4,* 609-612.

Ursin, H., Sundberg, H., & Menaker, S. (1969). Habituation of the orienting response elicited by stimulation of the caudate nucleus in the cat. *Neuropsychologia, 7,* 313-318.

Ursin, H., Wester, K., & Ursin, R. (1967). Habituation to electrical stimulation of the brain in unanesthetized cats. *Electroencephalography and Clinical Neurophysiology Journal, 23,* 41-49.

Ursin, R., Ursin, H., & Olds, J. (1966). Self-stimulation of hippocampus in rats. *Journal of Comparative and Physiological Psychology, 61,* 353-359.

Vanderwolf, C.H. (1987). Near-total loss of "learning" and "memory" as a result of combined cholinergic and serotonergic blockade in the rat. *Behavioral Brain Research, 23,* 43-57.

Vinogradova, O.S. (1975). Functional organization of the limbic system in the process of the registration of information: Facts and hypothesis. In R.L. Isaacson & K.H. Pribram (Eds.), *The hippocampus* (Vol. 2, pp. 3-69). New York: Plenum.

Walker, E.L., Dember, W.N., Earl, R.W., Fawl, C.L., & Karoly, A.J. (1955). Choice alternation: 3. Response intensity vs. response discriminability. *Journal of Comparative and Physiological Psychology, 48,* 80-85.

Walsh, R.N., & Cummins, R.A. (1976). The open-field test: A critical review. *Psychological Bulletin, 83,* 482-504.

Ward, A.A., Jr. (1949). The relationship between the bulbar-reticular suppressor region and the EEG. *Electroencephalography and Clinical Neurophysiology Journal, 1,* 120.

Welker, W.I. (1957). "Free" versus "forced" exploration of a novel situation by rats. *Psychological Reports, 3,* 95-108.

Werka, T., Skår, J., & Ursin, H. (1978). Exploration and avoidance in rats with lesions in amygdala and piriform cortex. *Journal of Comparative and Physiological Psychology, 92,* 672-681.

Wester, K. (1971). Habituation to electrical stimulation of the brain in unanesthetized cats. *Electroencephalography and Clinical Neurophysiology Journal, 30,* 52-61.

CHAPTER II.4

Two Characteristics of Surprise: Action Delay and Attentional Focus

Michael Niepel, Udo Rudolph, Achim Schützwohl, Wulf-Uwe Meyer

Surprise is a syndrome of reactions which is conceived by many authors as a primitive (Descartes, 1649/1911), primary (Plutchik, 1980), or fundamental (Izard, 1977) emotion. It consists of physiological changes, specific behavioral patterns (including a distinct facial expression), and subjective experience. Surprise is assumed to interrupt other ongoing processes, to direct attention to the eliciting stimulus, and thus to enable the organism to respond adaptively to sudden changes in its environment.

Surprise is elicited by events that deviate from the organism's schemata. A schema is part of a knowledge structure which is activated by a given stimulus (Rumelhardt, 1984; Rumelhardt, Smolensky, McClelland, & Hinton, 1986). Schemata form "a kind of informal, private, unarticulated theory about the nature of events, objects or situations. The total set of schemata we have available for interpreting our world in a sense constitutes our private theory about the nature of reality" (Rumelhardt, 1984, p. 166). Thus, schemata help us to "understand" situations, events, and objects. Furthermore, in order to provide an accurate account of our surroundings and internal states, schemata have to be continuously monitored as to their compatibility with the data presently available. As long as a schema is compatible with the situations and events that we experience, there is no need to revise it. However, if a discrepancy between schema and input occurs, surprise is elicited. The main function of surprise is to enable processes that help to remove this discrepancy. Interruption of ongoing activities and focusing of attention on the surprising event serve this purpose.

The interruption of other ongoing processes as a major characteristic of surprise has been emphasized by many authors. Tomkins (1962) for example describes surprise as "a general interrupter of ongoing activity. This mechanism is similar in kind and function to that in a radio or television network which enables special announcements to interrupt any ongoing program" (p. 498). Izard and Buechler (1980) stress that surprise replaces other emotions. Interruption obviously puts the organism in a position to attend to the surprising stimulus and thus enables the unhindered execution of processes that help to remove the discrepancy between event and schema.

Involuntary focusing of attention as a characteristic of surprise has already been stressed by Descartes (1649/1911) and Darwin (1872). Focusing one's attention on the surprising stimulus is a prerequisite for its analysis and evaluation. This analysis presumably starts with a verification of one's own perception of this stimulus. Obviously revision of a schema without such a verification would be inappropriate to adaption.

Surprising events often seem to give rise to exploration, that is to why-questions and a tendency for causal search in order to identify the causes of these events (Darwin, 1872; Isaacs, 1930; Meyer, 1988); an understanding of the causal structure underlying the events often seems necessary in order to decide whether and to what extent a schema is to be revised. Ideally, such search processes will lead to a subjectively valid attribution of the schema-discrepant event. In the light of this attribution an extension, correction, or restructuring of the relevant schema may be required. Such revisions of a schema contribute to effective action control. They enable the individual (a) to anticipate and control future occurrences of the previously discrepant event (i.e., to bring it about or to prevent it), (b) to avoid the event if it turns out to be negative and uncontrollable, or (c) to ignore the event if it proves to be irrelevant with respect to further action.

In sum, surprise in contrast to other emotions (see Scherer, 1984) does not require elaborate analyses of an event. It is rather based on a simple, continuously executed compatibility check of events and schemata. Surprise serves an effective action control: It consists of and causes processes that may result in a revision of our action guiding schemata.

Few people (neither the proverbial "man in the street" nor the many philosophers and psychologists who have been speculating about surprise) would disagree with these propositions. However, to our knowledge there exists with some few exceptions no systematic research on the interrelation between surprise and attention or surprise and performance (interruption). Among these exceptions is the work of Charlesworth (1964) on the relation between surprise and curiosity behavior. He showed that a surprising (unexpected) event can instigate curiosity behavior as manifested in the tendency to repeat such events. In general, however, these propositions seem to be based mainly on common sense or anecdotal observations.

The Experimental Paradigm

In the beginning of our research on surprise, we focused on developing an experimental paradigm that would allow us to perform a systematic analysis of this emotion. The paradigm that has emerged consists of two stages. The first stage serves the generation of a specific schema in our subjects. In the second stage, subjects are confronted with an event that does not correspond to this schema an event that is schema-discrepant. Interruption (action delay) is measured by reaction times (RT), and attentional focus by memory performance.

Throughout the experiment subjects perform a choice reaction time task. In each trial two words are simultaneously presented on a computer screen, one word above the other. Each word pair is displayed for 3 s. During the presentation of the words, a dot appears for 0.1 s either above the upper word or below the lower word. Subjects are instructed to press as quickly as possible one of two response keys depending on the position of the dot. Stage 1 of the experiment consists of a number of trials whose uniformity is expected to establish a schema with respect to the mode of presentation of the words. In the experimental group, both words are presented as black letters on a white ground (*normal video*, NV). In the last trial (Stage 2), one of the two words is presented in inverted typescript (*reverse video*, RV) that is, as white letters on a black ground. This event is supposed to be discrepant to the schema established in Stage 1 and thus to elicit surprise. In the control group, the same event occurs in the last trial. However, the event is not schema-discrepant in this group because in each of the preceding trials (Stage 1) one of the two words has been presented in RV mode.

Immediately after the last trial, subjects are given an unexpected memory test. They are asked to recall the two words and the position of the dot in the last trial. Reaction times for each trial are registered by a computer.

Using this prototype experimental procedure, we conducted a number of experiments, some of which will be summarized here.

Effects of Schema Discrepancy

In the first experiment (Meyer, Niepel, Rudolph & Schützwohl, 1991) we investigated the effects of introducing a RV word after 29 preceding trials (containing no RV word) on action delay and attentional focus. We expected that in Trials 1 to 29 a schema would be established with respect to the mode of presentation of the words (black letters on white ground, NV). The presentation of a RV word on Trial 30 was supposed to be schema-discrepant and to elicit two reactions: First, a delay in executing the required action (key pressing response) and, second, a focusing of attention on the discrepant event, which should lead to enhanced memory performance for the RV word.

Each word pair was presented for 3 s. Half a second after the onset of the words a dot appeared for 0.1 s either above the upper or below the lower word. Subjects were 51 male and female students (experimental group: $n=34$; control group: $n=17$).

Figure 1. Mean reaction times in Trials 21 - 29 and in Trial 30.

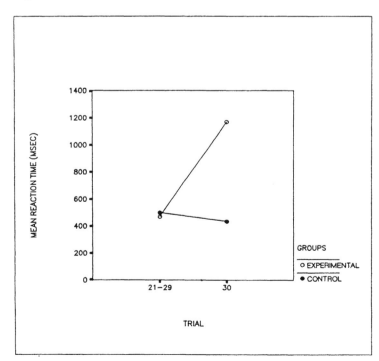

Figure 1 shows the mean reaction times (RT) for Trials 21 to 29 and for the critical Trial 30. The figure indicates a strong increase of RT on Trial 30 in the experimental group compared to the control group[1].

Apparently, the discrepant event occurring in Trial 30, causes a pronounced delay in executing the choice reaction time task in the experimental group.

Table 1. Percentage of Subjects who recall the RV Word and the Position of the Dot

	RV word	Position of dot
Experimental Group ($n = 34$)	56	68
Control Group ($n = 17$)	0	6

Note. Recall for the position of the dot was tested using a format offering three alternatives: "above", "below", and "don't know".

The results of the memory test performed immediately after Trial 30 are shown in Table 1. Most experimental subjects recall the RV word as well as the position of the dot, whereas most of the control subjects do not[2].

First of all, these findings support the assumption that the surprise reaction, elicited by a schema-discrepant event, causes a delay in the execution of an ongoing activity. Furthermore, the enhanced memory

---

[1] An analysis of variance (ANOVA) revealed a main effect of the within-group factor Trials, $F(1,48) = 11.02$, $p<.01$, and a main effect of the factor Groups, $F(1,48) = 12.09$, $p<.01$. However, these main effects were qualified by a significant two-way interaction of Trials X Groups, $F(1,48) = 15.45$, $p<.001$. A further detailed analysis of this interaction revealed that the difference between the two groups was only significant on Trial 30, $t(35) = 5.21$, $p<.001$. Furthermore, the differences between the RTs of Trials 21-29 and Trial 30 were significant in the experimental group, $t(32) = 5.10$, $p<.001$, and in the control group, $t(16) = 2.40$, $p<.05$.

[2] Memory performance of the experimental subjects was superior to the control subjects with regard to the RV word, $\chi^2 (1, N = 51) = 12.84$, $p<.001$, and the position of the dot, $\chi^2 (1, N = 51) = 14.96$, $p<.001$. The χ^2-tests are based on cell frequencies.

performance for the RV word in the experimental group is congruent with the proposition that the surprise reaction includes a focusing of attention on the schema-discrepant event (the RV word). Unexpectedly, memory for the position of the dot also differed significantly between the two groups. We shall discuss this finding later.

Duration of the Surprise Reaction

Many authors have suggested that the surprise reaction is of a relatively short duration (e.g., Tomkins, 1962). In order to find out how long the reaction would last under the circumstances created in our experimental paradigm, we varied the interval between the onset of the unexpected event (RV word) and the action-relevant dot. The surprise reaction was assumed to be finished when no increase in RT to the dot occurred.

Three experimental and three corresponding control groups were formed, using word-dot intervals of 0.5 s, 1 s, or 2 s, respectively. A total of 48 male and female students were randomly assigned to these six groups. Except for the word-dot interval, the same procedure as in the first experiment was employed. The first stage of the experiment consisted of 32 trials; in Trial 33 (the critical trial) a RV word was introduced in the experimental groups.

As expected there was a strong relationship between length of the word-dot interval and action delay in the experimental groups. RT differences were computed between the mean RT of 8 preceding trials and RT in the critical Trial 33. Mean RT differences were 288 ms in the 0.5-s condition, 229 ms in the 1-s condition and 88 ms in the 2-s condition[3].

In the control groups, the corresponding differences were -4 ms, -30 ms, and -116 ms. Increasing the interval between the onsets of words and dot up to 2 s obviously gives the experimental subjects the opportunity to "recover" almost completely from their surprise before the dot appears on the screen. Thus, in this condition, there is only a small delay in the required reaction to the dot.

---

[3] The 0.5-s condition differed significantly from the 2-s condition, $t(14) = 2.95$, $p<.05$. No other comparisons between experimental groups were significant at the 0.05-level. The data are taken from a larger design. For a complete description of the results and the design see Niepel, Rudolph, Schützwohl & Meyer (submitted, Experiment 2).

Generality of Effects

In order to test whether our findings are limited to the visual domain, we modified the experimental procedure by introducing auditory stimulus material. We expected that action delay and attentional focusing would also emerge in an auditory paradigm.

Subjects (48 male and female students) were placed in front of three loudspeakers. From the speaker in the middle, tape-recorded words were played one at a time (corresponding to the presentation of the word pairs in the visual paradigm). Either 0.2 s after the onset of a spoken word (i.e., during the word-display) or 1.5 s after the onset (i.e., following the word-display) a tone was presented either from the right or the left speaker. To indicate the location of the tone, subjects had to press the right or the left of two response keys. A total of 33 trials consisting of a word and a tone were presented. In the experimental groups, each word in Trials 1 to 32 (Stage 1) was spoken by the same male voice. This procedure was expected to establish a schema with respect to the mode of presentation of the words. In Trial 33 (Stage 2) the word was spoken by a female voice. We expected this event to be schema-discrepant and to elicit surprise. In the control groups, one half of the words was spoken by a male voice while the other half was spoken by a female voice (using a fixed random order). Thus, in the control groups, the female voice in Trial 33 was not schema-discrepant.

This procedure led to results that were very similar to those obtained with the visual paradigm (see Figure 2). When the interval was short (0.2 s), there was a pronounced increase of RT in the experimental group in Trial 33, as compared to the control group without a schema-discrepant event. When the interval between the onsets of the word and the tone was increased to 1.5 s, the surprise reaction did not inhibit the choice reaction time task[4].

---

[4] A two-way ANOVA on mean differences revealed a main effect of the factor Groups, $F(1,42) = 18.64$, $p<.001$, and of the factor Interval length, $F(1,42) = 19.72$, $p<.001$. Main effects were qualified by a significant Groups X Interval length interaction, $F(1,42) = 12.39$, $p<.001$. Separate t-tests showed that experimental and control groups differed significantly in the 0.2-s interval, $t(12) = 4.26$, $p<.001$, but not in the 1.5-s interval, $t(22) = 1.11$.

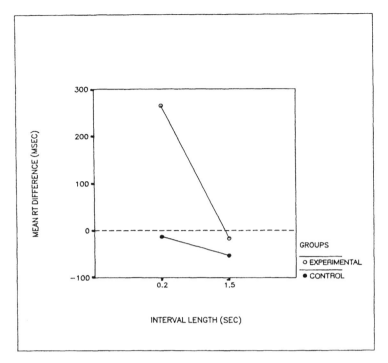

Figure 2. Mean difference between reaction times in the critical trial and in the preceding trials as a function of the interval length between the onsets of word and tone.

Number of Trials

The experiments reported this far clearly demonstrate that the presentation of an inverted word after 29 uniform trials (or 32 trials, respectively) leads to a pronounced delay in the execution of the required task and attentional focusing upon the inverted word. However, these numbers of preceding trials were chosen arbitrarily. Hence, we decided to examine the effect of the number of uniform trials before introducing an inverted word upon action delay and attentional focusing more systematically.

Utilizing the original visual paradigm, we conducted an experiment which included four experimental and four control groups (Schützwohl, 1993). In the four experimental groups, the RV word was introduced either in Trial 3, 13, 23, or 33, respectively. In the corresponding control groups, in each trial one of the two words was presented in RV mode. The onsets of words and dot were separated by a 0.5-s interval. Subjects were 130 male and female students.

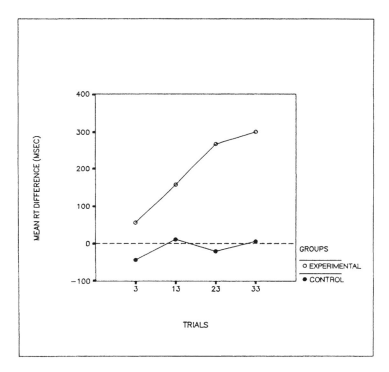

Figure 3. Mean difference between reaction times in the critical trial and in the preceding trials as a function of the number of preceding trials.

The effects of the number of preceding trials upon RT are shown in Figure 3. Obviously, within the experimental groups, the amount of RT increase is a function of the number of preceding trials: Action delay is more pronounced with a growing number of preceding trials[5].

---

[5] A two-way ANOVA on mean differences revealed a main effect of the factor Groups, $F(1, 82) = 60.5$, $p<.001$, and of the factor Number of preceding trials, $F(1, 82) = 5.40$, $p<.01$. These main effects were qualified by a significant Groups X Number of preceding trials interaction, $F(1, 82) = 3.0$, $p<.05$. Separate t-tests within the experimental groups showed significant differences between the 3-trials condition and the 23-trials condition, $t(27) = 3.90$, $p<.01$, and between the 3-trials condition and the 33-trials condition, $t(19) = 3.14$, $p<.01$. No other comparisons were significant at the 0.05-level.

However, there is no effect of the number of preceding trials upon memory performance in the experimental groups: Recall of the inverted word and of the dot's position does not differ between the four groups (see Table 2). In each of the experimental groups recall of the inverted word was better than in the corresponding control group. This is not true for the recall of the dot's position: Better recall in the experimental groups does not emerge until there are at least 23 preceding trials. With 3 and 13 trials memory performance does not differ between experimental and control groups[6].

Table 2. Percentage of Subjects who recall the RV Word and the Position of the Dot

	Trial			
	3	13	23	33
RV word				
Experimental Groups ($n = 65$)	38	44	35	38
Control Groups ($n = 65$)	1	13	0	6
Position of dot				
Experimental Groups	73	75	88	81
Control Groups	61	75	20	31

Note. Recall for the position of the dot was tested using a format offering three alternatives: "above", "below", and "don't know".

---

[6] The number of preceding trials had no effect upon memory performance for the inverted word within the experimental groups, $x^2(3, N = 65) = 0.27$, nor within the control groups, $x^2(3, N = 65) = 2.11$. More experimental than control subjects recalled the RV word, $x^2(1, N = 130) = 18.6$, $p<.001$. There was no difference in recall of the dot's position between experimental and control subjects in the 3-trials, $x^2(1, N = 33) = 0.14$, n.s., and the 13-trials condition, $x^2(1, N = 32) = 0.00$, n.s. However, recall differed significantly in the 23-trials condition, $x^2(1, N = 31) = 11.65$, $p<.001$, and in the 33-trials condition, $x^2(N = 32) = 6.22$, $p<.05$. All x^2-tests are based on cell frequencies.

The effect of varying numbers of uniform trials prior to the introduction of an inverted word on RT presumably reflects an increase in time consuming processes of schema revision.

Again, focusing of attention emerged as an integral component of surprise: Even after only a few trials the introduction of an inverted word led to better recall of this word in the experimental groups. Recall of the dot's position will be discussed in the next section.

General Discussion

The results of the experiments reported here strongly support our view on surprise: Events which conflict with a schema initiate processes that lead to an impairment of other ongoing processes (e.g. to a delay in the execution of an action) and to a focusing of attention on the surprising event.

The magnitude of the effect of a schema-discrepant event upon action delay obviously depends on two variables: First, with a relatively short interval between the onset of the schema-discrepant event and the action-relevant stimulus (0.5 s in the visual paradigm and 0.2 s in the auditory paradigm) the execution of the choice reaction time task is considerably impaired. However, increasing the interval between the schema-discrepant event and the action relevant stimulus heavily reduces (in the 1.5-s-interval condition in the auditory paradigm) the effect of the surprising event upon action delay. These findings suggest that the surprise reaction is of only short duration in our experiments.

Second, action delay depends on the number of uniform trials preceding the schema-discrepant event: With a growing number of uniform trials the effect of the critical event upon action delay steadily increases. We assume that at the beginning of our experiment there exists only a general, vague schema concerning the mode of presentation of the stimulus materials. As a consequence, the introduction of an inverted word is not or only to a small degree schema-discrepant, resulting in a small RT increase. However, with an increasing number of uniform trials, the schema gets specialized. Thus, the introduction of the inverted word deviates more and more from the schema. Because the revision of a specialized schema is presumably more time consuming than the revision of a general vague schema, a continuous increase in RT is observed with a growing number of trials preceding the schema-discrepant event.

Furthermore, memory performance for the inverted word indicates that focusing of attention on the schema-discrepant event is a constituent feature of the surprise reaction: Recall for the inverted word is higher in each of the experimental groups than in the corresponding control groups.

A result, which was unpredicted concerns the recall of the dot's position. Let us first consider the findings in the control groups. Recall is low in the control group of the first experiment which used 29 preceding trials (see Table 1). Recall is also low in the two control groups of the experiment on schema formation with similar numbers of preceding trials (23-trials condition and 33-trials condition, see Table 2). With smaller numbers of preceding trials (3-trials condition and 13-trials condition, see Table 2) the recall of the dot's position within the control groups is relatively high. In the experimental groups, on the other hand, memory performance for the dot's position is high, irrespective of the number of preceding trials (see Tables 1 and 2). This pattern of results presumably reflects the allocation of attentional resources for the successful execution of the choice reaction time task. At the beginning of the experiment the attentional demands of the task are high, because practice is low at this early stage. Consequently, recall for the dot's position is high in the control groups with a small number of trials (3 and 13 trials). With an increasing number of trials (increasing practice) the attentional demands decrease; as a result, memory performance for the position of the dot is low in the control groups with 23 and more trials.

Let us now consider the experimental groups. These groups do not differ with respect to the recall of the dot's position. One possible explanation of this finding is that the surprising event (at least in the 23-trials and 33-trials conditions) caused a change from automatic processing of the stimulus material to a "conscious" analysis involving a high degree of attention to the dot.

References

Charlesworth, W.R. (1964). Instigation and maintenance of curiosity behavior as a function of surprise versus novel and familiar stimuli. *Child Development, 35,* 1169-1186.

Darwin, C. (1872). *The expression of the emotions in man and animals.* London: Murray.

Descartes, R. (1911). The passions of the soul. In E.S. Haldane & G.R.T. Ross (Eds. and Trans.), *The philosophical works of Descartes* (Vol. 1). Cambridge: University

Press. Original work published 1649).
Isaacs, N. (1930). Children's "why" questions. In S. Isaacs (Ed.), *Intellectual growth in young children* (pp. 291-349). London: Routledge & Kegan Paul.
Izard, C.E. (1977). *Human emotions.* New York: Plenum Press.
Izard, C.E. & Buechler, S. (1980). Aspects of consciousness and personality in terms of differential emotions theory. In R. Plutchik & H. Kellerman (Eds.), *Emotion. Theory, research, and experience.* (Vol. 1, pp. 165-187). New York: Academic Press.
Meyer, W.-U. (1988). Die Rolle von Überraschung im Attributionsprozeß [The role of surprise in the process of causal attribution]. *Psychologische Rundschau, 39,* 136-147.
Meyer, W.-U., Niepel, M., Rudolph, U. & Schützwohl, A. (1991). An experimental analysis of surprise. *Cognition and Emotion, 5,* 295-311.
Niepel, M., Rudolph, U., Schützwohl, A. & Meyer, W.-U. (submitted). *Temporal characteristics of the surprise reaction induced by schema-discrepant visual and auditory events.*
Plutchik, R. (1980). A general psychoevolutionary theory of emotion. In R. Plutchik & H. Kellerman (Eds.), *Emotion. Theory, research, and experience.* (Vol. 1, pp. 3-33). New York: Academic Press.
Rumelhardt, D.E. (1984). Schemata and the cognitive system. In R.S. Wyer, Jr. & T.K. Srull (Eds.), *Handbook of social cognition.* (Vol. 1, pp. 161-188). Hillsdale, N.J.: Erlbaum.
Rumelhardt, D.E., Smolensky, P., McClelland, J.L., & Hinton, G.E. (1986). Schemata and sequential thought processes in PDP models. In J.L. McClelland, D.E. Rumelhardt, and the PDP Research Group (Eds.), *Parallel distributed processing. Psychological and biological models* (pp. 7-57). Cambridge, Mass.: The MIT Press.
Scherer, K.R. (1984). On the nature and function of emotion. A component process approach. In K.R. Scherer & P. Ekman (Eds.), *Approaches to emotion* (pp. 293-317). Hillsdale, N.J.: Erlbaum.
Schützwohl, A. (1993). *Schema und Überraschung: Untersuchungen zum Zusammenwirken von Kognition und Emotion* [Schema and surprise: Studies on the interaction between cognition and emotion]. Unpublished doctoral dissertation. University of Bielefeld, Bielefeld, FRG.
Tomkins, S.S. (1962). *Affect, imagery, consciousness: Vol. 1. The positive affects.* New York: Springer Publishing Co.

CHAPTER II.5

Interest and Curiosity. The Role of Interest in a Theory of Exploratory Action

Andreas Krapp

Research into curiosity and exploration concerns itself primarily with describing and explaining exploratory behavior in terms of general laws. To do this, researchers typically rely on formal categories that allow the formulation of general statements independent of the particular goals or specific content of behavior and experience. This results, however, in the inadequate treatment, if not the complete exclusion, of important research questions. This applies, for example, to special cases of exploratory behavior as well as to the analysis of coherent sequences of exploratory actions. For instance, why would some individuals leave a situation that most people find stimulating, to seek out a situation that most people find rather dull (mineral collectors in a stretch of rocky desert)? Or how can we explain the tenacity and regularity with which individuals pursue certain forms of exploration involving great exertion and resulting in a gradual extension of competence?

Explanations based on everyday experience often incorporate the notion of an active individual pursuing particular content-related goals. The individual is not simply motivated, he or she is "interested" in some matter of content. Everyday-type theories regard the content-orientation and the intensity of interests as important factors for explaining curiosity and exploration. The current scientific discussion, on the other hand, neglects the concept of interest. This is true not only for curiosity and exploration research, but also for more recent research into motivation (Krapp, 1989, 1992a; Krapp, Renninger & Hidi, 1992). This is especially surprising in light of the fact that, as early as 1949, Berlyne, the founder of modern curiosity research, dealt extensively with the relation between interest, curiosity, and exploratory behavior, and, in a statement made shortly before his death, identified the object of his scientific life's work with the words: "My interest is interest" (cited in Wohlwill, 1981, p. 7).

We believe that previous research into curiosity and exploration has overlooked and dealt inadequately with two related deficits: *firstly*, the inadequate treatment of the content and goals of exploratory behavior, and *secondly*, the inadequate treatment of content continuity in a coherent sequence of exploratory actions. In the following section, I attempt to show that this criticism applies equally to all of the theoretical approaches developed to date. Following this, I present the main ideas of a theoretical conception based on traditional views of interest-guided person-environment-engagements. In light of this conceptualization, I will then discuss why and how it could be useful to consider in more detail those aspects of exploratory behavior which traditionally have been essential components in a theory of interest and in our everyday thinking about interest.

The Neglect of Content in Curiosity and Exploration Research

Three research perspectives have especially influenced the research conducted up to now, namely, the perspectives stemming from general psychology, differential psychology, and a theory of action.

The Research Perspective of General Psychology

In interpreting curiosity, research approaches based on drive theories, theories of activation (Berlyne, 1960), or cognitive theories typically exhibit a general psychology perspective. Investigations center around phenomena that can be described and explained in terms of *general laws of human behavior*. Researchers are concerned first and foremost with the regularities or laws found to be equally valid for all individuals. Person-specific orientations, for example, content preferences within particular topic areas of curiosity, are ignored. Research into curiosity from a general psychology perspective also treats *changes* in exploratory behavior over time as a peripheral matter. It comes as no surprise, then, when the triggering conditions of exploratory behavior are assumed to lie in the stimulus properties of the situation (e.g., collative variables; cf. Berlyne).

The Research Perspective of Differential Psychology

Within the tradition of differential psychology, regularities of exploratory behavior that can be observed for all individuals are of less interest. Instead, researchers focus on differences in behavior and on the explanation or prediction of interindividual differences in light of personality characteristics. Researchers typically have assumed motivational factors to be the source of these differences in the field of exploratory behavior.

Indeed, numerous catalogs and diagnostic methods for the description of various degrees of curiosity, in the sense of an exploration motive, have been developed. Typical examples would be the theories, conceptions and methods pertaining to "seeking curiosity" (Livson, 1967), "sensation seeking" (Zuckerman, 1979), "variation motivation" (Fischer & Wiedl, 1981; Pearson & Maddi, 1966), "quest for knowledge" (Krieger, 1976, 1981; Lehwald, 1985), "cognitive orientation" (Kreitler & Kreitler, 1976, 1981), or "experience-producing tendencies" (Henderson, 1989).

However, the theoretical concepts and methods which have been developed by differential psychologists also exhibit shortcomings. To the extent that these theories and concepts relate to stable dispositions that characterize a person over a longer period of time, they are forced to neglect the process aspect. Over the past few years, however, progress has been made toward the development of approaches that abandon a global conceptualization of curiosity in favor of process variables (e.g., Henderson & Moore, 1979; Kreitler & Kreitler, this volume). But even these models are not able to escape the criticism that research from a differential psychology perspective neglects the content dimension of exploratory behavior.

An exception to this can be found in Day's (1971, 1981) multidimensional model for the differential measurement of curiosity. Similar to Guilford's cube-model of intelligence, Day postulates three orthogonal dimensions of curiosity behavior: (a) the preferred source of stimulation, (b) the preferred type of exploratory activity, and (c) a person's *specific interests*. This last dimension borrows from Kuder's (1960) theory of career interests in relying on ten content-related topic areas. These ten areas, however, are defined in very general terms (e.g., outdoors, scientific, musical). Consequently, they are insensitive to the great variability of content-related interests as the trigger for or goal of exploratory behavior.

The Theory of Action Research Perspective

A third research approach is based on a theory of action perspective. Research along these lines stresses typical human aspects, such as the conscious planning of a person-environment engagement. Actions represent organized sequences of ongoing behavior that correspond to "inner" cognitive processes and entities such as goals, values, and knowledge.

Within the context presented here, the goal-orientedness of an exploratory action is of special importance. "An exploratory action ... involves an active dealing with objects or events of the environment aimed at building up new structures of knowledge" (Voss & Keller, 1983, p. 159). At this point, the question arises as to *which* bits of knowledge are serving as the focus of the exploratory behavior. But this is a matter which this theoretical approach also fails to address. Instead, this research relies on a motivational tendency defined in purely formal terms. In accordance with Hunt (1961), researchers assume that experiences which induce cognitive conflict in a person represent a source of motivation for exploratory engagement.

One notes here again that nothing has been said about the content-orientation of exploratory behavior. This is somewhat disappointing, for if autonomous activity and conscious goal-orientation represent fundamental aspects of exploration, then an interpretation based upon a theory of action ought to say something about how individual value-orientations, interests or content-specific preferences influence exploratory behavior.

Unsolved Problems in Curiosity and Exploration Research

Conceivably, this problem in traditional curiosity and exploration research has gone unrecognized or been largely ignored due to the fact that the majority of work in the field has been done with animals and young children. At this level of development the problem of autonomous goal selection plays no predominant role in activity. The preference for experimental designs involving standardized stimulus conditions and preselected exploration objects only reinforces this tendency (e.g., Hutt-Box; cf. Keller, Schölmerich, Miranda, & Gauda, 1987).

If, though, one considers the phenomena of curiosity and exploration as they occur outside the laboratory, then the problem of the content-orientation of exploratory activity increases in importance with increasing willingness and ability to engage in autonomous behavior. A theory of exploration useful in describing everyday activity should, for instance, yield insight into whether and in what way specific exploration is influenced by the presence or absence of content preferences, or how such preferences determine stimulus-search behavior in the case of (diversive) exploration. I refer to this gap in exploration research as *inadequate analysis of the content aspect of exploratory behavior.*

A further deficit appears when exploratory behavior is regarded not as a one-time act, but rather as a series of actions undertaken in relation to one another. Indeed, the sum of all individual actions results not in some random pattern, but in a structure having subjective meaning. Individual actions can relate to one another in a number of ways. For instance, there are sequences of activity that relate to analogous objects or that pursue comparable content-related goals. In light of the cognitive effects of exploratory activity, the sequence involved in a topically structured series of actions merits special attention. For what comes to mind sooner than the idea that long-term preferences for objects, possibilities of action, and topics (in other words "interests" in the original sense of the word) exert a fundamental influence on both the orientation and nature of individual exploratory actions, and in doing so determine the very nature of "cognitive construction" as well?

This matter becomes of even greater importance when sequences of exploration, arranged along topical lines, are viewed from an ontogenetic perspective. Doing this brings into focus problems that have been largely ignored up to now, such as the development of exploratory styles as a function of object domain preferences, or the effects of content-specific explorations on the type of cognitive construction within the person. This brings into view a second group of questions and unsolved problems. These questions and problems arise in conjunction with the (all too long neglected) aspect of continuity of goals and content in exploratory behavior. This, too, represents a great deficit in traditional exploration research. In what follows, I refer to this deficit as *inadequate analysis of content continuity.*

In the following, I attempt to show that conceptualizations originating in a theory of interest can aid us in the theoretical and empirical treatment of these two problems. First, I outline a concept of interest, in which interest is regarded as a specific kind of "person-object-relationship" (Fink, 1991; Krapp, 1992b, 1993; Krapp & Fink, 1992; Prenzel, 1988). Following this, I introduce a few examples to show how this concept of interest can be used to clarify the content aspect of exploratory behavior.

A Concept of Object-specific Interests

The conceptualization of interest presented here is based on theoretical consideration of the origin and effect of person-environment relationships. It is assumed that individual development is determined largely by the quality and course of a person's relationship to the social and physical environment. Continual interaction with people, objects, events, and areas of subject matter (content) leaves behind traces in both the person and the environment. Each experience adds to and differentiates a person's store of knowledge. The person acquires cognitive representations about the "nature of things" (declarative knowledge) and about action possibilities (procedural knowledge). Person-environment engagements thereby shape a person's cognitive and motivational structure. This includes the development of values, attitudes, motivational orientations, and other components closely associated with cognitive processes.

In the course of an individual's development, experiences are organized into categories and classes of categories, which are themselves subject to reorganization depending on their meaningfulness to the person. Thus, each person builds unique, subjective cognitive structures. But different individuals also acquire very similar cognitive structures simply because similar realities and interpretations occur within the cultural context. The cognitive categories and structures reflecting these aspects of the environment are objective in the sense that they are the same across individuals which are members of a social system.

It seems safe to assume that individual categories vary in subjective importance across different situations and phases of development. In the course of development the person develops a special relationship to certain parts of the environment. An individual interest is a unique relationship between a person and an object, or object domain, found in that person's environment. This relationship must be of some duration, and does not refer to one-time, unrepeated forms of engagement[1].

---

[1] Person-object-relationships may be analyzed in terms of both process and structure. These represent two theoretical perspectives, and are associated with two levels of analysis. The first level of analysis deals with the internal and external interest-oriented actions related to an object. At this level, interest is understood as a state that represents the actualized relationship between a person and an object in a specific situation at a certain time. The second level of analysis interprets interest as a persisting disposition. As a dispositional category, interest may manifest itself in subjective representations that are value-related and cognitive in nature.

A more precise theoretical description of this interest-concept centers around three aspects: (a) the object of interest, (b) the structural components of interest, and (c) the characteristics of the interest-oriented person-object-relationship.

Interest Object

Three conceptual levels of objects are distinguished: object domains, objects of interest, and reference objects. The levels differ from one another in their degree of specificity The most general level involves *domains of interest*, such as "music", "sports", or "travel". School content areas, such as math, biology, or history are also interest domains.

At the next level are *interest objects*, also referred to as "objects of interest". An interest object consists of that part of an entire interest domain which a particular person at a particular time includes as a personal interest. Objects of interest are person-specific. Although two different people may enjoy the same things and action possibilities with a certain domain, and therefore have very similar interest objects, each individual will have had some unique experiences, thus excluding the possibility of their interest objects being identical. Someone who is interested in "sports", for instance, is very unlikely to be interested in or even aware of every aspect of every sport on earth. Rather, one person might be interested in performing non-team sports that require good physical condition (marathon running, cross country skiing), whereas another person might be interested in watching professional team sports.

Finally, the third level involves the particular, concrete things used when engaging in activity with the object of interest. These are referred to as *reference objects*. Going back to the above examples, a person interested in non-team sports might have running shoes, and a stopwatch as pertinent reference objects, whereas someone interested in watching professional team sports might have binoculars, season tickets, and subscriptions to sports magazines. It should be noted that reference objects are not the only elements found in a person's interest object.

It is important to clarify some further matters relating to these interest objects. Firstly, objects of interest are based on more than just concrete reference objects. Abstract or ideal elements, for example, symbolic representations of things, concepts, and events, information, and questions of a scientific nature can all be part of an object of interest. Secondly, insofar as they function as interaction partners, people are not regarded as objects of interest. Thirdly, even though one and the same domain can be involved in the formation of completely different person-object-relationships in different people, these different relationships can all be described and explained in a consistent manner on the bases of general rules.

Structural Components of the Interest-Oriented Person-Object-Relationship

Each interest has a more or less differentiated structure. The complexity of the structure can be modified by eliminating elements (substructures, components), by adding elements, or by processes of incorporation and exclusion (see Fink, this volume). Simple as well as complex person-object-relationships can be broken down into individual components. These basic structural components include reference objects, activities (i.e., action possibilities), and topics.

Reference objects. Concrete objects (reference objects) are essential elements for most person-object-relationships, for example, books for the object domain "literature"; or instruments, sheet music, or records for the object domain "music". In empirical studies, reference objects serve as landmarks for charting the boundaries and content of subjective domain perception. In many cases, they represent the primary content of an interest (e.g., collecting of certain things). Furthermore, the objective characteristics of an interest object often have a direct influence on a person's engagement with it. For example, an engagement may be limited by the material characteristics or the socio-culturally determined purpose of the interest object (e.g., artistic work that involves valuable materials).

Activities. A second structural component is the type of activity associated with the object of interest. Such activities include not only observing, perceiving, manipulating, and exploring the different characteristics of an object, but also changing the object or making the imagined object real (e.g., when a child paints or builds something that he or she thought of earlier). A further important activity is acquiring and processing information about the object or activities associated with it (e.g., searching for relevant sources of information). Social contacts also are frequently included within object-related activities, particularly when the person-object-relationship can be realized only within a group (e.g., games involving social interaction).

Topics. The present and future forms of activity that a person undertakes with an object depend, in large part, on a person's goals, topics, and questions regarding the object (Renninger, 1989, 1990). Topical categories often serve to guide interest-oriented engagement and thereby influence the specific course of events involved in an activity as well as its overall nature. A teenager's interest in computers will exhibit a completely different structure for activity with a prepackaged computer game than for activity with a home-made toy robot, even though the interest involves the same reference objects (e.g., hardware, programs) and competence (programming).

Special Characteristics of an Interest-Oriented Person-Object Relationship

Although the basic structural components of an interest-oriented person-object-relationship are important landmarks for the empirical reconstruction of an interest, they give only an incomplete and approximate picture of the theoretical construct. A more detailed picture emerges from investigation of the special theoretical characteristics of an interest-oriented person-object-relationship. These characteristics include: selective persistence, value orientation (self-intentionality), positive emotionality, and differential cognitive representation.

Selective persistence. The term "persistence" means that interest-oriented action is not a one-time or short-term affair. Rather, it is characterized by a certain amount of stability. Strong individual interest elicits repeated engagements with the object. Thus, a child's fleeting, curious attention toward an object in the environment cannot be classified as interest. Usually it is not difficult to recognize different degrees of persistence in intraindividual comparison of actions with respect to particular objects and topics (Renninger & Leckrone, 1991). In empirical investigations the degree of persistence may be operationalized as the *frequency* of engagement within a particular object domain. A more precise analysis might compare the frequency of actual person-object-engagement in a particular domain with the frequency of possible opportunities for engagement in that domain. Within a single incident of interest-based activity, the *duration* of engagement can serve as an indicator of persistence.

Defining persistence as repeated engagement with an object and individual willingness for long-term involvement with that object implies choices. From an array of action possibilities and objects, an individual must choose those which best match his or her interests. Thus, persistence of person-object-relationships develops only through a *selective* process (Prenzel, 1988, 1992).

Value orientation and self-intentionality. Objects, activities, and topics associated with an interest are experienced as important and meaningful because they are closely related to personal attitudes and values. Linking personal interests to values and attitudes is common to older and more recent theories of interest (e.g., Dewey, 1913; Hidi & Baird, 1986, 1988; Kerschensteiner, 1926; Lunk, 1927; Rathunde, 1992; Renninger, 1989, 1990; Renninger & Wozniak, 1985). This does not mean that a person is necessarily aware of personal value judgements, or that a reflective judgement is made prior to interest-based activity. But, greater personal maturity and increased differentiation in a person's set of values will increase the likelihood of the person developing interests on the basis of conscious value judgements.

The value orientation of interest can surface in diluted form, as a *preference* for particular objects, activities, and topics (see Fink, this volume). Empirically useful indicators of preference include relative desirability of the object in the eye of the individual, time spent with the object, and choice in favor of interest-relevant activities.

Another important characteristic of full-fledged interest is *self-intentionality*. This is somewhat comparable to both Csikszentmihalyi's (1975) "autotelic" behavior (which is based on Bühler's [1918] concept of "Funktionslust") and Deci's (1980) intrinsic motivation, both of which refer to activities that are conducted in the absence of external stimulation (e.g., sanctions or reinforcement), and thus are under subjective internal control (Deci & Ryan, 1985, 1991). One can speak of an activity as self-intentional only when the person can independently plan it and carry it out. Hence, the principle of self-determination must be involved in interest (Deci, 1992; Krapp, 1992b, 1993; Van der Wilk, 1991)[2].

Positive emotions. Interest-oriented engagement with an object is usually accompanied by positive or pleasant feelings. In extreme cases, the positive emotions that accompany activity can intensify to the point of total immersion in that activity. This results in a "flow" experience, as described by Csikszentmihalyi (1975, 1990). But positive emotion is a global evaluation, which certainly cannot describe every aspect of an interest-oriented activity. During the course of such engagement, some negative feelings, such as anger and discouragement, may also occur. However, the emotional balance is presumed to be positive when the activity is considered as a whole (Prenzel, Krapp, & Schiefele, 1986).

Cognitive aspects. Repeated engagement with an object of interest results in specific cognitive structures (Renninger, 1989, 1990). An individual tends to develop relatively differentiated knowledge about an object of interest. This includes knowledge about the object and knowledge about action possibilities (procedural knowledge). Action-oriented knowledge relates both to previous concrete experiences, and to object-based activities (experiences) that the person has not yet attempted but has learned about from watching others, or from other information sources (Prenzel, 1988, 1992). In its developed form, then, interest is characterized by a high level of object-specific cognitive complexity.

As Piaget (1974, 1981) suggested, a person working in a domain of interest is exceptionally willing to assimilate that which is experienced and to accommodate his or her thinking accordingly. The use of available schemata to set goals, make sense of experience, and store new information in memory (assimilation), as well as the modification of schemata and the resulting extended competence (accommodation), are based on interest to a great extent. Piaget (1974, p. 131) even goes so far as to say, "Every intelligent activity is founded upon an interest."

---

[2]Self-determination distinguishes the present view of interest from instrumental action models of motivation (cf. Heckhausen, 1980), which define the motivational basis of an action as a process of rational calculation. These theories presuppose that a person chooses activities on the basis of the perceived likelihood of desirable or undesirable outcomes. However, consideration of long-term "pay-offs" actually plays a subordinate role in actions guided by interest. Rather, participation itself and the immediate outcomes are considered to be sufficient reasons for performing the action.

Interest and Exploration

Of what significance are interests for exploratory activity, and how can the concept of object-specific interest be applied to the two problems mentioned above, namely the problems of inadequate analysis of content specifity and of content continuity in exploratory activity? This section attempts to answer these questions by presenting hypotheses about the influence of interests on exploratory behavior, along with the available empirical findings relating to these hypotheses. It will be seen that interests play an important role in both diversive and specific curiosity.

Interests Determine the Content and Direction of Diversive Exploration

Diversive exploration commonly occurs when a person has ended his or her specific exploration of an object and either begins to play around with the object or turns to a new object. In this stage, the choice of objects, actions, or topics for activity may appear rather arbitrary, seemingly dependent on the stimulus conditions of the situation. But experimental findings do not give a complete picture of this phenomenon, as methodological considerations often require a purposeful limiting of the action possibilities made available to study subjects. This limitation does not allow for a very common, everyday phenomenon, namely the purposeful, consciously controlled search for a subjectively preferred object or action possibility.

Our observations of kindergarten children, made during a study of interest development in natural settings (interest genesis project; cf. Fink, in this volume), revealed that, following a stressful situation, children often seek out play activities involving familiar objects in their areas of interest (Fink, 1991; Krapp & Fink, 1992).

Example: Michael was a loner in the kindergarten group. Coming from a modest background, he found it difficult to assimilate to the social group, and to participate in games with his peers in the way to which they were accustomed. Whenever these situations became too much of a strain for him, Michael turned away and sought out familiar toys that, as shown by our reconstructions, belonged to his most preferred person-object-relationship. This observation could also be interpreted to mean that Michael, in a state of diversive curiosity, tended to occupy himself within the framework of his personal interests.

Observations like this support the hypothesis that the motivational state of diversive curiosity must be viewed separately from a content-neutral willingness to engage in activity. Not even small children limit their play behavior to the opportunities and stimulus qualities of the objects that happen to lie within the immediate field of perception. Instead, they often conduct a purposeful search for favorite objects or play opportunities. The more an individual is able to recognize the alternatives that are available and to determine his or her activities, the greater will be the role of interests in shaping behavior in such situations. Studies involving adolescents and adults confirm that persons having a pronounced preference for certain object domains will very probably engage in activity with objects from those areas if the situation permits free selection of activity (Hoff, 1986; Schiefele & Krapp, 1991).

Interests Influence the Goal Orientation of Specific Curiosity and Exploration

From a theory of interest perspective, interest-oriented action is closely related to specific exploration. In both instances, a person examines portions of the environment in a goal-oriented and self-determined manner, in order to gain understanding and to eliminate cognitive uncertainty. In so doing, the person develops new knowledge, and changes his or her cognitive structure. Different approaches have been taken in the attempt to explain this form of behavior. Traditional curiosity theories emphasize stimulus aspects (collative variables). To the extent that personal motivational factors are considered, content-neutral explanatory factors occupy center stage, for example, cognitive incongruence (Hunt, 1963, 1965) or feelings of efficacy (White, 1959). Interest concepts, on the other hand, focus on matters of content. These concepts ascribe repeated and intensive engagement with topics or objects to personal preference.

Most forms of specific exploration are integrated into complex action sequences and allow for great flexibility in selecting the aspects to be investigated or in focusing a single exploratory action on a small portion of the entire range of possible activities. On the surface, such decisions may seem quite arbitrary. We suspect, however, that such content selection does not occur randomly, but rather is the result of previous experience, and of topic or object preferences. Although our research has yielded no definitive empirical proof of this, observations do support this hypothesis (cf. Fink, 1991).

Example: Over the course of the kindergarten study mentioned above, one child (Daniel) exhibited an especially noticeable interest in the object domain "animals". Whenever the opportunity to relate an activity to some area of content presented itself, for example, in role playing, crafts, or painting, Daniel related the activity to animals. This was also true of situations containing opportunities for specific exploration, for example, looking at books, or observing the surroundings while on field trips.

Interests Influence the Design of an Exploratory Action

All other things being equal, an exploratory act undertaken out of interest will involve a different course of events than an exploratory act triggered by other factors. This involves characteristics such as persistence (endurance), and the variability of forms of engagement.

Renninger (1990), whose theoretical conception of interest is similar to the person-object-conceptualization used here, investigated the effects of interest on the quantity and quality of play activity of children between the ages of 2;9 and 4;2 years. Albeit using a different terminology, she also analyzed diversive and specific forms of engagement involving play objects. Her findings suggest a powerful and consistent effect of interest on children's play actions involving exploratory activities. The children played with identified objects of interest for longer periods of time than with identified objects of noninterest. This suggests that the children may have seen more possibilities for action with identified objects of interest, and may have been setting challenges for themselves which vary from those they set for themselves with identified objects of noninterest.

Interests Serve as the Basis of Content Continuity in a Sequence of Exploratory Actions

As mentioned earlier, persistence is a fundamental characteristic of interest (Renninger & Leckrone, 1991). The repeated engagement and the willingness for long-term involvement with an object implies selective choices. From an array of possible actions and objects, an individual must choose those actions and objects which best match his or her interests. If one assumes that interests are often at the root and determine the content of freely selected forms of exploratory behavior, then it becomes impossible to regard exploratory actions inspired by interest as single, isolated phenomena. The goals of single actions, the nature of exploratory activities, expectations in terms of emotional and cognitive effects, and other aspects as well can be interpreted satisfactorily only by viewing the single action as an element in a coherent sequence of activity.

Our studies pertaining to the development of interests have shown that the preference structure of a particular interest can contain diverse possibilities for concrete realization in terms of both objects and activities (cf. Fink, 1991, p.194).

Figure 1: The preference-structure of a topicspecific interest

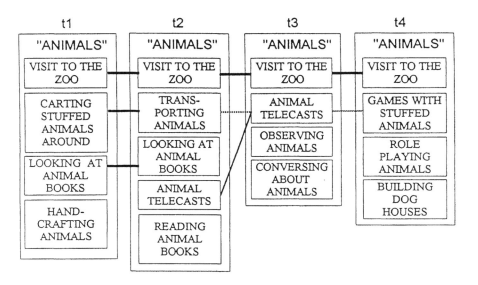

Figure 1 portrays the preference structure of a four-year-old child, who is interested in the topic "animals" and who, as a result of this interest, brings together a wide assortment of activities under the roof of this common topic. The markings extending from t1 to t4 refer to the four phases of data collection at the beginning of each kindergarten year. Even at this relatively general level of reconstructing a person's interest structure over the course of time, one can see that the internal structure of an interest can contain components that at first glance seem completely unrelated but which are in fact connected to one another, for example, looking at books (t1,t2), watching TV (t3) or building houses (t4). An absolutely rational connection can exist between different exploratory actions and objects that, to the adult, appear to be nothing more than randomly selected play objects and activities.

It seems significant that, over the course of time, a combination of highly preferred activities has developed out of repeated engagements with certain objects and topics, and that these engagements were, in many cases, coupled with exploratory actions. Any new exploratory action that the child undertakes exists and acquires meaning within this greater context. The orientation of exploratory acts, as well as the topic that the child selects for the next action during the course of an "exploration unit" will, to a certain extent, depend on the logical consequences of previous explorations.

Interests Determine the Nature and Orientation of Cognitive Construction

Extending this train of thought, and considering especially the long-term effects of explorations, one arrives back at the aspect of cognitive construction. Our study results reveal considerable individual differences in exploratory behavior involving novel objects and considerable differences in willingness to acquire more knowledge about the objects. In order to explain these interindividual differences, Voss (1981) hypothesized a "rate of cognitive construction" that differs from person to person. This explanation assumes a general trait-like disposition towards more demanding or less demanding forms of exploratory behavior, independent of the nature of the object and the content to be explored.

Viewing the situation from the perspective of a theory of interest leads to a simpler and probably more adequate explanation: In most situations, interest is the primary determinant of whether or not a certain object is even considered for exploratory activity, and of the extent to which the exploration is carried out. In this sense, the preferred "rate of cognitive construction" will vary with the situation and will fluctuate as a function of the presence or absence of interest.

Interests also influence the orientation of cognitive construction. The areas in which a person chooses to learn more, or, in other words, where discrepancies between the cognitive structures already available and the complexity of the "object structure" found in the object itself are sought out, discovered, and resolved, will depend to a great extent on a person's previous experiences, on the store of knowledge that is already available in the wake of previous engagements with the object, and last but not least on the wish for completeness in the knowledge structure pertaining to particular objects (Krapp, 1992b, 1993).

The following example may illustrate this point. Imagine two groups of teenagers who have the opportunity to explore a new computer system containing a large and varied assortment of computer software. The two groups have been told to take as long as they like to explore the new computer system. The first group is made up of teenagers who have developed highly differentiated personal interests in working with computers. The other group is made up of teenagers who, in the course of their schooling, have also learned how a computer works and how it can be used in certain situations, but who are not particularly interested in computers per se. Under otherwise identical conditions, the exploratory behavior seen in the two groups will be entirely different. One would expect to find, for instance, that the extent and duration of exploratory behavior are markedly more pronounced in the first group than in the second group. In addition, a greater variety in both the orientation and quality of exploratory behavior would be expected in the first group. For each of the teenagers in this first group would be likely to explore the possibilities offered by the new machine and to try out action possibilities in those areas where his or her special interests and abilities lie.

Numerous studies have explored the effect of interest-guided action on resulting knowledge structures (cf. Hidi, 1990; Krapp, 1992a; Krapp & Prenzel, 1992; Renninger, Hidi, & Krapp, 1992; Schiefele, 1992). Our own studies have shown that topic interest considerably affects the cognitive engagement with relevant text content when reading, and hence exerts considerable influence on knowledge representation. The knowledge structures acquired by high-interest subjects differ especially in *qualitative terms* from those acquired by low-interest subjects. The object-specific knowledge of high-interest subjects exhibits more depth, is based more firmly in the central concepts and relations presented in the text, and is, as a result, more useful when discussing principal problems or when attempting to transfer knowledge to a new situation (Hidi, 1990; Krapp, Renninger & Hidi, 1992; Schiefele 1991; Schiefele & Krapp, 1991).

Summary and Conclusion

Previous research into curiosity and exploration has largely ignored two problems: the analysis of both content specifity and content continuity in exploratory behavior. This article suggests that an object-related theory of interest offers a plausible basis for the description and explanation of the content aspect of exploratory behavior. Our thinking in this direction centers around what we call the person-object-theory of interest.

Interests, insofar as they represent a stable relationship between a person and an object domain in that person's environment, provide us with a possible explanation of the content dimension in diversive exploration. We believe that diversive curiosity and exploration are not directed randomly at whatever objects or action possibilities happen to be available, but instead often exhibit a goal-oriented character. According to this view, interests are an important component of this phase of stimulus-search behavior.

Interests also play a decisive role in the content orientation of specific curiosity. Findings from studies involving adults as well as children lead to the conclusion that engagement with an object in which a person is greatly interested proceeds in different fashion and exhibits a different qualitative character than does engagement with an object in which a person has little or no interest. Interests influence the goal-orientation, the course of events, and the result of single exploratory actions.

In addition, interests also influence the content and effects of a chain of exploratory actions separated from one another in terms of time or situation. Interest serves as the basis of the content continuity found in a sequence of exploratory actions. This content continuity especially affects cognitive construction. We believe that personal interests determine both the "rate of cognitive construction" and the quality and direction of the modifications associated with this construction.

This hypothetical position takes on a new dimension when one adopts an ontogenetic perspective in considering continuity of interest content over a longer sequence of exploratory actions. One can well assume, for instance, that during the course of development a close relationship exists between exploratory and interest-oriented behavior. For one thing, the ontogenetic changes in the overall structure of all a person's interests must have some effect on the content and form of a person's exploratory behavior. On the other hand, it seems reasonable to assume that general exploratory tendencies, which at the beginning of development have no specific content, influence either directly or indirectly the development of interests. Experiences of success or failure while exploring unfamiliar objects and other related psychic phenomena play an important role in the origin and change of object-specific preferences.

References

Berlyne, D. E. (1949). Interest as a psychological concept. *The British Journal of Psychology, 39*, 184-195.
Berlyne, D. E. (1960). *Conflict, arousal, and curiosity.* New York: Grove Press.
Bühler, K. (1918). *Die geistige Entwicklung des Kindes.* Jena: Fischer.
Csikszentmihalyi, M. (1975). *Beyond boredom and anxiety: The experience of play in work and games.* San Francisco: Jossey-Bass.
Csikszentmihalyi, M. (1990). *Flow.* New York: Harper & Row.
Day, H. I. (1971). The measurement of specific curiosity. In H. I. Day, D. E. Berlyne, & D. E. Hunt (Eds.), *Intrinsic motivation: A new direction in education* (pp. 99-112). Toronto: Holt, Rinehart & Winston of Canada.
Day, H. I. (1981). Neugier und Erziehung. In H.-G. Voss & H. Keller (Eds.), *Neugierforschung* (pp. 226-262). Weinheim: Beltz.
Deci, E. L. (1980). *The psychology of self-determination.* Lexington, MA: Heath.
Deci, E. L. (1992). Interest and intrinsic motivation. In K.A. Renninger, S. Hidi, & A. Krapp (Eds.), *The role of interest in learning and development* (pp. 43-70). Hillsdale, NJ.: Erlbaum.
Deci, E. L., & Ryan, R. M. (1985). *Intrinsic motivation and self-determination in human behavior.* New York: Plenum Press.
Deci, E. L., & Ryan, R. M. (1991). A motivational approach to self: Integration in personality. In R. Dienstbier (Ed.), *Nebraska Symposium on Motivation: Volume 38. Perspectives on Motivation.* Lincoln, NE: University of Nebraska Press.
Dewey, J. (1913). *Interest and effort in education.* Boston: Riverside Press.
Fink, B. (1991). Interest development as structural change in person-object relationships. In L. Oppenheimer & J. Valsiner (Eds.), *The origins of action: Interdisciplinary and international perspectives* (pp. 175-204). New York: Springer.
Fischer, M., & Wiedl, K.-H. (1981). Variationsmotivation. Empirische und theoretische Beiträge zur Weiterentwicklung eines persönlichkeitspsychologischen Konstrukts. In H.-G. Voss, & H. Keller (Eds.), *Neugierforschung* (pp. 109-143). Weinheim: Beltz.
Heckhausen, H. (1980). *Motivation und Handeln.* Berlin: Springer.
Henderson, B. B. (1989). Individual differences in exploration: A replication and extension. *Journal of Genetic Psychology, 149*, 555-557.
Henderson, B. B., & Moore, S. G. (1979). Measuring exploratory behavior in young children: A factor-analytic study. *Developmental Psychology, 15*, 113-119.

Hidi, S. (1990). Interest and its contribution as a mental resource for learning. *Review of Educational Research, 60*, 549-571.
Hidi, S., & Baird, W. (1986). Interestingness: A neglected variable in discourse processing. *Cognitive Science, 10*, 179-194.
Hidi, S., & Baird, W. (1988). Strategies for increasing text-based interest and students' recall of expository texts. *Reading Research Quarterly, 23*, 465-483.
Hoff, E. H. (1986). *Arbeit, Freizeit und Persönlichkeit.* Bern: Huber.
Hunt, J. McV. (1961). *Intelligence and experience.* New York: Ronald Press.
Hunt, J. McV. (1963). Motivation inherent in information processing and action. In O. J. Harvey (Ed.), *Motivation and social interaction: Cognitive determinants* (pp. 35-94). New York: Ronald Press.
Hunt, J. McV. (1965). Intrinsic motivation and its role in psychological development. In D. Levine (Ed.), *Nebraska symposium on motivation* (pp. 189-282). Lincoln: Nebraska University Press.
Keller, H., Schölmerich, A., Miranda, D., & Gauda G. (1987). The development of exploratory behavior in the first four years of life. In D. Görlitz & J. F. Wohlwill (Eds.), *Curiosity, imagination, and play* (pp. 130-150). Hillsdale, NJ.: Erlbaum.
Kerschensteiner, G. (1926). *Theorie der Bildung.* Leipzig: Teubner.
Krapp, A. (1989). Neue Ansätze einer pädagogisch orientierten Interessenforschung. *Empirische Pädagogik, 3*, 233-255.
Krapp, A. (1992a). Interesse, Lernen und Leistung. Neuere Forschungsansätze in der Pädagogischen Psychologie. *Zeitschrift für Pädagogik, 38*, 747-770.
Krapp, A. (1992b). Das Interessenkonstrukt. Bestimmungsmerkmale der Interessenhandlung und des individuellen Interesses aus der Sicht einer Person-Gegenstands-Konzeption. In A. Krapp & M. Prenzel (Eds.), *Interesse, Lernen, Leistung. Neuere Ansätze einer pädagogisch-psychologischen Interessenforschung* (pp. 297-329). Münster: Aschendorff.
Krapp, A. (1993). *The construct of interest: Characteristics of individual interests and interest-related actions from the perspective of a person-object-theory* (Studies in Educational Psychology). Munich: Institut für Empirische Pädagogik und Pädagogische Psychologie der Universität der Bundeswehr.
Krapp, A., & Fink, B. (1992). Continuity of interests between home and school. In K. A. Renninger, S. Hidi, & A. Krapp (Eds.), *The role of interest in learning and development* (pp. 397-429). Hillsdale, NJ: Erlbaum.
Krapp, A., & Prenzel, M. (Hrsg.). (1992). *Interesse, Lernen, Leistung. Neuere Ansätze einer pädagogisch-psychologischen Interessenforschung.* Münster: Aschendorff.
Krapp, A., Renninger, A. & Hidi, S. (1992). Interest, learning and development. In K.A. Renninger, S. Hidi & A. Krapp (Eds.), *The role of interest in learning and development* (pp. 3-25). Hillsdale, NJ: Erlbaum.
Kreitler, H., & Kreitler, S. (1976). *Cognitive orientation and behavior.* New York: Springer.
Kreitler, H., & Kreitler, S. (1981). Die kognitiven Determinanten des Neugierverhaltens. In H.-G. Voss, & H. Keller (Eds.), *Neugierforschung* (pp. 144-174). Weinheim: Beltz.
Krieger, R. (1976). *Determinanten der Wißbegier.* Bern: Huber.
Krieger, R. (1981). Ungewißheit und Wißbegier. Von der reizinduzierten Motivation zu einer Wert-Erwartungs-Theorie. In H.G. Voss & H. Keller (Eds.), *Neugierforschung* (pp. 80-108). Weinheim: Beltz.
Kuder, G. F. (1960). *Manual to the Kuder Preference Record, Form C (Vocational).* Chicago: Scientific Research Association.
Lehwald, G. (1985). *Zur Diagnostik des Erkenntnisstrebens bei Schülern. Beiträge zur Psychologie Band 20.* Berlin: Volk und Wissen Volkseigener Verlag.
Livson, N. (1967). Towards a differentiated construct of curiosity. *Journal of Genetic Psychology, 111*, 73-84.

Lunk, G. (1927). *Das Interesse*. Leipzig: Klinkhardt.
Pearson, P. H., & Maddi, S. R. (1966). The similes preference inventory: Development of a structured measure of the tendency toward variety. *Journal of Consulting Psychology, 30*, 301-308.
Piaget, J. (1974). *Theorien und Methoden der modernen Erziehung*. Frankfurt: Fischer.
Piaget, J. (1981). Intelligence and affectivity: Their relationship during child development. In T. A. Brown, & M. R. Kaegi (Eds.), *Annual Review Monographs*. Palo Alto, CA: Annual Reviews.
Prenzel, M. (1988). *Die Wirkungsweise von Interesse. Ein Erklärungsversuch aus pädagogischer Sicht*. Opladen: Westdeutscher Verlag.
Prenzel, M. (1992). Selective persistence and interest. In K. A. Renninger, S. Hidi, & A. Krapp (Eds.), *The role of interest in learning and development* (pp. 71-98). Hillsdale, NJ.: Erlbaum.
Prenzel, M., Krapp, A., & Schiefele, H. (1986). Grundzüge einer pädagogischen Interessentheorie. *Zeitschrift für Pädagogik, 32*, 163-173.
Rathunde, K. (1992). Serious play: Interest and adolescent talent development. In A. Krapp & M. Prenzel (Eds.), *Interesse, Lernen, Leistung. Neuere Ansätze einer pädagogisch-psychologischen Interessenforschung* (pp. 137-164). Münster: Aschendorff.
Renninger, K. A. (1989). Individual patterns in children's play interests. In L. T. Winegar (Ed.), *Social interaction and the development of children's understanding* (pp. 147-172). Norwood, NY.: Ablex
Renninger, K. A. (1990). Children's play interests, representation, and activity. In R. Fivush, & J. Hudson (Eds.), *Knowing and remembering in young children* (pp. 127-165). Emory Cognition Series (Vol. III). Cambridge, MA: Cambridge University Press.
Renninger, K. A., Hidi, S., & Krapp, A. (Eds.). (1992). *The role of interest in learning and development*. Hillsdale, NJ: Erlbaum.
Renninger, K. A., & Leckrone, T. G. (1991). Continuity in young children's actions: A consideration of interest and temperament. In L. Oppenheimer, & J. Valsiner (Eds.), *The origins of action: Interdisciplinary and international perspectives* (pp. 205-238). New York: Springer.
Renninger, K. A., & Wozniak, R. H. (1985). Effect of interest on attentional shift, recognition, and recall in young children. *Developmental Psychology, 21*, 624-632.
Schiefele, U. (1991). Interest, learning, and motivation. *Educational Psychologist, 26*, 299-323.
Schiefele, U. (1992). Topic interest and levels of text comprehension. In K. A. Renninger, S. Hidi, & A. Krapp (Eds.), *The role of interest in learning and development* (pp. 151-182). Hillsdale, NJ: Erlbaum.
Schiefele, U., & Krapp, A. (1991, April). *The effect of topic interest and cognitive characteristics on different indicators of free recall of expository text*. Paper presented at the AERA-Meeting in Chicago.
Van der Wilk, R. (1991). Interest and their structural development: Theoretical reflections. In L. Oppenheimer, & J. Valsiner (Eds.), *The origins of action: Interdisciplinary and international perspectives* (pp. 159-173). New York: Springer.
Voss, H.-G. (1981). Kognition und exploratives Handeln. In H.-G. Voss, & H. Keller (Eds.), *Neugierforschung* (pp. 175-196). Weinheim: Beltz.
Voss, H.-G., & Keller, H. (1983). *Curiosity and exploration: Theories and results*. New York: Academic Press.
White, R. W. (1959). Motivation reconsidered: The concept of competence. *Psychological Review, 66*, 297-333.
Wohlwill, J. F. (1981). Vorwort. In H. G. Voss, & H. Keller (Eds.), *Neugierforschung* (pp. 7-9). Weinheim: Beltz.
Zuckerman, M. (1979). *Sensation seeking*. Hillsdale, NJ.: Erlbaum.

CHAPTER **II.6**

Interest and Exploration: Exploratory Action in the Context of Interest Genesis

Benedykt Fink

Research into curiosity and exploration has tended to relate the initiation and maintenance of children's exploratory behavior primarily to the physical properties of objects (Berlyne, 1960; H. Keller, 1981; Keller, Föse, & Schölmerich, 1985). Berlyne (1960) referred to an object's exploration-inducing stimuli such as novelty or complexity, as "collative variables" and considered them to be at the root of children's purposeful exploratory behavior (Voss, 1985; Voss & Keller, 1983). The perception of these stimulus qualities, Berlyne proposed, leads to cognitive experiences, such as discrepancy, uncertainty, and conflict, which in turn guide exploratory behavior.

This emphasis on an object's collative variables in explaining purposeful exploratory behavior "breaks up" the material object into a number of distinct, structurally-based stimuli. This theoretical approach pays little attention to matters of content, such as the meaning or personal significance of a material object (Csikszentmihalyi & Rochberg-Halton, 1981; Fink & Forster, 1992, 1993). In other words, viewing the object of exploration in terms of its stimulus properties leads away from consideration of the object's meaning and of the role of exploratory behavior involving that object within the context of a larger content area.

It is possible, however, to adopt the opposite tack, away from the structural stimuli and towards the meanings of objects. This entails consideration not only of the meaningfulness of the objects in isolation from one another, but also of their topical relatedness to one another. This approach throws light of a different sort on exploratory behavior, revealing the role of that behavior within the context of engagement with a larger topical area. In the wake of such engagement, the child

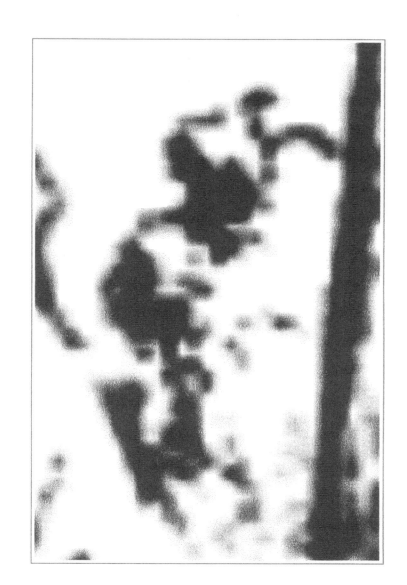

builds long-term, subjectively valued relations to the topical area, or object domain, involved. To a great extent, children's activity can be described in terms of the content of a valued object domain ("content orientation") and of the extent of repeated engagement ("content continuity") involving that domain (Fink, 1991, 1992; Gottfredson, 1981; Krapp & Fink, 1992; Oerter, 1987; Travers, 1978).

Krapp (in this volume) considers the lack of theoretical and empirical clarification of the content orientation and content continuity of exploratory behavior as a grave deficit in the research into curiosity and exploration. This deficit seems all the more peculiar when one considers that early childhood research has shown that young children are capable of establishing very close and long-term relationships to particular objects or object domains (Keller, Schölmerich, Miranda, & Gauda, 1987; J.A. Keller, 1987; Papousek, 1984; Papousek, Papousek, & Harris, 1987). These relationships can form the basis for important selection mechanisms employed in the exploration of various areas in the child's surroundings and in the choice of objects during the development of specific interests (Izard, 1977; Keil, 1979, 1983; Roe, 1957; Roe & Siegelmann, 1964; Travers, 1978). The period of early childhood witnesses the genesis of interests resulting from the exploration and structuring of the physical environment, accompanied by the successive differentiation of particular object domains (Gottfredson, 1981; Todt, 1985). Limiting the child's exploration to particular, preferred areas can even lead to the development of stable interests (Travers, 1978).

It seems sensible for workers in the field of curiosity and exploration research to consider the results of interest research to explain the content orientation and content continuity of exploratory activity. Based on a longitudinal case study of interest genesis in preschool and elementary school children (Fink, 1991, 1992; Krapp & Fink, 1992), the following contribution attempts to do exactly this. First, basic theoretical features of a conception of interest development in childhood are presented and explained using examples from the empirical research. The concluding discussion considers exploratory behavior from a perspective emphasizing the origins of childhood interests.

The Interests of Individuals

The "theory of interest" of H. Schiefele and colleagues (Fink, 1989; Prenzel, 1988; Prenzel, Krapp, & Schiefele, 1986) serves as the point of departure in portraying the earliest forms of childhood interest. According to this theory, interests represent an individual's personally meaningful relationship with a nonhuman object or object domain. Repeated engagement with subjectively meaningful objects leaves behind long-term, dispositional traces in the individual's personality as well as a cognitive representation of the object domain and of related action possibilities. These dispositional traces do not represent simple attitudes in the sense of long-term dispositions for certain areas of knowledge or types of activity. Compared to attitudes, these dispositional traces are more strongly combined with the individual's value structures. Also, they are more likely than attitudes to show themselves in forms of observable behavior involving specific objects (Krapp, 1989). A person acting within the framework of such a dispositional trace willingly engages in activity with different aspects of the object of interest, and, when presented with a choice, prefers items that are related to the object of interest over nonrelated items. An interest-oriented person-object-relation contributes to the realization of interests in particular situations and to the phenomenon of a particular object coming to be of especial worth to the individual.

The term "object domain" refers to some broad category, such as sports or nature. An "interest object" or "object of interest", on the other hand, refers to the structure of the cognitive representation that a particular individual has developed. Two different people might both say that they are "interested" in sports (common object domain), but if the one person likes to perform whereas the other person likes to watch, then their interest objects are quite different. It is hypothesized that no two individuals' objects of interest will ever be identical, even when a common object domain is involved.

Researchers working within the theory of interest rely on three characteristics in their assessments and descriptions of interest-oriented activity: selective persistence, value orientation, and positive emotionality (Prenzel, 1988; Krapp & Fink, 1992). The term persistence means that interest-oriented action is not a one-time or short-term affair, but rather exhibits a certain degree of stability. Defining persistence as the repeated engagement with an object and the individual's willingness for long-term involvement with that object implies selective choices. Individuals must choose actions and objects which correspond most to their interests. Value orientation is meant to indicate that an interest comes into existence when the individual attaches subjective value to an object domain. Objects and activities associated with a particular interest are experienced as being important and meaningful. Interest-oriented engagement with an object is usually accompanied by positive or pleasant feelings. In extreme cases, the accompanying positive emotions can lead the individual to lose him/herself completely in the activity (cf. flow-experience by Csikszentmihalyi, 1975).

The Characteristics of Beginning Interests

It should be noted that the theory of interests focuses on the interests of adults. Empirical findings indicate, however, that the interests of children in preschool and in the early elementary grades are not yet fully developed. Rather, young children's relationships with objects exhibit less pronounced forms of the three characteristics described above. These less pronounced characteristics are referred to here as duration and frequency, subjective valuing, and fondness (Travers, 1978; Fink, 1991). These three aspects, along with a fourth important aspect, that of object continuity, allow one to reconstruct a child's *preference structure*, and yield insight into the degree of complexity of the child's object of interest (Fink, 1991; Krapp & Fink, 1992).

Duration and frequency. The duration and frequency of repeated engagement with a particular object are empirical indicators of interest-oriented preference. They correspond to the *persistence* of a mature interest (Prenzel, 1988). This means that interest-oriented activity is neither a one-time affair nor a short-term matter, but stretches over time and situation, and, therefore, possesses some degree of constancy. The isolated incident, brought about by curiosity, wherein the child attends to an unusually conspicuous object does not fall under the heading of interest. In order to speak of a beginning interest, the child must attend to the object at different times and within different opportunities (e.g., "He plays with that every day.").

Subjective valuing. Interest-oriented person-object-relations, or rather their structural components (reference objects, activities, topics), are positively valued by the child. The child experiences an object of interest as meaningful and distinct from other objects. In this sense, precursor forms of interests are closely related to the system of subjective attitudes and value-related convictions. During childhood, before value orientations have stabilized, subjective valuations and emotional components of an interest-oriented person-object-relation show themselves in weakened form as preferences for certain reference objects, action possibilities, and topics in situations involving free choice (Fink, 1991; Oerter, 1987). The duration of a single activity can also serve as an indicator of the relative importance of this activity to the person (e.g., "He can spend all day doing that."). The subjective valuing of objects, activities, and topics is a relative criterion, involving a comparison with the extent to which something else is valued (e.g., "The stuffed animals are more important to him than any other toys.").

Fondness. The child's fondness for activities and objects is a further empirical indicator of interest-oriented preference. Fondness corresponds to the positive emotionality associated with mature interest. This means that interest-oriented engagements with an object area are usually accompanied by pleasant feelings (e.g., "Most of all he enjoys building houses with his blocks."). The element of fondness is not necessarily associated with every single interest-based action, but rather pertains to the entirety of an interest-oriented engagement. It is conceivable that negative feelings, such as discouragement or anger, can occur during the course of single interest-based actions. In toto, however, the entire interest-oriented engagement is experienced as being positive (Schiefele, Prenzel, Krapp, Heiland, & Kasten, 1983).

Object continuity. Object continuity is a further interest-specific aspect that is closely tied to the developmental dynamic of beginning interests. This refers primarily to the content-specificity and relative stability of the object domain over time. On the one hand, then, object continuity signifies that certain structural components of an interest-oriented person-object-relation can be observed repeatedly over time. On the other hand, however, one can speak of object continuity when new forms of activity, reference objects, or topics are integrated into the existing structure of a person-object-relation, and when this structural reorganization of the early interest exhibits a logical developmental course.

Structural Aspects of Early Interests

As mentioned earlier, an object of interest is the subjective representation of some part of the individual's surroundings. An object of interest can involve both concrete and abstract objects (Fink, 1989; Fink &

Forster, 1992, 1993). Bear in mind that an object of interest does not generally consist of one single, definable object and one single action possibility involving that object. Such a constellation is referred to as a *simple person-object-relation*. Though, technically speaking, an object of interest could consist of a single simple person-object-relation, it is usually thought of as more or less complex, hierarchical structure of interrelated simple person-object-relations. This structure results from (a) the individual's experiences from earlier engagements with the object domain that are stored in memory, and (b) the knowledge of possibilities that the individual has already carried out, plans to carry out, or at least knows of (Prenzel, 1988; Renninger, 1989, 1990). The subjective cognitive representation of an object domain generally consists of a number of interrelated, hierarchically organized simple person-object-relations, and can thus also be referred to as a *complex person-object-relation*. The basal components of interests, then, include (a) concrete objects, referred to as *reference objects*; (b) activities, or action possibilities; and (c) topics, which often serve as frameworks for the individual searching for interest-oriented engagements, thus influencing the type of activity selected and the course of events that results.

The distinction made in this paper between beginning and mature interests is based on the nature of their constituent person-object-relations and on the complexity of the interrelatedness of these person-object-relations. Beginning interests exhibit a relatively uncomplicated structure and consist of person-objects-relations that the adult world considers childlike, unproductive, innocent, focused on the here and now. Mature interests are not only more complex, but the person-object-relations themselves are more adultlike, more work-oriented, perhaps more abstract or having a longer planning horizon.

These two aspects, complexity and maturity, do not vary in discrete intervals, but rather slide along a continuous scale having no upper limit. Because of this, the labeling of an interest structure as beginning or mature is entirely arbitrary. An interest can be called "mature" only in a relative sense -- relative to its earlier stages of development, to the developmental stage of other interests, or to the developmental extent of the interests of other individuals.

The Interest Genesis Project

In conjunction with a longitudinal case study, theoretical concepts intended for use in describing early stages of interest development were defined more precisely. The entire project involved three basic elements of especial importance: (1) recording of descriptive characteristics of individuals' interest-oriented person-object-relations in preschool and kindergarten age children (Fink & Krapp, 1986; Kasten, 1985), (2) description and explanation of the developmental course of interests (Fink, 1991; Fink & Krapp, 1986; Krapp & Fink, 1986), and (3) description of the reciprocal influence of social and object-based relationships (Fink, Schiefele, & Krapp, 1985).

Information was gathered continuously from a small group of children (n=12), in some cases over a period of five years. The data collection began upon the children's entry into preschool or kindergarten. The study employed diverse techniques of data collection: interviews with parents and teachers, check-lists and questionnaires for the teachers, observations in the home and kindergarten.

Our developmentally oriented partial study relied primarily on data collected during interviews with parents. These data consist of 28 interviews with the parents of seven children, conducted at four points in time over the course of five years. The interview questions pertained to the type of games or activities that the child preferred; the duration of, valuing of, and child's fondness for these activities and the objects involved; the social contacts associated with the play activity, as well as any changes in terms of objects or social contacts. Idiographic case studies were conducted using qualitative content analysis. The findings presented here are based on the results of these analyses.

Empirical Findings Relating to the Interest Genesis Model

The interest genesis model contains several conceptual assertions meant to provide a thorough description of the developmental course of interests. The empirical findings from our developmentally oriented partial study can be used to formulate more precisely and to enrich the conceptual goals of the interest genesis model. In what follows, we address these points:

- highlight typical structures of beginning interests in children;

- indicate global principles of development useful in describing structural changes in beginning interests;
- identify specific organizational changes in the structure of the beginning interests;
- demonstrate hypothetical models of interest development.

Typical Structures of Interest-Oriented Person-Object-Relations

In the case of preschool and elementary school children, one finds both simple and complex person-object-relations (Fink, 1991; Fink & Krapp, 1986; Krapp & Fink, 1992). A simple person-object-relation is usually composed of a single, highly preferred component, which is at most unsystematically connected to other structural components. This involves either a preferred activity (e.g., drawing pictures), an important reference object (e.g., doll), or a dominant topic (e.g., behavior of animals).

Example of a simple person-object-relation:
At time t3, Sabine exhibited a highly to moderately preferred person-object-relationship centered on the activity of drawing pictures. Within the framework of the picture drawing activity, Sabine made use of a number of interchangeable materials (e.g., water colors, magic markers, crayons). The subject matter of the drawings varied greatly, including things Sabine had immediately experienced (e.g., sunsets with sailboats), and things that Sabine could remember having experienced (e.g., flowers and butterflies). Taken together, the material shows clearly that the activity of drawing pictures itself was more important than the specific reference objects and topics depicted. The observed variation in drawing materials, topics, and depicted objects underlines the importance of the drawing activity.

Complex person-object-relations contain several simple person-object-relations, grouped around a higher level, highly preferred anchor dimension (Wapner, 1981). An anchor dimension can be characterized as either a highly preferred activity (e.g., "smudging and smearing"), highly preferred reference objects (e.g., dolls), or a prominent topic for cognitive engagement (e.g., animals). Depending on its own character, an anchor dimension lends a corresponding accent to the complex person-object-relation associated with it. Thus, complex person-object-relations can be referred to as activity-centered, object-centered, or topic-centered interests.

Example of an activity-centered complex person-object-relation:
At time t1, and especially in her family environment, Katrin was very involved in an activity dimension that can be summarized as "smudging and smearing." Under this activity dimension fell the following: playing with sand and water, for example, throwing sand, building castles, baking sand cakes, plastering the playground slide with sand, finger painting, "cooking" cut vegetables.

Example of an object-centered complex person-object-relation:
At time t1, Sabine exhibited an object-centered person-object-relation typical for her age and gender, centered around dolls. The dolls were an important part of Sabine's mother-child roll playing: the dolls were dressed and undressed, taken in the stroller, fed, etc. They were also important for "story-telling," and for a version of roll-playing called "Dagmar," in which Sabine imitated her older cousin, and during which the dolls served as playmates in some undertaking (e.g., shopping). Although there was some variation in the selection of dolls for these games, they were not universally interchangeable.

Example of a topic-centered complex person-object-relation:
At time t4, Dirk exhibited a number of simple person-object-relations that fell under the topic of "animals." This included highly valued concrete as well as symbolic "animal objects": taking care of his pet bird (at this time his favorite play object), collecting toy animals (with which Dirk acted out various phantasy games and roll plays, such as "caring for sick animals," "zoo," "circus," or "animal doctor"), visiting the zoo and circus, looking at animal books and reference books, talking about animals with his parents and friends, and collecting tadpoles and sea shells.

Global Developmental Principles of Interest Development

In general, two basic structural aspects characterize all forms of development: differentiation and integration (Bronfenbrenner, 1979; Lewin, 1954; Oerter, 1977, 1981; Wapner, 1981, 1987; Werner, 1957, 1959). In keeping with these developmental principles, we distinguish between two global process components of interest development: incorporation and exclusion (Fink, 1991; Gottfredson, 1981; Todt, 1985; Travers, 1978). Using these two process components, it is possible to illuminate the structural modifications involved in the development process. This process proceeds at two levels, that of the *primary structure* and that of the *secondary structure*. Within our theoretical framework, activities, reference objects, and topics form the primary structure of simple and complex person-object-relations (Fink, 1991). The simple person-object-relations that are found within a complex person-object-relation and that are grouped under an anchor dimension comprise the secondary structure of a beginning interest.

Incorporation refers to the process whereby previously isolated elements are built into an already existing structure, for example, when a simple person-object-relation is built into an already existing complex relation. *Exclusion* refers to the opposite process. In other words, individual elements within a complex person-object-relation are separated from the greater structure. Incorporation and exclusion result in quantitative changes in the structure of a complex person-object-relation, as the number of component elements changes. In addition, qualitative changes are introduced, for example, the structure of the existing complex person-object-relation is modified or organized anew, in that some structural components gain or lose in importance, or a new anchor dimension emerges (see below).

Specific Components of Development

According to Fink's (in 1991) descriptive model, structural changes in beginning interests are possible at all structural levels of a person-object-relation, the primary and secondary levels of individual interests as well as the system level involving all of a person's interests. These changes can involve the basal components of individual person-object-relations (activities, objects, and topics) or a person's entire system of person-object-relations. Accordingly, the primary and secondary structures of beginning interests are subject to reorganization during the course of development. Reorganization can involve modification, but it can involve also developments in a completely new direction. Incorporation and exclusion result in the following structural changes during development:

Reorganization of existing complex person-object-relations. The incorporation and exclusion of both familiar and unfamiliar simple person-object-relations modifies the organization of complex person-object-relations.

Across all four episodes of data collection, Sabine exhibited a topic-centered complex interest in "animals." Notably, a few simple person-object-relations integrated in this complex interest could be pursued independently as well. For example, at time t1, the person-object-relations "being read to" and "read from books" (whose text Sabine had memorized) seemed to have little to do with the topic of animals. At time t2, however, there were numerous indications that both of these simple person-object-relations centered around books about animals and had become integral parts of the topic-centered interest. Parallel to this, however, Sabine also chose other types of books for reading. In addition, it became clear that the complex person-object-relation "animals" contained new structural elements at different points in time: "animal programs" at time t2, "observing animals" and "talk about animals" at time t3, and roll playing "animals" and "building dog houses" at time t4. Some integrated simple person-object-relations showed themselves repeatedly over a longer period of time (e.g., "visiting the zoo" and "playing with toy animals"), whereas others seemed to have been excluded.

Formation of new complex person-object-relations. A simple person-object-relation is incorporated into a newly created complex person-object-relation.

In the case of Dirk, the empirical evidence indicated that, at time t1, the familiar simple person-object-relation "building" was incorporated into a newly arising complex person-object-relation "animals." Dirk went beyond his original activity of building garages and castles, in that he began relating his building activity to his newly discovered preference for animals. The building of zoos was brought into connection with roll playing centered around the zoo, and became an important part of the topic-oriented person-object-relation "animals," which also involved other simple person-object-relations. This included, for example, collecting snails and earthworms, observing animals, racing snails, or making animals out of plasticine.

Structural connections between complex person-object-relations. Over the course of development, complex person-object-relations interlock, and exhibit common structural components (simple person-object-relations). The common structural components belong simultaneously to both of the interests involved, despite the fact that these interests differ qualitatively from one another.

In Dirk's case, several simple person-object-relations were involved in a structural connection between two complex person-object-relations. For example, at time t4, three simple person-object-relations (collecting seashells, collecting tadpoles, and collecting toy animals) were integral parts of both the highly preferred, topic-centered person-object-relation "animals" and the activity-centered complex person-object-relation "collecting things."

Hypothetical Models of Structural Change

Further theoretical consideration leads to hypotheses regarding typical courses of interest development during childhood. Using examples from our study's case material, it is possible to demonstrate the validity of these hypotheses, and to state them in more specific terms. In doing so, three ideal developmental models are especially characteristic: a growth model, a canalization model, and an overlap model.

Growth model. The structural reorganization of a complex person-object-relation tends toward greater differentiation. Instances of incorporation, on both the primary and secondary levels of person-object-relations, are more common, and therefore of greater importance than instances of exclusion. However, the degree of complexity increases not only through the incorporation of new elements in the already existing structure, but also through increased differentiation of components already present.

During all episodes of data collection, Dirk exhibited a topic-centered interest, "animals," which, as it developed, became increasingly differentiated at both the primary and secondary levels. For instance the highly preferred animal role plays grew constantly in extent and variety. The original role play topics, such as "tour through the zoo" or "observation and feeding of animals," were supplemented by new topics, such as "circus act" or "caring for sick animals." Later, these role plays came to include "taking care of one's own pet." The subtopic "collecting animals" was also extended continually. Beginning with a collection of toy animals, this later came to include the collecting of live animals, for example, snails, earthworms, and tadpoles. This served as the basis for an attempt to breed frogs. Finally, the activities of looking at books about animals and of "talking about animals" became increasingly important. Over the duration of the study, a few simple person-object-relations were excluded from the greater structure (e.g., pony riding), but most of the integrated person-object-relations remained and, under the anchor dimension "animals," increased in structural importance.

Canalization model. The reorganization of a complex person-object-relation occurs by way of the increasing differentiation of one area at the expense of other areas. Both of the processes of exclusion and incorporation are subject to the effects of selection. In other words, the range of activities, objects, and topics involved in the processes of exclusion and incorporation becomes increasingly narrow. Differentiation comes to be concentrated in one particular area of the interest structure.

At time t1, Sabine exhibited the object-centered interest "playing with water," which contained several simple person-object-relations as integral parts, such as "playing in a plastic swimming pool," various forms of "playing with water in the kitchen," or "running through the sprinkler." Over the course of development, some of these integrated components were excluded. At the same time, playing in the plastic swimming pool sometimes involved swimming and diving, activities of which Sabine was not yet capable. At a later point in time, following swimming lessons, Sabine increasingly associated the acts of swimming and diving with the highly preferred act of splashing around. "Swimming" and "diving" became increasingly differentiated forms of the previously undifferentiated "splash around in water." The later integration of "water slide" and "high diving" seems to indicate that the relation "playing with water," which at first was very broad, became canalized into a subarea that could be referred to as "athletic-oriented engagement with water."

Overlap model. Qualitatively distinct interests overlap with one another at both the primary and secondary levels, in the sense that they share common elements. Out of this repeated overlapping of individual person-object-relations can emerge a new anchor dimension, which in turn can serve to integrate additional person-object-relations. This new anchor dimension is the starting point for qualitative developmental change, which is accompanied by increasing differentiation.

During all episodes of data collection, Sabine exhibited the simple person-object-relation "drawing pictures." At first, her interest in drawing pictures extended to activities involving a small range of draw-

ing materials and fantasy topics chosen predominantly at random. With time, the drawing activity became more intense and grew to include new forms of engagement. During this phase of pronounced developmental change, Sabine's picture drawing became increasingly associated with other, distinct person-object-relations. The forming of this tie between "drawing pictures," "visiting an art exhibition," and "collecting art postcards" implies a discovery of common structural components. The case material seems to indicate that, based on the overlapping of these person-object-relations, Sabine created a new anchor dimension. This enabled Sabine to expand a simple drawing activity, and to develop a new quality in her interest in drawing, which could be referred to as "artistically oriented drawing activity."

According to the model presented here, the development of interests can be monitored and described in terms of both processes of modification (incorporation and exclusion) and developmental results. It should be noted that the process of incorporation is not necessarily associated with increased complexity, and that the process of exclusion does not always result in reduced differentiation. The reorganization of complex person-object-relations over the course of development is itself a complex matter. Its nature is a function of the reciprocal relationship between the processes of incorporation and exclusion as well as the level at which these processes occur.

Discussion - Interest Genesis as both the Result and a Conditional Factor of Exploratory Behavior

Researchers working in the fields of curiosity and exploratory behavior and basing their work on conceptions of intrinsic motivation have pointed to the relatedness of interest genesis and exploration (e.g., Berlyne, 1949, 1960; Hunt, 1965, 1971; White, 1959, 1960). Of special relevance for a theory of interest are these authors' investigations into the factors and processes found in the relationship between a person and an object.

Berlyne (1949) regarded interest as a form of motivation anchored primarily in the relationship between the organism and a stimulus. He distinguished between three aspects of interest: a striving for novelty, a search for new experiences and knowledge, and esthetic experiences. Within this framework, Berlyne (1960) proposed a class of intrinsically motivated, exploratory behavioral forms (orientation reaction, locomoting exploration, exploratory manipulation, epistemic curiosity) which, he believed, require no additional inducement to that already found in the relationship between organism and stimulus. Collative properties (novelty, complexity, surprisingness, and cognitive conflict) lead to specific exploration, which reduces the degree of arousal. Lack of novelty, complexity, and contrast results in an arousal potential that is too low, and also leads to an increase in arousal. The lack of stimulation results in diversive exploration, whereby the individual searches for a more stimulating situation (Day & Berlyne, 1971). The nonspecific search for stimuli having greater arousal potential leads to a reduction in arousal.

While Berlyne saw no difficulty in tying exploration to the acquisition of new information about surrounding objects, Hunt (1965, 1971) assumed that incongruent information is necessary to trigger intrinsically motivated, exploratory activity. White (1959), on the other hand, sometimes compared a feeling of efficacy with interest, which finds expression in exploratory activities. In this case, the exploration is of a specific nature, and continues for a longer period of time. It is related to the directive, selective, and persistent engagements of the child with certain objects found in the surroundings.

All of these authors attempted to explain exploratory behavior in terms of intrinsic motivation, a concept closely related to interest. On the one hand, these theoretical approaches regard tendencies in exploratory behavior as a constituent part of interest genesis. On the other hand, they suggest that exploratory behavior in certain object domains is based on an already existing interest structure. In determining to what extent this assumption is justified, one can draw upon the model of interest genesis presented here, which is based on the empirical findings of our longitudinal study of the beginnings of interests during childhood.

Exploratory Behavior as a Factor in Interest Genesis

The model presented here for describing interest genesis assumes that, at any given time over the course of individual development, interests come into existence, undergo modification, or disappear entirely. Single components or substructures can extend themselves, whereas others lose in importance or are excluded. Occasionally one finds evidence of a more extensive restructuring process, which is referred to as reorganization.

Within the framework of developing interests, structural reorganization is largely either object- or content-specific. The basic processes referred to as incorporation and exclusion correspond to and grow out of exploratory tendencies at the level of activity or behavior. Interest-specific engagements, therefore, involve object-specific exploration. From the spectrum of action possibilities, material objects, and topics available, the child selects only those things that, in terms of content, are compatible with the preferred anchor dimension of a complex person-object-relation. Exploration involves incorporation and exclusion at both the primary (exploration of activities, reference objects, and topics) and secondary (exploration of simple person-object-relations) structural levels.

The "experiencing of discrepancies" (Voss & Keller, 1983), "seeking novelty" (Livson, 1967; McReynolds, Acker, & Pietila, 1961), "search for structure" (Travers, 1978), or "experiencing of success and failure" (Krapp, in this volume) represent various complementary mechanisms that, within the entire system of interest-oriented person-object-relations, prove favorable to or maintain the child's exploration activities. Although the overlap model of interest development shows that the child creates a new anchor dimension upon discovery of common structural elements in qualitatively distinct person-object-relations, the diversive exploratory behavior based on this new anchor dimension involves objects with which the child is already familiar, and therefore is not unspecific with regard to content. The exploratory behavior of the child exhibits a content or topical orientation in that this behavior serves the newly created anchor dimension (cf. Fink, 1991).

The Importance of Interests in the Development of Exploratory Behavior

The model of interest genesis yields insight into the content orientation of exploratory action tendencies. More specifically, the model identifies constituent elements of the object area that the child explores in the course of interest development. However, structural reorganization during interest development implies not only object specificity. It involves a continuity of object domain. In other words, general aspects of an interest-oriented person-object-relation, involving a particular object domain, remain constant, whereas more specific aspects of that person-object-relation may change through the processes of incorporation and exclusion.

The course of structural development in beginning interests is not the result of chance. Rather, the specificity and continuity of the object domain of interest determine the direction of development. This seems reasonable, when one considers that the individual does not experience structural reorganization as a breaking with something, or as a completely new beginning, but as a pleasant variation and extension of engagement with a familiar object domain. The structure that results contains familiar and content-specific components that, over the course of development, have either remained a part of the structure or have been added anew. The continuity of the interest-oriented object domain makes clear that the exploratory action tendencies that follow upon one another within the framework of interested engagements must necessarily exhibit a content-related continuity.

Conclusions

The theoretical conceptualization of the origins of interest during childhood is itself still at an early stage of development. However, work in this area has already witnessed the development of empirically supported perspectives which yield insight into the central questions of the present paper concerning the importance of exploratory action tendencies in the context of developing interests. Extensive clarification leading to a theoretical understanding of exploration remains to be done, and involves the following:
  - The theory of interest centers on activities that are self-determined, goal-oriented, and directed toward specific objects or object domains. It is hypothesized that our theory, oriented around objects and object domains, is also relevant to early forms of goal orientation and exploratory action tendencies in childhood. The object specificity of interested engagement gives an exploratory action sequence its content-related orientation or specificity.
  - The models of interest development (growth, canalization, and overlapping) form a theoretical basis for describing interest-oriented action structures. Parallel to this, the goal orientation of the explorations involved in the development of a child's interest-oriented structures become evident. The development and restructuring of a child's action goals entail similar changes in that child's observable interests.
  - The development of the goal structures associated with the interests also involves qualitative changes, in the sense of a reorganization of the

structural components of an interest-oriented person-object-relation (e.g., emergence of a new anchor dimension, around which new goals may be grouped). These changes do not occur at random, but instead exhibit continuity in the interest-related object domain. Accordingly, the sequence of events within exploratory activity exhibits continuity in terms of content. In such a case, the interest-oriented object domain determines the direction of exploratory behavior.

References

Berlyne, D.E. (1949). 'Interest' as a psychological concept. *British Journal of Psychology, 39*, 184-195.
Berlyne, D. E. (1960). *Conflict, arousal, and curiosity.* New York: McGraw-Hill.
Bronfenbrenner, U. (1979). *The ecology of human development.* Cambridge, MA: Harvard University Press.
Csikszentmihalyi, M. (1975). Beyond boredom and anxiety. San Francisco: Jossey-Bass.
Csikszentmihalyi M., & Rochberg-Halton, E. (1981). *The meaning of things. Domestic symbols and the self.* Cambridge: Cambridge University Press.
Day, H.I., & Berlyne, D.E. (1971). Intrinsic Motivation. In G.S. Lesser (Ed.), *Psychology and educational practice.* Glenview, Illinois: Scott,Foresman and Company.
Fink, B. (1989). *Das konkrete Ding als Interessengegenstand.* Frankfurt/Main: Peter Lang.
Fink, B. (1991). Interest development as structural change in person-object relationships. In L. Oppenheimer & J. Valsiner (Eds.), *The origins of action: Interdisciplinary and international perspectives* (p. 175-204). New York: Springer.
Fink, B. (1992). Interessenentwicklung im Kindesalter aus der Sicht einer Person-Gegenstands-Konzeption. In M. Prenzel & A. Krapp (Eds.), *Neuere Ansätze einer pädagogisch-psychologischen Interessenforschung* (p. 53-83). Münster: Aschendorff.
Fink, B., & Krapp, A. (1986). *Komponenten eines Modells zur Beschreibung der Interessenentwicklung als Veränderung von Person-Gegenstands-Beziehungen: Theoretische Überlegungen und empirische Befunde aus Fallanalysen.* Beitrag zur Tagung der Arbeitsgruppe für Empirische Pädagogische Forschung (AEPF) in Fribourg, Schweiz, vom 17. bis 19. September 1986.
Fink, B., & Forster, P. (1992). Die Bedeutung materieller Dinge in der häuslichen Lebensumwelt: Eine Pilotstudie. *Zeitschrift für Pädagogische Psychologie, 6*, 115-131.
Fink, B., & Forster, P. (1993). The meaning of things in home environments: a pilot study. *The German Journal of Psychology, 17*, 81-83.
Fink, B., Schiefele, U., & Krapp, A. (1985). *Zur wechselseitigen Abhängigkeit von sozialen und gegenständlichen Bezügen im Kindesalter.* Beitrag zur Arbeitstagung Pädagogische Psychologie, Trier, 25./26. September 1985.
Gottfredson, L. S. (1981). Circumscription and compromise: A developmental theory of occupational aspirations. *Journal of Counseling Psychology, 28*, 545-579.
Hunt, J. McV. (1965). Intrinsic motivation and its role in psychological development. In D. Levine (Ed.), *Nebraska Symposium on Motivation* (pp. 189-282). Lincoln: Nebraska

University Press.
Hunt, J. McV. (1971). Toward a history of intrinsic motivation. In H. Day, D.E. Berlyne, & D.E. Hunt (Eds.), *Intrinsic motivation: A new direction in education* (pp. 1-32). Toronto: Holt, Rinehart & Winston.
Izard, C. E. (1977). *Human emotions*. New York: Plenum.
Kasten, H. (1985). *Beiträge zu einer Theorie der Interessenentwicklung*. München: Habilitationsschrift.
Keil, C. F. (1979). *Semantic and conceptual development: An ontological perspective*. Cambridge, Mass.: Harvard University Press.
Keil, C. F. (1983). On the emergence of semantic and conceptual distictions. *Journal of Experimental Psychology: General, 112*, 357-385.
Keller, H. (1981). Entwicklung explorativen Verhaltens im ersten Lebensjahr. In H.G. Voss & H. Keller (Eds.), *Neugierforschung* (pp. 56-79). Weinheim: Beltz.
Keller, H., Föse, B., & Schölmerich, A. (1985). Materialanalyse explorationsinduzierender Objekte. In W. Einsiedler (Ed.), *Aspekte des Kinderspiels* (pp. 109-126). Weinheim: Beltz.
Keller, H., Schölmerich, A., Miranda, D., & Gauda, G. (1987). The development of exploratory behavior in the first four years of life. In D. Görlitz & J.F. Wohlwill (Eds.), *Curiosity, imagination, and play* (pp. 127-150). Hillsdale, N.J.: Lawrence Erlbaum.
Keller, J. A. (1987). Motivational aspects of exploratory behavior. In D. Görlitz & J.F. Wohlwill (Eds.), *Curiosity, imagination and play* (pp. 24-42). Hillsdale, N.J.: Lawrence Erlbaum.
Krapp, A. (1989). Neuere Ansätze einer pädagogisch orientierten Interessenforschung. *Empirische Pädagogik, 3*, 233-255.
Krapp, A. & Fink, B. (1986). *The transition from family to kindergarten and its impact on person-object-relations*. Paper presented at the 9th Int. Conference of the IAPS; Haifa, Israel, 7.-10.7.1986.
Krapp, A. & Fink, B. (1992). The development and function of interests during the critical transition from the home to preschool. In K. A. Renninger, S. Hidi, & A. Krapp (Eds.), *"Interest" in learning and development* (p. 397-429). Hillsdale, NJ: Erlbaum.
Lewin, K. (1954). Behavior and development as a function of the total situation. In L. Carmichael (Ed.), *Manual of child psychology* (pp. 918-970). New York: Wiley.
Livson, N. (1967). Towards a differentiated construct of curiosity. *Journal of Genetic Psychology, 111*, 73-84.
McReynolds, P., Acker, M., & Pietila, C. (1961). Relation of object curiosity to psychological adjustment in children. *Child Development, 32*, 393-400.
Oerter, R. (1977). *Moderne Entwicklungspsychologie*. Donauwörth: Ludwig Auer.
Oerter, R. (1981). Entwicklung. In H. Schiefele & A. Krapp (Eds.), *Handlexikon zur pädagogischen Psychologie* (pp. 100-107). München: Ehrenwirth.
Oerter, R. (1987). Entwicklung der Motivation und Handlungssteuerung. In R. Oerter & L. Montada (Eds.), *Entwicklungspsychologie* (pp. 637-695). München/Weinheim: Psychologie Verlags Union.
Papousek, M. (1984). Wurzeln der kindlichen Bindung an Personen und Dinge: Die Rolle der integrativen Prozesse. In C. Eggers (Ed.), *Bindungen und Besitzdenken beim Kleinkind* (pp. 155-184). München: Urban & Schwarzenberg.
Papousek, M., Papousek, H., & Harris, B. J. (1987). The emergence of play in parent-infant interaction. In D. Görlitz & J.F. Wohlwill (Eds.), *Curiosity, imagination, and*

play (pp. 215-246). Hillsdale, N.J.: Lawrence Erlbaum.
Prenzel, M. (1988). *Die Wirkungsweise von Interesse. Ein Erklärungsversuch aus pädagogischer Sicht*. Opladen: Westdeutscher Verlag.
Prenzel, M., Krapp, A. & Schiefele, H. (1986). Grundzüge einer pädagogischen Interessentheorie. *Zeitschrift für Pädagogik, 32*, 163-173.
Renninger, K. A. (1989). Individual patterns in children's play interests. In L.T. Winegar (Ed.), *Social interaction and the development of social understanding* (pp. 147-172). New York: Ablex.
Renninger, K. A. (1990). Children's play interests, representation, and activity. In R. Fivush & J. Hudson (Eds.), *Knowing and remembering in young children*. Cambridge, MASS: Cambridge University Press.
Roe, A. (1957). Early determinants of vocational choice. *Journal of Counseling Psychology, 4*, 212-217.
Roe, A. & Siegelmann, M. (1964). *The origin of interests*. Washington: American Personnel and Guidance Association.
Schiefele, H., Prenzel, M., Krapp, A., Heiland, A., & Kasten, H. (1983). *Zur Konzeption einer pädagogischen Theorie des Interesses*. Arbeiten zur Empirischen Pädagogik und Pädagogischen Psychologie, Gelbe Reihe, Nr. 6. München: Universität München und UniBw.
Todt, E. (1985). *Theorie der Interessenentwicklung*. Manuskript für Workshop "Interesse", 8 - 10.3.1985, Gießen.
Travers, R.M.W. (1978). *Children's interests*. Kalamazoo: Western Michigan University.
Voss, H.G. (1985). Zum Handlungssystem Exploration-Spiel in der frühen Kindheit. In W. Einsiedler (Ed.), *Aspekte des Kinderspiels* (pp. 97-108). Weinheim: Beltz.
Voss, H.G. & Keller, H. (1983). *Curiosity and exploration. Theories and results*. New York: Academic Press.
Wapner, S. (1981). Transactions of persons-in-environments: Some critical transitions. *Journal of Environmental Psychology, 1*, 223-239.
Wapner, S. (1987). A holistic, developmental, systems-oriented environmental psychology: Some beginnings . In D. Stokols & J. Altman (Eds.), *Handbook of Environmental Psychology* (pp. 1433-1465). New York: Wiley & Sons.
Werner, H. (1957). The concept of development from a comparative and organismic point of view. In D. B. Harris (Ed.), *The concept of development* (pp. 125-148). Minneapolis: University of Minnesota Press.
Werner, H. (1959). *Einführung in die Entwicklungspsychologie*. München: Barth.
White, R. W. (1959). Motivation reconsidered: The concept of competence. *Psychological Review, 66*, 297-333.
White, R. W. (1960). Competence and the psychosexual stages of development. In M. R. Jones (Ed.), *Nebraska Symposium on Motivation* (pp. 97-141). Lincoln: University of Nebraska Press.

CHAPTER III

Development and Interindividual Differences

CHAPTER **III.7**

The Relationship Between Attachment, Temperament, and Exploration

Dymphna C. van den Boom

Classical theories of exploration and play predict or explain why these behaviors exist and why they are a significant force in development (Groos, 1901; Hall, 1920; Lazarus, 1883; Schiller, 1954). Most early twentieth-century theorists dealt with exploration and play as a secondary topic of interest, because each of these psychologists made a mark in other areas of developmental psychology (Freud, 1959; Piaget, 1962; Vygotsky 1967). In recent years there have been three major threads of thought concerning exploration and play. Berlyne (1960, 1966) has introduced an arousal theory based on behavioral learning theory. Bateson (1955, 1956) has proposed a theory that focuses on the communicative features of exploration and play using anthropological-systems theory. Finally, Sutton-Smith (1966, 1967) and Bruner (1972) have emphasized exploration and play as a source of behavioral variability based on a cognitive adaptation framework (Rubin, Fein, & Vandenberg, 1983).

Several researchers have studied exploration in the context of attachment theory (e.g. Ainsworth, 1967; Bowlby, 1969). A basic assumption of attachment theory is that a dynamic balance exists between the attachment and exploration behavior systems. This notion is captured by the secure base phenomenon stating that the infant's tendency to explore the environment varies inversely with its tendency to maintain proximity and contact with the caregiver. Given this frequently documented phenomenon, it is surprising how little work is available examining associations between security of attachment and the quantity and quality of exploration (Belsky, Garduque, & Hrncir, 1984). In this chapter a study is reported which explicitly examines the relationship between attachment and exploration in a population of irritable infants.

The Balance between Attachment and Exploration

Although Bowlby's work *Attachment and Loss,* of which *Attachment* is the first volume, is an effort to re-examine the nature of the child's tie to his mother, Bowlby takes care to point out that attachment behavior is not the only determinant of the mother-child interaction. According to Bowlby (1969) the spatial relationship between a mother and a child is the result of the dynamic equilibrium of four classes of behavior: the attachment behavior of the child; the child's exploratory behavior and play, which is antithetic to attachment; caretaking behavior of the mother; and behavior of the mother that is antithetic to mother-child interaction (e.g. household duties). Hence, a child's attachment behavior is seen as only one class of four separate classes of behavior that make up mother-child interaction. Behavior in each of these four classes varies greatly in intensity from moment to moment, and behavior from any one class may, for a time, be absent altogether. Moreover, each class of behaviors is likely to be affected by the presence or absence of the others, because the consequences of behavior of any one class are likely either to elicit or to inhibit behavior of the other three classes. Thus, when mother is called away, a child's attachment behavior is likely to be elicited and his exploratory behavior inhibited. Conversely, when a child explores too far, a mother's caretaking behavior is likely to be elicited and whatever else she is doing to be inhibited. In a happy dyad these four classes of behavior are likely to occur and progress together in harmony (Bowlby, 1969).

Bowlby considers exploratory behavior to be a class of behavior in its own right. By taking him away from mother, a child's exploratory behavior and play is seen as antithetic to his attachment behavior. In Bowlby's theorizing, then, the term attachment is restricted to that component of a relationship which has to do with the regulation of security and protection. The model he proposed to explain the nature of the child's tie to his mother is a security-regulating model, in which exploratory behavior as such is not incorporated. It was Ainsworth (1973; Ainsworth, Blehar, Waters, & Wall, 1978; Stayton, Ainsworth, & Main, 1973) who incorporated the infant's exploratory behavior in her view of the development of attachment, paying attention to the balance between attachment and exploration. Later Waters (1981) and Bischof (1975; Gubler & Bischof, 1990) elaborated this view, incorporating exploratory behavior into the arousal system as the antipode of fear.

The Influence of Infant Irritability on Exploration

Attachment behavior is elicited by a too distant and too prolonged withdrawal of the mother. Irritable infants can be expected to resort more to distress signals as a means of reducing this distance than do nonirritable infants. Distress signals, however, are less effective than more positively toned attachment signals because they operate through negative feedback. That is, distress signals decrease with increasing proximity of the mother, thereby reducing the likelihood of a response from the mother that lasts until the child's need for security is met. The mechanism through which irritability exerts its influence on the security system is reported more fully in van den Boom (1988) and will not be considered further in the present report.

Irritability may also be expected to influence arousal. The overemission of distress calls reduces the likelihood that the child will receive sufficient care from the mother (van den Boom, 1988). If the level of felt security provided by the mother is insufficient to meet the child's need for security, the child's arousal level will be elevated. A high level of arousal will lead to fearful behavior, therefore reducing the child's exploratory tendencies.

There is some empirical evidence that corroborates this notion. It has been shown that susceptibility to distress hinders cognitive development. Difficult infants, who have been shown to emit distress signals with a high frequency (Bates, Freeland, & Lounsbury, 1979), show slower cognitive development (Dunst & Lingerfelt, 1985; Field et al., 1978; Sostek & Anders, 1977; Wachs & Gandour, 1983). The infant with a lower mental test score is likely to receive higher ratings on negative emotion or mood (Matheny, 1989). This offers support for the notion that infant irritability affects the exploration side of the attachment-exploration balance in a negative way. After all, exploration is a direct avenue to the child's cognitive functioning in the infancy period. Possible irritability effects on the cognitive functioning of infants have also been demonstrated in laboratory procedures involving habituation. According to two recent studies, there may be meaningful differences between infants who complete testing and those who do not. The noncompleters are more difficult (Trieber, 1984; Wachs & Smitherman, 1985). In addition, infants who are fearful as assessed in the laboratory tend to score lower on tests of infant development (Lamb, 1982). Fear of exploration of the environment and fear of orienting to interesting features of it can result in behavioral disorganization. Such disorganization may prevent learning in the immediate situation. These results support the possibility of differential reactivity to the environment by easy versus difficult babies.

Thus, irritability may negatively influence the arousal system by activating the fear component more often than the exploratory component. This, in turn, may negatively influence the quality of the child's exploratory behaviors because the child will have fewer opportunities to practice his exploratory capacities.

Exploratory Behavior in the First Year

Despite the fact that exploratory behavior gained gradually more prominence in attachment models, none of the major theorists went into detail regarding the content of the infant's exploratory behavior. In attachment models, exploration simply means being busy with something other than the mother or the relationship (Keller, Schölmerich, Miranda, & Gauda, 1987). This is in contrast to attachment behaviors which have been described in detail (Bowlby, 1969; Ainsworth, 1963, 1967, 1973). It can be easily inferred from attachment theory that the use of an adult as a secure base supports the quality of exploration and thus supports cognitive and social development as much as it supports survival (Waters & Deane, 1982). Exploratory behavior is conceived of by Bowlby (1969) as being mediated by a set of behavioral systems evolved for the special function of extracting information from the environment. As in the case of attachment behaviors, a different meaning can be attributed to every instance of phenotypically similar behaviors. The meaning of exploratory behavior is also acknowledged to be a function of context. Sucking, for instance, is a clear example of this. Bowlby (1969) identified nutritive and nonnutritive sucking as behaviors that promote infant-adult proximity and contact; and yet sucking is also used extensively in exploration throughout infancy (Waters, 1981).

The study of exploration is a field of investigation in its own right. Psychologists, anthropologists and other behavioral scientists agree that young children spend lengthy time exploring their environments (Rubin et al., 1983). However, exploration is difficult to define. The results of several studies indicate that it is important to take into account the distinction between exploration and play when painting a picture of a child's developmental status. Historically, a distinction has been drawn between specific exploration, play and diversive exploration. Exploratory behavior is followed by playful manipulation and play activities can lead to diversive exploration (Berlyne, 1960). In a similar way, Weisler and McCall (1976) offer a contrast between exploration and play in the course of integrating and summarizing in general terms how scholars have characterized these two classes of behaviors. Exploration has been described as a relatively stereotyped perceptual-motor examination of an object, situation, or event, the function of which is to reduce subjective uncertainty. Exploration is seen when a child is confronted with a novel object and intends to ask, "What is this and what can it do?" Furthermore, it is characterized by visual investigation, active examination, manipulation and a prolonged attention to the object and activities with it. There is considerable evidence that through exploration of novel objects the child develops a fuller understanding of these objects. The principal feature of exploration is that it is dominated by and focused on the current stimulus situation (Weisler & McCall, 1976). Hence, exploratory behavior is behavior directed toward objects in the environment for the purpose of extracting information from them through visual investigation, active examination, handling, and prolonged attention to the objects and the effects of the activities that can be performed with them. In the remainder of this text these behaviors will be captured in the term exploration.

Play, in contrast, consists of behaviors and behavioral sequences that are organism- dominated rather than stimulus-dominated. Play behaviors appear to be intrinsically motivated and performed for "their own sake" and are conducted with relative relaxation and positive affect (Weisler & McCall, 1976). In play the child intends to ask "What can I do with this object?" In contrast to exploration, in which the child examines the properties of an object, in play, the child investigates different ways of using the object. Play, then, is a more advanced activity following specific investigation of an object. Play involves trial and error and the chance combinations of responses and is a form of "variation-seeking" with an object and a child's own behaviors (Sutton-Smith, 1975). Thus, exploration transforms the novel into the familiar, while play transforms the familiar into the novel (Miller, 1974).

Nunnally and Lemond (1973) presented a temporal scheme that specifies the various components of exploratory behavior in relation to one another. These investigators identified observable behaviors and correlated covert processes that are intended to cover the domain of exploratory behavior. It is assumed that throughout life there is an endless cycle of encountering a stimulus that initiates the proposed sequence. In encountering a stimulus, the first behavior to be noted typically is orienting behavior which provides an overall, immediate heightening of attention to the presence of the object. The stage of orienting behavior is followed by a stage of perceptual investigation. The covert processes corresponding to perceptual investigation are continued attention and, more importantly, the beginning of a process of encoding. In this stage an individual gives meaning to the stimulus in the sense of identifying and categorizing the object. After perceptual investigation leads to a partial encoding of the stimulus, a stage of manipulatory behavior frequently ensues. In this stage the covert process of encoding continues leading to a mental speculation about the object in terms of its origin, its usefulness, its relation to other objects in the environment, and so forth. This type of encoding is referred to as transformational thinking. Blending into the end of the stage of manipulatory behavior there is a set of activities which is best referred to as play. Nunnally and Lemond (1973) point out that no hard and fast distinction can be made between these two phases, but that, play activity appears to be less directed toward learning about the object in any sense, and more toward using the object for some pleasant activity. In play activity the covert process is referred to as autistic thinking, because the organism is pleasantly distracted from thinking about the object at all or the object enters into fantasies. Typically, in this stage of exploratory behavior, boredom begins before the play activity ends. As boredom grows in intensity, restlessness leads to the last stage in the temporal scheme, namely, searching behavior. Searching behavior leads an encounter with another stimulus which has the properties to evoke exploratory behavior, and the sequence is reinstigated. It is important to keep in mind, however, that only some objects in the environment induce exploratory behavior. Many investigations of exploratory behavior have concerned the particular stimulus characteristics that initiate this sequence of activities. Furthermore, not all objects that initiate the initial stages elicit subsequent stages of manipulatory behavior or play activity. And, finally, the amount of time spent in each of the successive phases and the probability of moving from one phase to the next depend upon the nature of the stimulus, the other stimulus impingements in the environment, organismic states, maturity of the individual, and possibly many other factors (Nunnally & Lemond, 1973).

A number of theorists have suggested that the structural properties of exploration and play change with and reflect development (McCall, Eichorn, & Hogarty, 1977; Piaget, 1962; Vygotsky, 1967). The distinction between exploration and play and the temporal relationship between the two has been examined empirically by Hutt and her colleagues (Hughes, 1978; Hughes & Hutt, 1979; Hutt, 1967, 1970, 1979). Over time exploration decreases, whereas the amount of time spent playing with toys increases (Hutt, 1966). Because the study to be reported here covers the first year of life, we will restrict ourselves to exploratory behavior, which is a central aspect of a child's behavioral repertoire in this developmental phase (McCall et al., 1977). Perhaps during no other era is exploration as direct an avenue to the child's cognitive functioning as during the infancy period, when a large component of thought is externalized (Fenson, Kagan, Kearsley, & Zelazo, 1976).

Much of the empirical research on exploration has been descriptive and as such has yielded provocative information about the changing pattern of behaviors that infants produce in "free play" settings. In the typical study, infants are presented with a set of attractive toys and are invited to play. In some studies, children are observed in their own homes (Belsky & Most, 1981; Ruff, 1984), and in others they are observed in a special playroom (Fenson et al., 1976; McCall, 1974; Zelazo & Kearsley, 1980). The purpose of these studies has been to record the way children respond to these objects. Although investigators differ concerning their particular observational coding schemes, a number of common features have emerged. An infant's interest in objects is apparent from the first week of life, when he will focus and then follow a brightly colored object dangled in his line of vision (Brazelton, 1973; Packer & Rosenblatt, 1979). *Visual fixation* is the predominant mode of exploration in the first few months of life. That this is an information-getting activity has consistently been reported in studies showing cardiac deceleration while fixating on a visual array (Lewis, Kagan, Campbell, & Kalafat, 1966). Even though Bower and his collaborators have argued that *reaching* behavior may be demonstrated in neonates (Bower, 1972, 1974) - a claim which was contradicted by more recent findings (Di Franco, Muir, & Dodwell, 1978; Field, 1976; Ruff & Halton, 1978) - the fine and complete modulation of this behavior is only apparent around 5 months (Pomerleau & Malcuit, 1980). A 6-month-old baby typically explores one object at the time (Fenson et al., 1976). There is considerable agreement that from 7 months on, behaviors involving a single object decrease.

The infant, then, begins to combine or *relate two separate objects* and shows an interest in similarities among objects (Fenson et al., 1976; Kagan, Kearsley, & Zelazo, 1978; McCall, 1974; Rosenblatt, 1977; Zelazo & Kearsley, 1980). The downward trend in single-object use, as well as the percentage of these behaviors produced at a given age level, is strikingly consistent across studies, in which different toys and slightly different response definitions are used, and in which children from different cultural backgrounds are observed (Rubin et al., 1983). Another consistent finding is that simple two-object combinations occur infrequently at any age (Fenson et al.,1976; Kagan et al., 1978). From about 8 months on, actions in keeping with the *specific functions* and social usages for the object increase (Rosenblatt, 1977; Zelazo & Kearsley, 1980). This implies that the infant has developed the capacity to deal with materials in a discriminatory fashion, extracting unique information that is inherent in them. Again, the appearance and timing of these behaviors seems little affected by situational or cultural variations. These patterns have been observed in French children in a crèche (Inhelder, Lezine, Sinclair, & Stambak, 1972), in Guatemalan children reared in an impoverished village (Kagan et al., 1978) and in American and English children from a variety of social class groups (Fenson et al., 1976; Rosenblatt, 1977; Fein & Apfel, 1979). By 12 months the infant is interested in a great number of toys, which he subjects to a number of instrumental activities, such as *banging,waving*, and *pushing* (Fenson et al., 1976).

Thus, infants' exploratory behaviors in relatively unstructured "free play" situations seem to demonstrate systematic changes during the first year of life. In the next section we will turn to parental behavior that may influence the quality of the infant's exploratory behavior.

Infant Exploration: Relations to Home Environment and Security of Attachment

Two factors that have often been postulated to influence exploratory behavior are child-rearing and parental factors. It is possible that the relationship between child-rearing and exploration is influenced *indirectly* by systems that facilitate or restrict children's behaviors in general. Attachment could be such a system. Some evidence indicates that the infant's tendency to explore the environment varies inversely with its tendency to maintain proximity and contact with the caregiver. There is good reason to expect that securely attached babies should be more competent explorers, because of their assumed freedom to attend to the environment beyond the attachment figure. Still, it is especially surprising that the limited data available on this issue are remarkably inconsistent, particularly with respect to performance under low-stress circumstances.

In many of the studies initiated from attachment theory, the Strange Situation (Ainsworth & Wittig, 1969) is used to classify infants into three molar categories: A, or Insecure Avoidant; B, or Secure; and C, or Insecure Resistant. Recently Main (Main & Solomon, 1986) has added a fourth category, D, or Disorganized. These molar categories are further divided into subcategories. Tracy, Farish, and Bretherton (1980) reported that securely attached infants spent more time manipulating toys than did insecure (A or C) infants. In another study using a low-stress lab situation, Jennings, Harmon, Morgan, Gaiter, & Yarrow (1979) reported the opposite results, though it should be noted that, unlike Tracy et al., these investigators did not employ the Strange Situation to assess security of attachment. Finally, Bretherton and her colleagues reported that when subcategories of attachment classifications measured at 1 year are scaled in terms of their "distance from the normative B_3 group," this continuous measure of quality of attachment is consistently related to the breadth, variety, and level of symbolic play observed during the last 2 months of the infant's first year. These results are consistent with Matas, Arend, and Sroufe's (1978) longitudinal data linking 18-month attachment with mean duration of bouts of symbolic play at 24 months. The findings are inconsistent with Main's (1973) results, which revealed no link between security of attachment at 12 months and level and duration of symbolic play at 20 months. Sorce and Emde (1981) explored the issue further, investigating the relationship between a mother's signalling of her unavailability and infant exploration in an ambiguous situation. The study was designed in such a way that confounding of the mother's physical presence and her sensitive responsiveness could be avoided. Maternal unavailability was simulated by having the mothers read a newspaper during toy presentations to the infant. In this way the mothers remained unresponsive to their infants' requests for attention. Other mothers served as a nonreading contrast group by watching their infants during the same toy presentations and responding sensitively to infant requests. Infants in the maternal reading condition exhibited less pleasure and less exploration, suggesting that the critical feature of the mother's availability is her willingness to respond to her infant's signals on an emotional level. Belsky et al. (1984) conducted a study representing another effort to identify contemporaneous patterns of covariation between individual differences in attachment and in exploration. The exploratory categories used were similar to the ones presented in the previous section. The results of this study support the contention that infants whose attachments to their parents provide a secure base from which to explore experience more freedom to attend to the environment and are thus more able to engage in cognitively sophisticated exploration. Infants who feel less secure, and presumably spend more time being concerned about the world surrounding them, appear less able to engage in cognitively sophisticated play and to engage freely in exploratory activity.

The studies by Sorce and Emde (1981) and Belsky et al. (1984) especially underscore the construct validity of the exploratory behaviors presented in the previous section as measures that tap the infant's inclination to spontaneously deploy, in free play, the cognitive capacities he or she has available. Exactly these exploratory behaviors are related to attachment in a way predicted by contemporary attachment theory. The samples of the latter two studies consisted of infants aged 12 months and beyond. However, the results clearly specify that the kinds of exploratory behaviors that should be taken into consideration when studying the attachment-exploration balance at earlier ages are those exploratory activities representing the child's developmental level. In addition, Sorce and Emde's study suggests that the mechanism responsible for the relationship between attachment and exploration is maternal sensitive responsiveness.

Taken together, the data suggest that security of attachment, for the most part brought on by maternal sensitive responsiveness, may affect the quality of exploration. As mentioned before, a basic assumption is that a balance exists between the attachment and exploratory behavioral systems. This seems to be supported by the results presented above, at least for the age of 12 months and beyond. However, given this frequently documented phenomenon, it is surprising how little work is available examining associations between security of attachment and the quantity and quality of exploration *during* the first year of life.

From an evolutionary perspective, it is clear that lack of inhibition in exploration, or ineffective strategies of exploration will adversely affect an individual's prospects of survival. Exploration is the activity through which a child acquires information about novelty, thereby reducing informational conflict (Nunnally & Lemond, 1973). Exploration is the outcome of a decision between either fleeing or investigating a novelty in the environment. However, if a stimulus is both novel and relevant, it is more *adaptive* to explore than to retreat because of fear which is antithetic to exploration. Thus, both the mode of decision and the strategy of exploration have a certain survival value. If a fear response is evoked, attachment behavior will be displayed. In case of an excessive deviation from the arousal level in either direction, the appetence component of the arousal system will be disinhibited.

Method

The study presented here is an intervention study in which sensitive responsiveness was experimentally manipulated in dyads with irritable infants. This offers the possibility to examine the relationship between sensitive responsiveness and quality of attachment and exploration in a more causal way. So far the majority of studies initiated from attachment theory have been correlational in nature. The following hypotheses were put forward:

1. Infant of mothers who participated in the intervention program will explore more than infants of control group mothers.

2. The quality of exploratory behavior of infants of intervention mothers will be higher than that of infants of control group mothers.

Subjects

The sample consisted of 100 irritable infants, that is, 47 girls and 53 boys, and their mothers. All infants were first-born children of lower socioeconomic status and for the most part from intact families. All infants were carried to term and had weighted more than 2,500 grams at birth. The infants were born to women who had uneventful pregnancies, no major delivery complications and who had not received more than routine medication during delivery. Apgar scores were at least 7 and 8 at 1 and 5 minutes, respectively. All mothers and infants were in good physical health and were not considered to be medically at risk. To determine the degree of infant irritability, the Neonatal Behavioral Assessment Scale (Brazelton, 1973) was administered on the tenth and the fifteenth day of life. Scores from the *peak of excitement, rapidity of buildup* and *irritability* items were combined and averaged across the two administrations. Infants whose irritability scores were six or higher were considered irritable; those with scores below six were considered nonirritable. The Brazelton had to be administered to 588 infants to find the predetermined number of 100 irritable infants.

Design

In order to study intervention effects on maternal and infant behavior, two treated and two control groups were used (Solomon & Lessac, 1968). One experimental and one control group were pretested to estimate the subjects' capacities at the start of the experiment. However, certain elements of pretests, such as, for instance, the presence of an observer, may contain aspects that have an influence on maternal behavior. Such pretests may have an interaction effect with the actual intervention procedures and/or main effect on the outcome variables. On the other hand, without a pretest, the investigator has limited knowledge as to whether the intervention advanced development or prevented a more rapid retardation relative to the control group. To overcome such shortcomings, in a four-group design the other experimental and control groups are not pretested. The pretested control group was observed as often as the treatment group received intervention visits to control for the possibility that getting attention alone would have a positive effect on maternal behavior. The four-group design requires random assignment of subjects to groups. Since studies of irritable infants are often limited to a small population, randomization may be insufficient to assure group equivalence. In such a case a more adequate procedure is to block infants on preselected variables of known importance (e.g., birthweight, labor complications, medication, etc.) and then to randomly assign each cohort to one of the four study groups, which is the procedure followed in this study.

Variables and Data Collection Procedures

The first group of dependent variables in this intervention study related to the quality of maternal responses to infant signals. The second dependent variable was the quality of infant exploratory behavior. We decided to index the quality of exploration in a free play situation for the following reasons. The validity of such an approach is suggested by Belsky, Goode, and Most (1980). These investigators not only showed that infant free play behavior, if appropriately measured, is sensitive to ontogenetic change, but also that the quality of such activity can be influenced by the infant's early environment in a theoretically predicted manner.

The choice of behavioral measures to index exploratory quality was suggested by Sorce and Emde (1981) and Belsky et al. (1984) and descriptive empirical research on exploration. Categories reflecting the influence of maternal sensitive responsiveness were preferred. The observational system consisted of looking, undifferentiated exploratory behaviors (mouthing, simple manipulation, rotation), and functional behaviors, (i.e., behaviors that fit the specific features of the play objects). In addition to the variables included in the observational system, several summary and derived measures were employed which represented indices of tempo, level of exploration, attention span, and exploration time. Several investigators noted vast individual differences in tempo of exploration - the extent to which an infant dashes from one toy to the next as opposed to displaying long sustained contact with a single object (Kagan, 1971; Maccoby & Feldman, 1972; McCall, 1974). Wenckstern, Weizmann, and Leenaars (1984), who studied the relationship between temperament and tempo of play, suggested that rapid-tempo infants are more intense and less sensitive to environmental stimulation and change. The stability of play tempo and relationships with temperament found support the idea that there is behavioral consistency in infancy and that broader stylistic patterns may partially underlie these consistencies. Because our sample consisted of irritable infants we found it important to include a measure on behavioral tempo. *Perseverance time* served as an index of tempo of exploration and was derived separately for each infant. Perseverance time was computed in the following way. First, the total amount of time spent in physical contact with a toy (total contact time) was calculated. Then, the total number of times an infant made contact with a different toy (act changes) was computed. Multiple contacts with the same toy were counted as act changes when the contacts were separated by intervals of three seconds or more except for an interruption caused by social contacts. In the latter case, continuation of play with the same toy was not counted as an act change because almost every infant continued to explore the toy it already played with after social contact. Finally, perseverance time was computed by dividing the contact time scores by the number of act changes.

In their study, Wenckstern et al. (1984) did not find a relationship between tempo and the ability to sustain attention, at least not when attention span was measured by means of a questionnaire as a temperament dimension. The temperament dimension attention span/persistence did not contribute to the discrimination of the tempo groups. However, the investigators suspect that there does exist a relationship between attentional factors and tempo, but that the attentional indices relevant to tempo are more subtle than the relatively gross measure of overall attention captured by temperament questionnaires. We therefore coded two measures of *attention span* that were directly derived from the infant's behavior during the free play session to be able to examine tempo/attention relationships in more detail. Previous research had shown that in exploration sessions, in which several toys are offered at the same time, infants become distracted by one toy while they are playing with another. In addition to that, infants spend time sorting through the objects before finding one that eventually catches their attention (Power, Chapiesky & McGrath, 1985). Therefore, as Power et al. (1985) did, two measures of attention span were coded: total amount of time spent exploring toys during the entire session, and mean length, in seconds, of the two longest periods of uninterrupted object involvement. Attention span was defined as the tendency to focus attention on an object or objects for prolonged periods, especially in case suitable object(s) for extended exploration had been found. Therefore the second measure of attention span was chosen instead of the average length of object involvement calculated over all instances. A measure using all periods of object involvement would be inappropriate because it would include much of the infant's sorting behavior. The mean and median are thus inadequate, because brief periods of object involvement are included in these scores. Moreover, unlike the mean of the two longest periods of object involvement, Power et al.(1985) found that the overall mean and median did not show significant levels of stability across testing sessions.

Another measure used was *level of exploration*, the highest level of exploration observed, which had been identified by Belsky and Most (1981) as an important individual difference measure. Thus, the summary and derived measures included in the analyses, in addition to the variables included in the observational system, were: perseverance time, two measures of attention span (total amount of time exploring toys [I], and mean length of the two longest periods of uninterrupted object involvement [II]), and level of exploration.

The infant played with three toy sets for 5 minutes, and each toy set was presented in a fixed order. Every 3 seconds, exploratory behaviors and the toy they were performed with were registered.

The third dependent variable was the quality of attachment. Between the ages of 12 and 13 months a follow-up measure was conducted for every mother-child dyad. This follow-up consisted of observation of the dyads in the Strange Situation procedure. Instructions for scoring the Strange Situation were followed exactly as outlined by Ainsworth et al. (1978). Infants were classified into the three main groups: secure, insecure avoidant, and insecure resistant.

The independent variable was the presence or absence of the intervention aimed at enhancing maternal sensitive responsiveness. The intervention was implemented between the sixth and the ninth month of life. Intervention took place during everyday interactions that were immediately relevant to the process of intervention. The advantage of this procedure is that the infant's behavior in this natural environment is likely to be similar to - although not identical with - that which the mother experiences most of the time. It also gives the intervenor a more realistic perspective of what the mother is coping with in the home. Three intervention sessions were scheduled for the dyads of the experimental groups. Sessions took place once every three weeks. The model adopted for parent education in this investigation centered entirely around the concept of sensitive responsiveness. The intervention was deliberately restricted to manipulating this one variable to facilitate evaluation of the intervention's effect. Many infant intervention programs are so multifaceted that it is difficult to ascertain which aspect of the intervention is responsible for which change in a participant's behavior.

Mothers were assisted to adjust their behaviors to the infant's unique cues. In doing so we elaborated on what we learned from the results of an observational study (van den Boom, 1990). Because the majority of mothers of irritable infants were uninvolved or ineffectively involved with their infants, intervention began with manipulations to affect maternal attentive behavior (imitation, repetition of vocalizations, and silencing during gaze aversion). It often happened that the infant fussed or cried during the sessions. These instances were used as behavior change targets regarding sensitive responsiveness to negative infant signals. For mothers with irritable babies it is also important to show that interaction can be positive and rewarding. Hence, during playful interactions, attention was devoted to recognizing infant signals, responding appropriately and observing the infant's subsequent response. In between discussions with mothers, when they were interacting with their infants, maternal and infant behaviors were observed using the same observational system that was used for pre- and posttest observations. In this way, interactive sequences were led in a more objective way and these protocols were analyzed before the next intervention session took place to see if changes occurred and what should be the focus of the next intervention session.

Results

Effect of the Intervention on Infant Exploratory Behavior

A major purpose of observing the infant's exploratory behavior was to examine increases in the quality of this behavior after intervention.

In order to reduce the data base, the variables of the observational system were divided into six groups based on prior analyses. Active and other mouthing, and alternating constituted the *mouthing* set. Detailed manipulation and rotating composed the *undifferentiated manipulation* set, whereas *simple manipulation* constituted a set of its own with the same name. Vigorous manipulation and throwing with visual guidance composed the *vigorous exploration* set. The *relational* set consisted of the categories "simple relational" and "group." And, finally, functional, functional-relational, and pretense acts constituted the *specific exploration* set. Perseverance time, the two measures of attention span (I and II), and level of exploration were the other variables entered into the analysis. To circumvent the problem of positively skewed distributions, principal components analysis was performed on logarithmic transformations of the original variables. Complete data were available for pretest sessions of 50 subjects.

Table 1 displays the principal components solution.

Table 1. Principal Components Analysis of Infant Exploratory Behavior (pretest)

Exploratory variable		Component loading			
		I	II	III	IV
I.	Sophisticated exploration (27%)				
	Relational	.85	-.22	.12	.13
	Level of exploration	.84	.03	-.20	-.10
	Specific exploration	.73	.31	-.11	-.19
II.	Tempo of exploration (26%)				
	Perseverance time	.03	.92	.01	.18
	Attention span (II)	-.04	.91	-.01	-.09
III.	Nonspecific manipulation (15%)				
	Simple manipulation	-.09	-.11	.91	-.23
	Attention span (I)	.28	.49	.64	.36
	Vigorous manipulation	.45	-.02	-.62	-.34
IV.	Mouthing (12%)				
	Mouthing	.09	.45	.09	.79
	Undifferentiated manipulation	.38	.31	.12	-.67

Four components accounted for 80% of the variance. Infants who score high on the first component are high in relational and specific exploration. This component appears to tap sophisticated exploration. The second component extracted is tempo of exploration. Infants who score high on this component are high in perseverance time and high in attention span as measured by the mean length of the two longest periods of uninterrupted object involvement. Infants who score high on the third component are high in simple manipulation and attention span as measured by the total amount of time the infant spent exploring toys during the entire session, and low in vigorous manipulation. Hence, this component seems to tap nonspecific manipulation. And, finally, infants scoring high on the fourth component are high in mouthing and low in undifferentiated manipulating. Thus, the fourth component extracted pertains to mouthing.

The statistical analysis of the exploratory data was further carried out in the following way. First, analyses of variance were applied to the pretreatment observations to establish the initial equality of means of the groups. Second, differences among the groups during the posttreatment observations were examined by analyses of (co)variance (using the pretest scores as covariate). Again no sex differences were obtained in separate analyses centered on the variables. Means and standard deviations of the components are reported in Table 2.

Table 2. Means and standard deviations for pre- and posttest components of infant exploratory behavior (logarithmic transformations)

	Experimental groups		Control groups	
	M	SD	M	SD
pretest sophisticated exploration	-0.77	2.44	0.55	2.57
posttest sophisticated exploration	0.67	1.78	-0.67	2.10
pretest tempo of exploration	-0.04	2.06	-0.28	3.22
posttest tempo of exploration	-0.08	1.80	0.08	2.16
pretest nonspecific manipulation	1.02	1.08	-1.43	2.55
posttest nonspecific manipulation	0.23	1.34	-0.23	1.86
pretest mouthing	0.40	1.55	-0.50	1.60
posttest mouthing	-0.31	1.42	0.31	1.42

At the time of the pretreatment observations, infant exploratory behavior did not differ significantly on seven of the ten variable sets among experimental and control conditions, all F-values < 3.25, p's $> .10$. Experimental infants showed more simple manipulation, $F(1,48) = 12.94$, $p < .001$, and less vigorous exploration, $F(1,48) = 14.19$, $p < .01$ when compared to the control infants.

Next, completely randomized 2 x 2 (intervention - no intervention versus pretest - no pretest) factorial analyses of variance were performed on posttest scores. Main effects for treatment condition were found for sophisticated exploration, $F(1,96) = 12.64$, $p < .001$, and mouthing, $F(1,96) = 4.61$, $p < .05$. Experimental infants showed significantly more sophisticated exploration and less mouthing when compared to control group infants. No significant differences were found for either tempo of exploration or nonspecific manipulation. Significant main effects for pretest condition were found for sophisticated exploration $F(1.96) = 7.66$, $p < .01$, and nonspecific manipulation $F(1,96) = 5.04$, $p < .05$. Post hoc comparisons revealed that infants who were pretested received higher scores on sophisticated manipulation than did infants who did not receive a pretest. The opposite was found for nonspecific manipulation. In this case infants who did not receive a pretest had higher scores than pretested infants. No significant interactions were discerned.

In univariate analyses of covariance, we further obtained significant differences for sophisticated exploration, $F(1,95) = 16.58$, $p < .001$, and mouthing, $F(1,96) = 8.17$, $p < .01$, indicating again that program infants scored higher on the first infant summary variable and lower on the second when compared to the control infants. A significant main effect for pretest condition was found for sophisticated manipulation, $F(1,95) = 7.48$, $p < .01$. Pretested infants scored higher than infants who did not receive a pretest. For sophisticated manipulation we also found a significant interaction of treatment x pretest condition $F(1,96) = 3.95$, $p < .05$.

A completely randomized analysis of variance on the experimental and control groups that were not pretested yielded only marginally significant effects of sophisticated manipulation, $F(1,48) = 7.92$, $p < .10$, and mouthing, $F(1,48) = 3.49$, $p < .10$, again indicating higher scores on sophisticated manipulation and lower ones on mouthing for program infants.

Because it might be conceived that irritable infants would be easily distracted in a play session in which a number of toys are available because of their high paced behavior, we tested for differences in crying behavior between the groups. None of the tests performed was significant, indicating that groups did not differ in unfocused (crying) behavior during the play sessions.

Results from the analyses of the data gathered during the free play assessment of exploratory competence provide additional support for the conclusion that the intervention was effective in enhancing the quality of infant exploratory behavior. While the control infants were significantly more likely to engage in the least cognitively sophisticated kinds of play, the experimental infants engaged in functional and pretend play significantly more frequently, and scored significantly higher on the composite exploratory competence score. No group differences were discerned for the two measures of tempo of exploration and nonspecific exploration.

Effect of Intervention on the Quality of Attachment

The follow-up measure assessed was security of infant-mother attachment at 12 months. The quality of attachment was coded from videotapes of the Strange Situation according to the standard classification scheme specified by Ainsworth et al. (1978) for the three major groups and the eight subgroups. Only the three major classifications were included in the analyses. Classification of the 100 infants according to the Strange Situation is shown in Table 3.

Table 3. Attachment classifications at 12 months of infants in each treatment group

Attachment group	A*	B	C
Experimental group I	9	15	1
Experimental group II	4	19	2
Subtotal	13	34	3
Control group I	14	6	5
Control group II	14	8	3
Subtotal	28	14	8

* A = Insecure Avoidantly Attached
  B = Securely Attached
  C = Insecure Resistantly Attached

In order to examine the impact of infant irritability on the proportion of A, B, and C attachment classifications, a test for multinomial distributions was applied to the data of both control groups. The proportions found by Kellenaers (1984) were used as the expected values. The results indicate that the distribution over the attachment classifications for the control group differed from that of Kellenaers' sample, c^2 (2, N = 50) = 49.43, p < .001. The difference between the distributions seems to be the result of more A- and C-categories in the control group. This is a replication of the results obtained in the observational study. The same test was done in both experimental groups. Here the distribution found did not differ from the one to be expected, c^2 (2, N = 50) = 1.73, p = .42.

There was an overall effect of treatment on the distribution of attachment patterns when the two experimental groups were compared with the two control groups by Mann Whitney U statistics, U = 677.5, p < .001. When the experimental and control group that were not pretested were compared, to control for contamination of the pretesting procedures, the overall effect of treatment on the distribution of attachment classifications again was found to be significant, U = 157.0, p < .01. Thus, there was a significant overall effect of treatment on attachment classifications. Seventy-two percent of untreated infants were classified in the insecure categories, compared to 32% in the treated groups. Only 28% of all untreated irritable infants were assigned the traditional secure classification.

Discussion

The intervention aimed at enhancing maternal sensitive responsiveness in dyads with irritable infants clearly had a positive impact on the mothers and children who participated in the intervention. Differences between program and control mothers were evident immediately after the intervention. The home intervention strategy produced marked behavioral changes in maternal and infant behavior during interactions. Data on mother-infant interaction show that experimental infants explored more than control infants (Van den Boom, 1988). In addition, assessments of the infants' free play behavior indicate that experimental infants are more able to initiate, maintain, and engage in cognitively sophisticated exploration and are less engaged in the least cognitively sophisticated kinds of exploration (i.e. mouthing). Hence, the hypothesis that maternal sensitive responsiveness fosters a sense of security in the infant which motivates subsequent engagement of the environment receives support from the results of both the mother-infant interaction data and the infant exploratory data.

The findings on the attachment-exploration balance found in this study also show that "felt security" is the functional goal of infant attachment behavior. The substitution of Bowlby's "regulation of proximity seeking" as the set goal by an affective condition "felt security" recognizes differences in infants in the behavior required of mothers to foster feelings of security in most situations (Thompson, 1986). Our results suggest that irritability is an important contributor to individual variations in felt security. Irritability seems to influence the infant's understanding of, interpretation of, and adaptation to social events. The results of the experimental manipulation of maternal sensitive responsiveness in dyads with irritable infants lends further support to this notion. Enhancing maternal responsiveness seems to foster the infant's sense of felt security, since it leads to an increase in exploratory behavior which is also of a higher quality compared to infants receiving a low level of maternal responsiveness.

Explanatory Models

Explanatory models of the attachment process that have been proposed differ sharply in complexity and inclusiveness. What seems to be needed is a model which encompasses both attachment and exploration as bidirectional processes *and* individual differences between infants. A model recently proposed by Gubler and Bischof (1990), aimed at uncovering the motivational dynamics of the distance regulation that can be observed in mother and infant, seems to be the most advanced attachment model to date. Basically, the model postulates three interacting motivational subsystems: the *security* system, the *arousal* system, and the *autonomy/sex* system. Of special interest in Bischof's model is the incorporation of individual difference variables such as "dependency" (security system) and "enterprise" (arousal system) as codeterminants of the action of behavioral systems. In this model, exploration is part of the *arousal* system. This system is activated during an encounter with a strange object or an adult human stranger. The arousal system releases either withdrawal *(fear behavior)*, or approach enabling the child to *explore* the source of arousal. Very familiar objects, like mother, have difficulties in activating the arousal system; they can, however, do so by behaving in an unpredictable way. The arousal system, like the security system, is regarded as being *homeostatic*. The arousal level is matched against the reference variable *enterprise*. Aversion against arousal, and hence, *fear* behavior, results from arousal exceeding the setpoint of enterprise. If, however, arousal falls short of enterprise, an appetence for arousal develops, leading to *exploratory* behavior. Gubler and Bischof's (1990) model is an elegant integration of current theories of attachment, fear, and exploration.

In the present study no group differences were discerned in tempo of exploration and nonspecific manipulation. The absence of differences in tempo of exploration between experimental and control group children supports the notion that tempo is a stylistic aspect of the infant's behavior. Because the total sample of the intervention study consisted of irritable infants, differences in this aspect of behavior during exploration would not be expected to occur. It has been established that infants differ in tempo. Kagan's (1965, 1971) concept of "conceptual tempo" reflects a general style or predisposition to attend to and interact with environmental stimuli in characteristic ways. Kagan suggested that fast-tempo (impulsive) children have much more difficulty sustaining attention than do slow-tempo (reflective) children. However, the search for attentional and behavioral consistencies in infancy that may serve as the antecedents of conceptual tempo prompted by this work has revealed inconsistent results. In some studies (Fenson, Sapper, & Minner, 1974; Kagan, 1971; Messer, Kagan, & McCall, 1974), it was found that fixation time is a relatively stable attribute of individual babies, determined not only by the properties of the stimulus and the state of the subject, but also by the characteristic tempo of the individual baby. McCall (1974) attempted to clarify matters by investigating the role of a number of factors, including individual predispositions in infant play, and found that play tempo apppeared to depend more on the child's age and sex and the specific stimulus attributes than on stylistic predispositions. The populations of these studies consisted of normal infants. In the present study, the population under investigation was a group of irritable infants known to have a high activity level (van den Boom, 1990). Although it could be expected that fast-tempo infants would be bored rapidly with each of a series of toys, running the risk of developing attentional deficits, the results clearly show an independence of exploratory tempo and quality. In addition, our results do not support the notion put forward by Wenckstern et al. (1984) that rapid-tempo infants would be less sensitive to environmental stimulation and change. Our findings indicate that the quality of exploratory behavior of irritable (fast-tempo) infants is susceptible to changes in maternal behavior.

References

Ainsworth, M. D. S. (1963). The development of infant-mother interaction among the Ganda. In B. M. Foss (Ed.), *Determinants of infant behaviour* (Vol. 2). London: Methuen.

Ainsworth, M. D. S. (1967). *Infancy in Uganda: Infant care and the growth of love*. Baltimore: John Hopkins University Press.

Ainsworth, M. D. S. (1973). The development of infant-mother attachment. In B. M. Caldwell & H. N. Riciutti (Eds.), *Review of child development research* (Vol. 3). Chicago: University of Chicago Press.

Ainsworth, M. D. S., Blehar, M. C., Waters, E., & Wall, S. (1978). *Patterns of attachment: A psychological study of the Strange Situation*. Hillsdale N. J. Lawrence Erlbaum Associates.

Ainsworth, M. D. S., & Wittig, B. A. (1969). Attachment and exploratory behavior of one year olds in a strange situation. In B. M. Foss (Ed.), *Determinants of infant behavior* (Vol. 4). London: Methuen.

Bates, J. E., Freeland, C. A. B., & Lounsbury, M. L. (1979). Measurement of infant difficultness. *Child Development, 50,* 794-802.
Bateson, G. (1955). A theory of play and fantasy. *Psychiatric Research Reports, 2,* 39-51.
Bateson, G. (1956). The message "This is play." In B. Schaffner (Ed.), *Group processes.* New York: Josiah Macy.
Belsky, J., Garduque, L., & Hrncir, E. (1984). Assessing performance, competence, and executive capacity in infant play: Relations to home environment and security of attachment. *Developmental Psychology, 20,* 406-417.
Belsky, J., Goode, M. K., & Most, R. K. (1980). Maternal stimulation, and infant exploratory competence: Cross-sectional, correlational, and experimental analyses. *Child Development, 51,* 1163-1178.
Belsky, J., & Most, R. K. (1981). From exploration to play: A cross-sectional study of infant free play behavior. *Developmental Psychology, 17,* 630- 639.
Berlyne, D. E. (1960). *Conflict, arousal and curiosity.* New York: McGraw-Hill.
Berlyne, D. E. (1966). Curiosity and exploration. *Science, 153,* 25-33.
Bischof, N. (1975). A systems approach towards the functional connections of fear and attachment. *Child Development, 46,* 801-817.
Bower, T. G. R. (1972). Object perception in infants. *Perception, 1,* 15-30.
Bower, T. G. R. (1974). *Development in infancy.* San Francisco: Freeman.
Bowlby, J. (1969). *Attachment and loss: Vol. 1. Attachment.* New York: Basic Books.
Brazelton, T. B. (1973). *Neonatal behavioral assessment scale.* Philadelphia: J. B.Lippincott Co.
Bruner, J. S. (1972). The nature and uses of immaturity. *American Psychologist, 27,* 687-708.
Di Franco, D., Muir, D. W., & Dodwell, P. C. (1978). Reaching in very young infants. *Perception, 7,* 385-392.
Dunst, C. J., & Lingerfelt, B. (1985). Maternal ratings of temperament and operant learning in two- to three-month-old infants. *Child Development, 56,* 555-563.
Fein, G. G., & Apfel, N. (1979). The development of play: Style, structure and situation. *Genetic Psychology Monographs, 99,* 231-250.
Fenson, L., Kagan, J., Kearsley, R. B., & Zelazo, P. R. (1976). The developmental progression of manipulative play in the first two years. *Child Development, 47,* 232-235.
Fenson, L., Sapper, V., & Minner, D. G. (1974). Attention and manipulative play in the one-year-old child. *Child Development, 45,* 757-764.
Field, J. (1976). Relation of young infants' reaching behavior to stimulus distance and solidity. *Developmental Psychology, 12,* 444-448.
Field, T., Hallock, N., Ting, G., Dabiri, C. & Shuman, H. H. (1978). A first year follow-up of high-risk infants: Formulating a cumulative risk index. *Child Development, 49,* 119-131.
Freud, S. (1959). Creative writers and daydreaming. In J. Strackey (Ed.), *The standard edition of the complete psychological works of Sigmund Freud* (Vol. 9). London: Hogarth.
Groos, K. (1901). *The play of man.* New York: Appleton.
Gubler, H., & Bischof, N. (1990). A system's perspective on infant development. In M. E. Lamb & H. Keller (Eds.), *Infant development: Perspectives from German speaking countries.* Hillsdale: Lawrence Erlbaum.

Hall, G. S. (1920). *Youth*. New York: Appleton.
Hughes, M. (1978). Sequential analysis of exploration and play. *International Journal of Behavioral Development, 1,* 83-97.
Hughes, M., & Hutt, C. (1979). Heart-rate correlates of childhood activities: Play, exploration, problem-solving and day dreaming. *Biological Psychology, 8,* 253-263.
Hutt, C. (1966). Exploration and play in children. *Symposium of the Zoological Society of London, 18,* 61-81.
Hutt, C. (1967). Temporal effects on response decrement and stimulus satiation in exploration. *British Journal of Psychology, 58,* 365-373.
Hutt, C. (1970). Specific and diversive exploration. In H. Reese & L. Lipsitt (Eds.), *Advances in child development and behavior* (Vol. 5). New York: Academic Press.
Hutt, C. (1979). Exploration and play. In B. Sutton-Smith (Ed.), *Play and learning*, New York: Gardner Press.
Inhelder, B., Lezine, I., Sinclair, H., & Stambak, M. (1972). Les debuts de la fonction symbolique. *Archives de Psychologie, 41,* 187-243.
Jennings, K. D., Harmon, R. J., Morgan, G. A., Gaiter, J. L., & Yarrow, L. J. (1979). Exploratory play as an index of mastery motivation: Relationships to persistence, cognitive functioning, and environmental measures. *Developmental Psychology, 15,* 386-394.
Kagan, J. (1965). Impulsive and reflective children: The significance of conceptual tempo. In J. D. Drumboltz (Ed.), *Learning and the educational process*. Chicago: Rand McNally.
Kagan, J. (1971). *Change and continuity in infancy*. New York: Wiley.
Kagan, J., Kearsley, R. B., & Zelazo, P. R. (1978). *Infancy: Its place in human development*. Cambridge Mass.: Harvard University Press.
Kellenaers, C. J. J. (1984). *Temperament en hechting*. [Temperament and attachment]. Unpublished master's thesis, University of Leiden.
Keller, H., Schölmerich, A., Miranda, D., & Gauda, G. (1987). The development of exploratory behavior in the first four years of life. In D. Görlitz & J. F. Wohlwill (Eds.), Curiosity, imagination, and play. Hillsdale,N. J.: Lawrence Erlbaum Associates.
Lamb, M. E. (1982). Individual differences in infant sociability: Their origins and implications for cognitive development. In H. W. Reese & L. P. Lipsitt (Eds.), *Advances in child development and behavior* (Vol.16). New York: Academic Press.
Lazarus, M. (1883). *Die Reize des Spiels*. Berlin: Ferd, Dummlers Verlagsbuchhandlung.
Lewis, M., Kagan, J., Campbell, H., & Kalafat, J. (1966). The cardiac response as a correlate of attention in infants. *Child Development, 37,* 63- 71.
Maccoby, E. E., & Feldman, S. S. (1972). Mother-attachment and stranger-reactions in the third year of life. *Monographs of the Society for Research in Child Development, 37,* (Whole No. 146).
Main, M. (1973). *Play, exploration and competence as related to child-adult attachment*. Unpublished doctoral dissertation, John Hopkins University, MD.
Main, M., & Solomon, J. (1986). Discovery of an insecure disorganized/disoriented attachment pattern: Procedure, findings and implications for the classification of behavior. In M. Yogman & T. B. Brazelton (Eds.), *Affective development in infancy*. Norwood N. J.: Ablex.

Matas, L., Arend, R.A., & Sroufe, L. A. (1978). Continuity of adaptation in the second year: The relationship between quality of attachment and later competence. *Child Development, 49,* 547-556.

Matheny, A. P. (1989). Temperament and cognition:Relations between temperament and mental test scores. In G. A. Kohnstamm, M. K. Rothbart, & J. E. Bates (Eds.), *Temperament in childhood.* New York: Wiley.

Messer, S. B., Kagan, J., & McCall, R. B. (1970). Fixation time and tempo of play in infants. *Developmental Psychology, 2,* 240-246.

McCall, R. B. (1974). Exploratory manipulation and play in the human infant. *Monographs of the Society for Research in Child Development, 39,* (Whole No. 155).

McCall, R. B., Eichorn, D. H., & Hogarty, P. S. (1977). Transitions in early mental development. *Monographs of the Society for Research in Child Development, 42,* (Whole No. 171).

Miller, S. N. (1974). The playful, the crazy, and the nature of pretense. *Rice University Studies, 60,* 31-51.

Nunnally, J. C., & Lemond, L. C. (1973). Exploratory behavior and human development. In H. Reese (Ed.), *Advances in child development and behavior* (Vol. 8). New York: Academic Press.

Packer, M., & Rosenblatt, D. (1979). Issues in the study of social behaviour in the first week of life. In D. Shaffer & J. Dunn (Eds.), *The first year of life.* New York: Wiley.

Piaget, J. (1962). *Play, dreams, and imitation in childhood.* New York: Norton.

Pomerleau, A., & Malcuit, G. (1980). Development of cardiac and behavioral responses to a three-dimensional toy stimulation in one- to six-month-old infants. *Child Development, 51,* 1187-1196.

Power, T. G., Chapieski, M. L., & McGrath, M. P. (1985). Assessment of individual differences in infant exploration and play. *Developmental Psychology, 21,* 974-981.

Rosenblatt, D. (1977). Developmental trends in infant play. In B. Tizard & D. Harvey (Eds.), *The biology of play.* Philadelphia: Lippincott.

Rubin, K. H., Fein, G. G., & Vandenberg, B. (1983). Play. In P. H.Mussen (Ed.), *Handbook of child psychology: Vol. 4. Socialization, personality and social development.* New York: Wiley.

Ruff, H. A. (1984). Infants' manipulative exploration of objects: Effects of age and object characteristics. *Developmental Psychology, 20,* 9-20.

Ruff, H. A., & Halton, A. (1978). Is there directed reaching in the human neonate. *Developmental Psychology, 14,* 425-426.

Schiller, F. (1954). *On the aesthetic education of man.* New Haven, Conn.: Yale University Press.

Solomon, R. L., & Lessac, M. S. (1968). A control group design for experimental studies of developmental processes. *Psychological Bulletin, 70,* 145-150.

Sorce, J. F., & Emde, R. N. (1981). Mother's presence is not enough: Effects of emotional availability on infant exploration. *Developmental Psychology, 17,* 737-745.

Sostek, A. M., & Anders, T. F. (1977). Relationships among the Brazelton Neonatal Scale, Bayley Infant Scales, and early temperament. *Child Development, 48,* 320-323.

Stayton, D. J., & Ainsworth, M. D. S., & Main, M. B. (1973). The development of separation behavior in the first year of life: Protest, following and greeting. *Developmental Psychology, 9*, 213-225.

Sutton-Smith, B. (1966). Piaget on play: A critique. *Psychological Review, 73*, 104-110.

Sutton-Smith, B. (1967). The role of play in cognitive development. *Young Children, 22*, 361-370.

Sutton-Smith, B. (1975). The useless made useful: Play as variability training. *School Review, 83*, 197-214.

Thompson, R. A. (1986). Temperament, emotionality, and infant cognition. In J. V. Lerner, & R. M. Lerner (Eds.), *New directions for child development: Temperament and social interaction during infancy and childhood.* San Francisco: Jossey-Bass.

Tracy, R., Farish, G., & Bretherton, I. (1980). *Exploration as related to infant-mother attachment in one year olds.* Paper presented at the International Conference on Infant Studies, New Haven, CT.

Trieber, F. A. (1984). Temperament differences between infants who do and do not complete laboratory testing. *Journal of Psychology, 116*, 95-99.

van den Boom, D. C. (1988). *Neonatal irritability and the development of attachment: Observation and intervention.* Doctoral dissertation. University of Leiden.

van den Boom, D. C. (1990). The influence of infant irritability on the development of the mother-infant relationship in the first six months of life. In J. K. Nugent, B. M. Lester & T. B. Brazelton (Eds.), *The cultural context of infancy* (Vol. 2). Norwood N. J.: Ablex.

Vygotsky, L. S. (1967). Play and its role in the mental development of the child. *Soviet Psychology, 12*, 62-76.

Wachs, T. D., & Gandour, M. J. (1983). Temperament, environment, and six-month cognitive-intellectual development: A test of the organismic specificity hypothesis. *International Journal of Behavioral Development, 6*, 135-152.

Wachs, T. D., & Smitherman, C. H. (1985). Infant temperament and subject loss in an habituation procedure. *Child Development, 56*, 861-867.

Waters, E. (1981). Traits, behavioral systems, and relationships: Three models of infant-adult attachment. In K. Immelman, G. W. Barlow, L. Petrinovich, & M. Main (Eds.), *Behavioral development.* Cambridge: Cambridge University Press.

Waters, E., & Deane, K. E. (1982) Infant-mother attachment: Theories, models, recent data, and some tasks for comparative analysis. In L. W. Hoffman, R. Gandelman, & H. R. Schiffman (Eds.), *Parenting: Its causes and consequences.* Hillsdale, N. J.: Lawrence Erlbaum Associates.

Weisler, A., & McCall, R. (1976). Exploration and play. *American Psychologist, 31*, 492-508.

Wenckstern, S., Weizmann, F., & Leenaars, A. A. (1984). Temperament and tempo of play in eight-month-old infants. *Child Development, 55*, 1195-1199.

Zelazo, P. R., & Kearsley, R. B. (1980). The emergence of functional play in infants: Evidence for a major cognitive transition. *Journal of Applied Developmental Psychology, 1*, 95-117.

CHAPTER **III.8**

Interindividual Differences in the Development of Exploratory Behavior: Methodological Considerations

Clemens Trudewind, Klaus Schneider

Almost all studies of exploratory behavior in preschoolers have noted the high amount of individual variation in different curiosity-oriented behaviors. Only a small amount of this variation can be explained by the age and sex of the children. In one of our own earlier studies in which we had confronted preschoolers with Corinne Hutt's novel box (see Schneider, Moch, Sandfort, Auerswald, & Walther-Weckman, 1983), interindividual differences in the total time the 99 children (ages 3 to 6) dealt with the novel box (exploration and playing) were determined only partially (about 20%) by the individual difference variables "age" (3 age groups) and "sex" of the children.

For the single categories of exploratory behavior observed in this situation, the percentage of individual variance explained by these biological variables was still lower: 13% for the most interesting single variable, the manipulation of the lever of the box, and between 6 and 8% for the remaining categories: looking, visual inspection, touching and question asking.

It seems, therefore, worthwhile to search for other individual difference measures in order to explain the high amount of variation in these behaviors. As students of motivation we believe that in addition to cognitive ability variables, *motives* as enduring behavioral dispositions determine the quality and quantity of manifest curiosity behavior in preschoolers.

We gratefully acknowledge the assistance of our students and research assistants in the collection of the data: A. Hillebrandt, H. Hungerige, L. Janczyk, B. Krüger, D. Lange, K. Mackowiak, M. Massie, E.-M. Matip, C. Neuhaus, E. Purmann, A. Ruge. The authors also wish to thank the children and parents of several nursery schools in Bochum and Marburg and Dr. H. Lugt-Tappeser for her cooperation organizing the data collection in Marburg.

Following Harlow (1953a, b), Lorenz (1943), White (1959) and others, we postulate an original motive to explore one's physical and social environment. The ultimate function of this behavioral disposition is the acquisition of knowledge or the assimilation of objective structures, whereas the immediate cause of exploration is assumed to be a state of subjective uncertainty created by certain aspects of the environment (Berlyne, 1960, 1966; Weisler & McCall, 1976).

However, as the pioneers of curiosity motivation already have conceded (James, 1892; McDougall, 1908; Montgomery, 1955; see Voss, 1984), differences in manifest exploratory behavior is always the expression of a compromise between instigating and inhibiting forces, for example, between the motivation to explore a novel object and to avoid this unknown object out of fear.

Another example from our own laboratory might illustrate this (Lugt-Tappeser & Schneider, 1987): 91 children between 3;0 and 6;8 years were again confronted with Hutt's novel object. Three individual difference variables - *age* (2 groups: 3;0 - 5;0 and 5;1 - 6;8), *sex* and dispositional *anxiety* - explained 23% of the variation of the time children were dealing with the novel object. The most important individual difference variable was anxiety, assessed with a check list of the free play situation of the children (see Lugt-Tappeser & Schneider, 1986). In this study, anxiety alone contributed more than half of the explained amount of the individual variation in total exploration.

What about the high amount of the unexplained individual variation? Can this be explained by a unique curiosity motive? Probably not, as Henderson and Moore (1979) as well as Kreitler, Zigler, and Kreitler (1975) have argued on the basis of factor analytic studies. Curiosity is not an unidimensional construct. Henderson and Moore (1979) found two modes of exploration, manipulation and question asking, and two styles, breadth and depth. Even with the restricted range of object related behaviors we observed in our studies in which the children were confronted with Hutt's novel object, exploration turned out to be a multidimensional behavioral construct. Looking at the novel object from a far distance, visual inspection of parts of the object, verbal remarks related to the object and the most prominent behavior, the manipulation of the lever, could only be organized in a two-dimensional schema (data from Schneider et al., 1983, study 1).

A principal component analysis of the intercorrelations of the frequencies of these behaviors over subjects, including three categories of playing with the object (transposition of function, unconventional

manipulation and incorporation of another play object in dealing with the novel box) suggested, according to Cattell's scree-test (Cattell, 1966), a two-dimensional solution, which explained 40% of the total variation (data from Schneider et al., 1983, study 1).
Figure 1 presents the loadings of these behavioral categories in a two-dimensional axis system after a varimax rotation.

Figure 1. The two dimensions found in a factor analysis of the exploratory behaviors and play behaviors of preschoolers observed in a first short encounter with a novel object (Hutt's "novel box"). Exploratory behaviors: 1. looking at the box (LOOK); 2. touching the box or parts of it (TOUCH); 3. manipulating the lever of the box (MANIP); 4. inspecting visually the box or parts of it (VIS INSP); 5. asking questions or making comments pertinent to the box (VERB R). Play behaviors: 1. transposing the function of the box (TRANSP); 2. manipulating the box and the lever in an unconventional way (UNC MAN) and 3. incorporating other play objects in playing with the box (INCORP). (data from Schneider et al., 1973, exp.1).

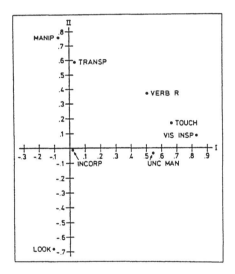

It can be seen that touching, visual inspection and verbal reactions related to the object on one hand, and manipulation and looking on the other hand have to be organized on two orthogonal dimensions. Manipulation and looking represent a bipolar dimension.

There is another source of individual variation in observed exploratory behavior in a given situation in children. The way children explore objects, the modalities they prefer in information intake and the persistence in exploratory behavior seem not only to be dependent on the situation, but also on the developmental status of the child - most prominently on the child's cognitive developmental level (Henderson & Moore, 1979; Keller & Boigs, 1989). The cognitive level determines the situational features which are new, complex, incongruent, etc. for the child, and determines via these qualities the amount of arousal of subjective uncertainty and resulting exploratory behavior.

To summarize, in estimating the degree of the determination of manifest exploratory behavior by stable dispositions or motives, we have to consider (a) the multi dimensionality of exploratory behavior; (b) its dependence on situational features and object qualities; (c) its dependence on the instigation of other motive systems in the same situation, and, finally, (d) its dependence on the developmental level of the children. Therefore, we need concurrent assessment procedures which enable us to estimate the individual strength of the curiosity motive as a behavioral disposition.

Generally, there are two different approaches to the assessment of dispositional variables in children's exploratory behavior. In the first approach, latent behavioral dispositions are inferred from observed curiosity-oriented behaviors, which are influenced by other motive systems as well and many situational constraints (norms, rules etc.). In the second approach, individual differences in the strength of the curiosity motive are inferred from attentional and emotional manifestations of the activation of this motive system by controlled curiosity incentives, which are not exploratory (information seeking) behaviors themselves.

The Behavior Assessment Approach

Following the first approach, the best way to assess the strength of individual disposition would be the observation of children's everyday behavior in their natural environment. One would have to make sure that in all settings in which children normally interact with their physical and social environment, a sufficient number of observations could be made. By counting the frequencies of prototypical exploratory behaviors, by measuring the duration of these behaviors, and the latency and the speed of approaching different objects, behavioral parameters could be obtained. These parameters would allow the examination of the dimensionality of exploratory behaviors in different periods of development, as well as the individual strength of the behavioral disposition for exploratory behavior under ecological conditions. In addition to the observations of children's behavior, parameters of the incentive structure of the child's environment could be assessed by the same procedure.

However, probably due to the enormous costs and manpower necessary for such observational studies, nobody has tried this so far. Presumably following this philosophy, Henderson and Moore (1979), among others, observed children confronted with specific curiosity instigating tasks. These authors found substantial correlations between two of their four different tasks only in regard to the frequency of question asking. All other parameters of exploratory behavior were specific for the different tasks. Nevertheless, Moore (1985) concluded: "Despite individual differences in mode and style of exploration, it was possible to identify children who overall showed high levels or low levels of exploration compared with their respective age and sex groups" (p. 4). The problem with such a selection of specific tasks is that the selected tasks might not be representative and therefore, the assessed parameters might be determined to a high degree by task-specific features. Hence, it is an open question whether individual difference measures in exploratory behavior when dealing with these tasks remain stable in other situations and with other tasks.

Another possibility for making sure that the selection of tasks is more representative without performing a total observation of children's behavior is the observation of children's exploratory behavior in specific "ecological key situations" (Trudewind & Husarek, 1979). We define ecological key situations as situations in which all children of the given age group are confronted, more or less often, in a stereotypical manner in their natural environment. Ecological key situations are characterized by the fact that the context instigates the interesting behavior in a prototypical way while still leaving room for the expression of individual differences. Examples of such key situations in exploratory behavior could be a visit to a museum or an exhibition (Henderson, Charlesworth, & Gamradt, 1982), a visit to a shopping market (Schneider & Unzner, this vol.), or confrontation with a novel object in a familiar environment (Görlitz, 1987; Schneider & Flohr, 1989). Other examples are: the child's entrance to an unfamiliar play room or to an unfamiliar house, walking or hiding in a strange environment, a repairman coming to the house, a construction area in the neighborhood, the presentation of a wrapped present, visiting a puppet-show, or looking at picture books or at the family photo collections. In such selected key situations, behavioral observation could give us hints to the dimensionality and consistency over situations of the exploratory behavior of children while keeping costs (time, manpower) low. The critical point in this approach is the selection of relevant key situations.

Another possible way to keep the costs of the data gathering low would be to question parents and/or teachers who had sufficient opportunities to observe their children in such ecological key situations. It should not be too difficult, especially for parents, to report if a specific prototypical behavior is shown regularly and frequently by their child in everyday situations.

The Development of a Parents' Questionnaire

So far, several approaches to assess behavioral dispositions for exploratory behavior in pre- and grade-school children via ratings of teachers have been critically evaluated (Henderson & Moore, 1979; Henderson, 1984b; Kreitler, Zigler, & Kreitler, 1975; Maw & Maw, 1966, 1975; Coie, 1974). In particular, low or absent correlations with behavioral measures and moderately positive correlations with measures of intelligence have been reported (Henderson, 1984b). However, because teachers observe their students' exploratory behavior mainly in learning and problem solving situations, shared individual variance between curiosity and intelligence is not surprising. Therefore, the existence of these correlations does not invalidate the assessment of individual differences in children's exploratory behavior with parents and teacher ratings. The aim of the present study was to develop a parents' questionnaire with which the stable disposition to explore in the prototypical everyday situations of preschoolers could be assessed. A second goal was to examine the dimensionality of exploratory behavior using information on individual patterns of preference for specific curiosity-instigating situations and for specific modes of exploratory behavior. Finally, we hoped that we would get information from this procedure about ecological key situations for preschoolers' exploratory behavior at home which could be used for further behavioral observations.

The curiosity motive is defined as a behavioral disposition to gather information on new, complex, surprising, and incongruent features in the environment in order to reduce the subjective uncertainty created by these features. It is closely associated with the concepts of competence, mastery, or effectance motivation (Harter, 1978a, 1981b; Morgan & Harmon, 1984; White, 1959; Yarrow, Morgan, Jennings, Harmon, & Gaiter, 1982). Everyday situations in the child's environment which instigate this curiosity motive are usually problem solving situations (Charlesworth, 1979), that motivate the child to master the task or situation. In order to find out if parents can differentiate curiosity-motivated behaviors from competence-motivated behaviors, we included 41 items which might be thought of as prototypical descriptions of preschoolers' competence-striving at home (Matip, 1989; Purmann, 1988).

The *contents* of parental assessment asked for in this questionnaire, therefore, were (a) children's curiosity behaviors, those guided by the presumed goal to gather information about aspects of the environment, and (b) behavior in which children presumably try to probe their competence and skills in everyday situations typical for preschool children. The *dimension of the evaluation* of these behaviors was the *typicalness* of the described behaviors for the observed child in these specified situations. The assessment was made on a four-point Lickert scale with "not at all typical" and "very typical" as end points.

The first version of the curiosity questionnaire consisted of 41 behavioral descriptions which characterized curiosity behavior of preschool children in everyday situations. These descriptions were organized into 5 categories, which also guided our search for these behaviors. Table 1 presents these a priori categories.

Table 1. Behavioral domains represented in the curiosity items of the questionnaire

Behavioral domains	Number of items
1. Exploration and manipulation of objects Example: My child often has a close look at new things I bought for the household or for the garden.	8
2. Exploration and orientation in strange environments Example: Going for a walk, my child often asks where the various ways lead to.	7
3. Looking for new experiences, focussing attention on new stimulation Example: My childs wants to have a taste of new foods or drinks without hesitation.	9
4. Exploring the functional or causal relationships in social or physical events Example: My child absolutely wants to watch whenever something is repaired.	9
5. Epistemic behaviors like observing, thinking about and questioning Example: My child frequently asks where particular natural events, such as lightning, thunder and snow, originate.	8

Analysis of the Dimensionality of the Parents' Questionnaire

We received completed questionnaires from parents of 103 boys and 88 girls, aged 2;5 to 6;8 (median = 4;10) years. The first version of the questionnaire consisted of 82 items.

Five curiosity and five competence items were excluded from the further analysis as parents agreed with them as "typical" or "very typical" for their children's behavior in more than 90% of the cases. For the remaining items, a factor analysis (principal components procedure with iterations of the communalities) was computed in order to determine if parents differentially evaluate their children's curiosity and competence-striving behaviors. According to Cattell's scree test (Cattell, 1966), a two- or four-dimensional solution could be accepted. An unambiguous separation of the two a priori scales was found for the two-dimensional solution after an orthogonal (varimax) rotation. For 20 of the 36 items from the curiosity scale, but only for one item of the competence scale, we found significant loadings on the first factor ($a > .35$). The opposite picture is found for the second factor: 20 of the original 36 items of the competence scale, but only one item of the curiosity scale had significant loadings on this second factor ($a > .35$). We conclude, therefore, that parents are able to differentiate in their descriptions of their children's behavior between curiosity and competence-motivated behaviors.

Because the two factors together explain only 21.7% of the total variance, and only 60% of all items have significant loadings on these first two dimensions, we computed additional factor analyses for the curiosity and competence items separately. A varimax factor analysis of the curiosity items for the whole sample yielded an interpretable three-factor solution. Computing this analysis separately for girls and boys or for two age groups (2;5 - 4;10 versus 4;11 - 6;8 years), demonstrated the stability of this three-factor structure. Based on a consistency analysis of the items and the factorial structure for the total sample as well as for the four subsamples, three scales were defined, representing mainly the factorial structure of the total sample.

The first scale characterizes the strength of "epistemic curiosity," the second scale measures the strength of "perceptive and manipulative curiosity," the third one we preliminarily named "searching for stimulating events."

In order to examine whether parents can discriminate on the manifest behavioral level between low curiosity and anxiety or behavioral inhibition, 37 items of the revised curiosity scale were mixed together with 36 items of an anxiety scale. This anxiety scale was constructed according to the same principles as the curiosity scale. Parents (mother or father) of 334 preschool children (165 boys and 169 girls) aged 3;0 to 6;8 (median = 5;1) years responded to this questionnaire. A factor analysis of all 73 items (anxiety and curiosity) revealed a two-factor solution. After an orthogonal (varimax) rotation, this two dimensional solution shows an unambiguous separation of the two a priori scales. For 29 of the 36 items of the anxiety scale, but only for one item of the curiosity scale, we found significant loadings on the first factor ($a > .35$), whereas 25 of the 37 items of the curiosity scale had significant loadings on the second factor ($a > .35$).

Additionally, we computed factor analyses for the curiosity items and the anxiety items separately. For the curiosity items we again found the same dimensional structure as for the whole sample where we had given the curiosity items mixed in one questionnaire with the competence items. Obviously, in their evaluation of the children's curiosity parents use the same latent dimension no matter whether the items are presented in the context of competence items or anxiety items. Therefore, we suppose that the latent dimension inherent in the parent's evaluation of the children's curiosity behavior is not only a dimension of this evaluation process but also of the children's behavior. It follows that we probably can use children's scores on the curiosity scale, based on the factor analysis, as indices of components of children's dispositional curiosity.

In Table 2 the 11 items of the first scale are presented.

Table 2. "Epistemic curiosity" - scale

Items	Loading a	Proportion of agreement	Corrected Item / Total correlation
My child frequently asks what the reasons for certain rules and customs are.	.70	.60	.58
My child frequently asks what the meaning of a word is.	.68	.72	.57
My child often asks where particular natural events, such as lightning, thunder and snow, originate.	.64	.74	.54
My child often asks what the reasons of my own or other persons' acting are.	.62	.71	.57
My child frequently tries to find out the origin of an unusual sound.	.58	.70	.49
My child always wants to know at once the designation of things which are unknown to him.	.55	.74	.48
My child often asks about persons we met by chance.	.53	.81	.51
My child always asks how things work.	.51	.79	.46
My child often tries to find out the reasons for unusual intensive smells.	.47	.54	.35
When there is music from outdoor, my child urgently wants to know what it means.	.41	.81	.36
Going for a walk, my child often asks where the various ways lead to.	.41	.52	.37

Cronbach's alpha: .82 (11 Items)
N = 279

These items are arranged according to their loadings on the first factor of the factor analysis for the total sample. Table 2 presents, in addition, the corrected item/total score correlations and the proportions of agreement for the individual items.

Scale 1 was interpreted as an "Epistemic curiosity" scale (see Berlyne, 1960). The behaviors specified here indicate a desire to gain insight, knowledge, and understanding. The most prominent behavior is question asking. However, the search for reasons and causes for unusual social and perceptual events are also included. The internal consistency of the whole scale (Cronbach's alpha) is high: alpha = .82.

Table 3. "Perceptive and manipulative curiosity" - scale

Items	Loading a	Proportion of agreement	Corrected Item / Total correlation
My child is particularly pleased to explore unfamiliar sheds, garages, huts or other uninhabited buildings.	.61	.36	.46
Very often my child refuses passing by a building site and insists on watching everything.	.59	.65	.47
In another childs' play-room, my child always tries to test the unknown toys immediately.	.50	.84	.42
My child often has a close look at new things I bought for the household or the garden.	.49	.86	.46
When my child discovers light switch anywhere, he/she tries to test it frequently.	.46	.39	.29
My child prefers walks or excursions to unfamiliar surroundings.	.46	.56	.37
When shopping, my child has a close look to all unusual things.	.46	.79	.40
Being in a strange house or in an unfamiliar building, my child likes to see as many rooms as possible.	.46	.49	.39
Whenever my child notice an unfamiliar plant or a small animal, he/she inspects it very carefully.	.45	.61	.37
When my child investigates new things, he/she can't be diverted by anything.	.45	.60	.40
When my child can't watch whenever some-thing is repaired, he/she is very angry.	.44	.65	.38
My child cannot be hindered rummaging in boxes, drawers, attics, and cellars.	.43	.74	.36
When we visit a fairy, my child always wants to try all the different merry-go-rounds.	.42	.78	.29
My child examines new things often systematically and with tenacity.	.41	.66	.37

Cronbach's alpha: .77 (14 Items)
N = 279

The 14 items of the second scale, presented in Table 3 specify behaviors of information gathering via looking at others, manipulation, and trial and error behaviors. This scale presumably measures the strength of perceptive and manipulative curiosity. The internal consistency (Cronbach's alpha) is satisfactory: alpha = .77.

Table 4. "Searching for stimulating events" - scale

Items	Loading a	Proportion of agreement	Corrected Item / Total correlation
When I telephone, my child always listens attentively.	.61	.68	.32
Every time the bell rings, my child runs immediately with me to the door.	.53	.89	.31
When I return from shopping, my child immediately wants to see what is in the shopping bag.	.53	.70	.46
When someone in our family gets a present, my child bothers us until it is unwrapped.	.52	.84	.40
When there is any "secret", my child wants to know it immediately.	.34	.76	.35

Cronbach's alpha: .61 (5 Items)
N = 279

Table 4 presents the 5 items of the third scale which presumably measure tendencies to look for interesting, exceptional, or surprising events and for hidden objects and secrets. We preliminarily named the scale "Searching for stimulating events." We suppose that this scale is related conceptually to diversive exploration (Hutt, 1970) and to the concept of "sensation seeking" (Zuckerman, Kolin, Price, & Zoob,

1964). The internal consistency of this scale, however, is lower than that for the two other scales and needs to be improved (alpha = .61).

Two of the three curiosity scales are significantly correlated with the age of the children. The Pearson-coefficients between the "Epistemic curiosity" scale and age were $r = .30$ ($n = 314$; $p < .001$) for the whole sample and $r = .37$ ($n = 157$) for boys and $r = .22$ (n = 157) for girls, respectively. The "Searching for stimulating events" scale was significantly correlated with age for the whole sample ($r = .10$, $n = 304$, $p < .03$) and for boys ($r = .14$, $n = 162$, $p < .04$). The "Perceptive and manipulative curiosity" scale is not correlated with age. Only for this scale we did find a significantly higher mean for boys than for girls. Perceptive and manipulatory exploration are, in the sight of the parents, more typical for boys than for girls. In the other dimensions of the curiosity questionnaire, parents evaluated boys and girls alike.

Summarizing, we conclude that parents seem to be able to evaluate the curiosity behavior of their children of preschool age when they are given a description of prototypical behaviors for their rating in ecological key situations. The dimensions found are comparable to dimensions extracted from behavioral observations and reported in the literature (see Henderson, 1984a; Henderson & Moore, 1979). We hope that these scales not only assess the preferred *modes* of information search and gathering in this age period, but also represent the *latent structure* of the curiosity motive. However, whether this holds true or not has to be examined in further validation studies.

The Incentive-Reactivity Assessment Approach

In the introductory remarks, we differentiated the behavioral approach from a second approach to assess the dispositions for exploratory behavior, which we call here the incentive-reactivity assessment approach. In this second approach, a window to the behavioral disposition is sought through the assessment of experiential and behavioral indices of the aroused motive under controlled conditions. Schneider and Schmalt (1981) paraphrase McDougall's (1908) definition of a motive (instinct) as an inborn or, as they term it, learned psychophysical disposition. Such a disposition enables the person to perceive features of the environment in a specific and attentive way, to experience a specific emotional arousal instigated by this perception, and to behave in a specific direction or experience at least behavioral intentions to do so. In the 1950's, David McClelland and his group based their assessment of individual differences in motives on the expression of motivational states, aroused by standardized TAT-pictures. With this assessment procedure, the strength of the motive arousal was not inferred from responsive, motive-dependent behaviors, but from mental states manifested in the production of the person's fantasies (McClelland, 1958, 1980; McClelland, Atkinson, Clark & Lowell, 1953). In this research program, it became necessary to differentiate between two dimensions of motive strength: intensity and extensity. The intensity dimension describes the strength of the activation of the motive system through environmental incentives; extensity describes the variety of environmental features which are able to instigate the motive system. Building on these theoretical insights, we planned to tap children's motive systems via the assessment of behavioral manifestations of experiential and emotional components of aroused motivation.

Development of a Puppet-Show Instrument Assessment Procedure

In ongoing research, we have attempted to assess individual differences in the strength of the curiosity motive by assessing children's reactions to the presentation of a puppet-show. We believe that the puppet-show fulfills several prerequisites for allowing a more direct approach to the assessment of individual motive strength, as the assessed indices are not mediated by information-gathering strategies, preferences, or socialized rules in the child's observable curiosity behavior. A puppet-show is a situation in which preschoolers expect new and surprising events. Even before the show starts, the curiosity motive might be aroused in young spectators to a degree. During the play it is possible to present curiosity incentives to spectators in a controlled way, for example, by presenting strange noises, strange or hidden objects, by introducing unexpected events, or by having the figures of the show manipulating objects and asking questions. Depending on the individual strength of the curiosity motive, these incentives should or should not instigate a sharp or mild increase of the already aroused curiosity motivation. The person-specific degree of this increase can be interpreted as an additional index of the strength of the curiosity motive. The special advantage of presenting such a puppet-show can be seen in the fact that a puppet-show facilitates children's expression of instigated curiosity motivation in a variety of their expressive, gestural, orienting, and verbal behaviors which are not as strongly influenced by rules, norms, and social constraints as manifest exploratory behaviors. The orienting reaction, manifested in looking behavior, head and trunk movements, facial and gestural expressions of surprise or tension, questions, suggestions and hints given by the children either verbally or behaviorally, could be taken as unbiased indices of the aroused curiosity system. Individual differences in the intensity of these reactions should reflect individual differences in the strength of the assumed underlying disposition.

Based on these considerations, we created a puppet-show in which curiosity-instigating events are presented in 23 episodes. The story deals with a preschool boy, who stays at home by himself and gets bored. One after another, different fantasy figures appear: a big tomato which wants to play with the child, a cleaning cloth, which complains about the harsh chemicals which create health problems for it, and finally, a brush which paints the noses of the other figures. As soon as the mother comes back, these fantasy figures fall back in their unanimate status and the main actor, the preschool child, asks the spectators not to reveal his secrets (Lange, Massie, & Neuhaus, 1990).

In the first study, we presented this puppet-show to 24 preschoolers (12 boys and 12 girls) aged 3;5 to 6;5 years in the gymnastics room of their nursery school. Two children, who knew each other, watched the show sitting close to each other at a distance of 3 meters in front of the stage. Both children were videotaped during the show.

As a first step in the data analysis, we computed the time children watched the show attentively. As expected, most of the children (91%) attend the show for more than 90% of the showtime.

In a second step, we defined behavioral categories for the 23 curiosity instigating episodes, which indicated a more or less strong arousal of the curiosity motive system.

Table 5 presents, as an example, the three main categories and the behaviors coded in these categories for those three situations in which a strange noise of unknown origin was to be heard.

Table 5. Coding system for children's reactions in those situations in which a noise of unknown origin was presented

No reaction (0):

In case children demonstrated no reaction related to this event.

Weak reaction (1):

This level was coded when at least one of the following behaviors was shown:

a) The child presents a surprised face (one of the following features: rised eyebrows, wide open eyes, jaw drop).

b) The child looks at least at two different places on the stage or in the room, at which the strange noise could have been presented.

c) The child looks to the other child as a reaction to this noise (not coded if shown as a response to the other child's behavior).

Strong reaction (2):

Child's behavior is coded at a strong reaction if at least one of the following behaviors was observed:

a) The child scans with eye and head movements the stage and the room.

b) The child stretches the neck, bows his trunk, erects her- or himself and/or stands up.

c) The child asks for the cause of this noise (e.g. "Where is that coming from?").

These categories are organized in 3-level rating scales, which represent the intensity of the child's reactions in the different episodes. The sum of these scale-values in the 23 episodes represents a confounded measure of the extensity and the intensity dimensions of the curiosity motive.

For the total score as well as for the scale-values in the different episodes, we ran consistency and item analyses. In several steps, 9 episodes were excluded from the computation of the total score. For this new total score, based on the remaining 14 episodes, the distribution is approximately normal. The total score ranges between 0 and 24 with a median of 10. The internal consistency (Cronbach's alpha) is only intermediate (alpha = .51 for the unstandardized and alpha = .77 for the standardized coefficient). This is not a terrific result, but it encouraged us to follow this line of investigation. One of the advantages of this instrument might be that with a relatively economical procedure we could get an intensity as well as an extensity index for the curiosity disposition. In order to get such an extensity index, we counted the numbers of episodes to which children react with curiosity-oriented behaviors without taking into account the intensity of the reactions.

However, for the extensity measure, the internal consistency was still lower. Therefore, we revised the puppet-show in order to render the curiosity instigating cues more salient. We also revised the coding system for children's reactions. The new puppet-show was presented to 64 children (31 boys and 33 girls) between the ages of 3;1 and 6;8 (median = 5;0) years. On the basis of an item analysis and an analysis of the internal consistency of the total scale, 19 episodes of the revised 22 episodes were selected. The total score was computed by summing up the scaled intensities of the children's reactions in these 19 episodes. This total score ranges between 4 and 31 with a median of 13. The internal consistency (Cronbach's alpha) is satisfying (alpha = .76, standardized and unstandardized coefficients). In addition to the total score two extensity measures were computed. In the first measure all episodes were counted in which children showed any reaction (weak or strong). In the second measure only those episodes were counted in which the children showed a strong reaction. For the first extensity measure we found internal consistency (Cronbach's alpha) of alpha = .68, for the second one of alpha = .74. Finally, a score for the intensity of the curiosity motive was computed by dividing the sum of the scaled reactions (1 = weak reaction; 2 = strong reaction) by the number of episodes in which children showed any reaction at all. In Table 6 the correlation coefficients between these different measures are presented.

Table 6. Pearson's correlation between the different curiosity measures of the puppet-show (n = 64)

	Extensity measure I	Extensity measure II	Intensity measure
Total score	.93	.91	.73
Extensity measure I		.71	.47
Extensity measure II			.90

Both extensity measures and the intensity measure correlate highly with the total score. Only the first extensity measure is not related strongly to the intensity measure. Whether the extensity and intensity measures represent different dimensions of curiosity motivation has to be examined in further validation studies.

First Steps in Validating the Curiosity Motive-Scores

First signs of the validity of the parents' questionnaire and the indices derived from children's reactions to the puppet-show were found in several studies pursuing different goals. In a pilot study we tried to develop a behavioral coding system for preschoolers' exploratory behavior in dealing with different curiosity objects (Banta-box; drawer-box; 10 pairs of objects presented to the child: "preference for the unknown") based on the work of Henderson & Moore (1979). We had also given a questionnaire the parents in order to ask their evaluation of the children's dispositional curiosity and anxiety. Although the number of observations is still low we found some promising relationships between parents' evaluation of curiosity and children's manifest exploration in this test situation. The total score of exploratory behaviors with all three tasks with the total score of the curiosity questionnaire was $r = .40$ ($n = 13, p < .10$). This total score was computed by summing up the scores of the three curiosity scales. The strongest relationship between children's manifest behavior and parents' evaluation of children's curiosity was found for the frequency of manipulation of the children (summed over the three tasks). This score correlated with the total score of the curiosity questionnaire, $r(13) = .51$, $p < .04$, with the "Epistemic curiosity" scale, $r = .40$ and with the "Searching for stimulating events" scale, $r = .41, p < .08$. Furthermore, a significant correlation was found between the "Searching for the stimulating event" scale and the frequency of playing, $r(12) = .59$, $p < .03$. This confirms the assumption that this scale measures a motive component which is stronger related to the concept of sensation seeking and diversive exploration (Hutt, 1970) than to specific curiosity.

We received further support for the concurrent validity of this assessment procedure in a study in cooperation with H. Lugt-Tappeser. Sixty-four children (31 boys and 33 girls) aged 3;1 to 6;8 years were observed in their nursery school in the free play situation. Their behaviors were coded according to a behavioral anxiety check list (Lugt-Tappeser & Schneider, 1986). In addition, children were observed watching the puppet-show and for 51 children we obtained the parents' curiosity and anxiety questionnaire.

The categories of the anxiety behavioral checklist, based on a factor analysis, have been named (1) "immobile observation", (2) "active defense", and (3) "solitary behavior" (Lugt-Tappeser, 1991). The "Epistemic curiosity" scale of the questionnaire correlates with the total score based on children's behavior watching the puppet-show, $r(47) = .32$, $p < .02$, with the extensitiy measure, $r(47) = .28$, $p < .03$ and with the intensity measure, $r(47) = .25$, $p < .05$. No relations were found between measures of children's reactions to the puppet-show and the two other dimensions of the curiosity scale of the parents' questionnaire. However, these correlations are based mainly on the relations found for girls. Separately tested for boys and girls, the relations are only significant for girls.

Though not predicted we also found negative relationships between the curiosity scores based on children's reactions to the puppet-show and the anxiety scores of the parents' questionnaire as well as anxiety scores based on the behavioral check list (see Table 7).

Table 7. Pearson's correlations between the curiosity scores of the puppet-show and the anxiety scores of the parent's questionnaire and the Marburger Anxiety Scale.

	Total score	Measure of extensity I	Measure of intensity
Social anxiety (n = 50)	-.23**		-.29**
Fear of physical impairment (n = 51)	-.21*	-.22*	
Cognitive anxiety (n = 50)			
Total anxiety (n = 49)	-.25**	-.21*	-.21*
Immobile observation (n = 64)	-.19*	-.22**	
Active defence (n = 64)	-.37***	-.36***	-.33***
Solitaire behavior (n = 64)	-.20*	-.17*	
Summed anxiety (n = 64)	-.34***	-.34***	-.20*

(*p < .10; **p < .05; ***p < .01, one-tailed)

With the exception of cognitive anxiety all anxiety scores correlated significantly and negatively with the total score of the curiosity scale.

Conclusions

Due to the many facets of the manifestation of the curiosity motive in exploratory behavior, we encountered some difficulties in the pursuit of the goal to measure individual differences in the curiosity motive through concurrent assessment procedures. We found that parents are able to differentiate between basic dimensions of their children's curiosity behavior in everyday situations and evaluate them consistently. In addition, parents seem to be able to differentiate between their children's behavior motivated by curiosity, by competence striving and by anxiety. The originally obtained dimensions of the curiosity scale could be replicated with a second sample. The derived scales "Epistemic curiosity" and "Perceptive and manipulative curiosity" have satisfying internal consistencies. Only the scale "Searching for stimulating events" needs to be revised.

The second behavioral approach, to measuring the curiosity disposition with a checklist of manifest exploratory behavior is still at the beginning. Nevertheless, we found, in a small sample, predicted relations between these behavioral curiosity scores and scores from the parents' questionnaire.

The puppet-show assessment procedure was conceived with the intention to create a situation in which the activation of the curiosity motive becomes manifest in facial and gestural expression, and in attentive and verbal behavior without being inhibited to the same degree as exploratory behaviors by constraints of the situations or by simultaneously activated motives. After the revision of the puppet-show and of the coding system we found for the total scale based on 19 episodes satisfying internal consistency. The positive correlations we found between measures from children's reactions to the puppet-show and the "Epistemic curiosity" scale of the parents' questionnaire supported our theoretical assumptions. However, the consistent negative correlations between curiosity scores from children's reactions to the puppet-show and the different anxiety scores were not predicted and contradicted our assumptions. Children's reactions to the puppet-show seem not be pure manifestations of curiosity but are also determined by children's anxiety dispositions. It is unclear at the moment and cannot be solved on the basis of the data we have whether instigated curiosity motivation is inhibited in this situation by anxiety, instigated by features of the puppet-show and/or the total situation, or whether anxious children have developed habits to inhibit the expression of curiosity and emotionality and demonstrate this habitual inhibition also in a situation in which no anxiety-instigating stimuli are present. These different hypotheses can only be examined in experimental procedures. However, the fairly good test characteristics of our assessment procedures encourage us to follow this program in order to clarify the interaction of the two motive systems, curiosity and anxiety, and their expression in manifest behavior.

References

Berlyne, D. E. (1960). *Conflict, arousal and curiosity.* New York: McGraw-Hill.
Berlyne, D. E. (1966). Curiosity and exploration. *Science, 153*, 25-33.
Cattell, R. B. (1966). The scree test for the number of factors. *Multivariate Behavioral Research, 1*, 245-276.
Charlesworth, W. R. (1979). Die Beobachtung adaptiven Verhaltens: Eine ethologische Ergänzung zur Entwicklungspsychologie in ökologischer Sicht. In H. Walter & R. Oerter (Eds.), *Ökologie und Entwicklung* (pp. 185-199). Donauwörth: Auer.
Coie, J. D. (1974). An evaluation of the cross-situational stability of children's curiosity. *Journal of Personality, 42*, 93-116.
Görlitz, D. (1987). Process orientation and research dilemmas. In D. Görlitz & J. F. Wohlwill (Eds.), *Curiosity, imagination, and play* (pp. 107-125). Hillsdale, NJ: Erlbaum.
Harlow, H. F. (1953a). Mice, monkeys, men and motives. *Psychological Review, 60*, 23-32.
Harlow, H. F. (1953b). Motivation as a factor in the acquisition of new responses. In J. S. Brown et al. (Ed.), *Current theory and research in motivation: A symposium* (pp. 24-49). Lincoln, NE: University of Nebraska Press.
Harter, S. (1978a). Effectance motivation reconsidered: Toward a developmental model. *Human Development, 21*, 34-64.
Harter, S. (1978b). Pleasure derived from challenge and the effects of receiving grades on children's difficulty level choices. *Child Development, 49*, 788-799.
Harter, S. (1981a). A model of mastery motivation in children: Individual differences and developmental change. In W. A. Collins (Ed.), *Aspects of the development of competence. The Minnesota Symposia on Child Psychology, Vol. 14* (pp. 215-255). Hillsdale, NJ: Lawrence Erlbaum.
Harter, S. (1981b). A new self-report scale of intrinsic versus extrinsic orientation in the classroom: Motivational and informational components. *Developmental Psychology, 17*, 300-312.
Henderson, B. B. (1984a). Social support and exploration. *Child Development, 55*, 1246-1251.
Henderson, B. B. (1984b). The social context of exploratory play. In T. D. Yawkey & A. D. Pellegrini (Eds.), *Child's play: Developmental and applied* (pp. 171-201). Hillsdale, NJ: Erlbaum.

Henderson, B. B., Charlesworth, W. R., & Gamradt, J. (1982). Children's exploratory behavior in a novel field setting. *Ethology and Sociobiology, 3*, 93-99.
Henderson, B., & Moore, S. G. (1979). Measuring exploratory behavior in young children: A factor-analytic study. *Developmental Psychology, 15*, 113-119.
Hutt, C. (1970). Specific and diversive exploration. In H. W. Reese & L. P. Lipsitt (Eds.), *Advances in Child Development and Behavior (Vol. 5)* (pp. 119-180). New York: Academic Press.
James, W. (1892). *The principles of psychology (Vol. 2).* New York: Holt.
Keller, H., & Boigs, R. (1989). Entwicklung des Explorationsverhaltens. In H. Keller (Ed.), *Handbuch der Kleinkindforschung* (pp. 443-464). Berlin: Springer.
Kreitler, S., Zigler, E., & Kreitler, H. (1975). The nature of curiosity in children. *Journal of School Psychology, 13*, 185-200.
Lange, D., Massie, M., & Neuhaus, C. (1990). *Entwicklung eines Puppenspielverfahrens zur Erfassung des Neugiermotivs bei Vorschulkindern* (Diplomarbeit). Bochum: Ruhr-Universität.
Lorenz, K. (1943). Die angeborenen Formen möglicher Erfahrung. *Zeitschrift für Tierpsychologie, 5*, 235-409.
Lugt-Tappeser, H. (1991). *Überprüfung der Marburger Angstzeichenliste zur Erfassung der Ängstlichkeit bei Kindern* (Unpublished paper). Marburg: Philipps-Universität.
Lugt-Tappeser, H., & Schneider, K. (1986). *Die Entwicklung einer Zeichenliste für die Erfassung von Ängstlichkeit bei Vorschulkindern.* Ruhr-Universität Bochum: Institutsbericht der Fakultät für Psychologie, Nr. 5.
Lugt-Tappeser, H., & Schneider, K. (1987). Ängstlichkeit und das Erkunden eines neuen Objektes bei Vorschulkindern. *Zeitschrift für Entwicklungspsychologie und Pädagogische Psychologie, 19*, 300-313.
Matip, E.M. (1989). *Entwicklung eines Elternfragebogens: Kompetenzstreben und Ängstlichkeit* (Diplomarbeit). Bochum: Ruhr-Universität.
Maw, W. H., & Maw, E. W. (1966). An attempt to measure curiosity in elementary school children. *American Educational Research Journal, 3*, 147-156.
Maw, W. H., & Maw, E. W. (1975). Note on curiosity and intelligence of school children. *Psychological Reports, 36*, 782.
McClelland, D. C. (1958). Risk taking in children with high and low need for achievement. In J. W. Atkinson (Ed.), *Motives in fantasy, action, and society* (pp. 306-321). Princeton, NJ: Van Nostrand.
McClelland, D. C. (1958). The importance of early learning in the formation of motives. In J. W. Atkinson (Ed.), *Motives in fantasy, action, and society* (pp. 437-452). Princeton, NJ: van Nostrand.
McClelland, D. C. (1980). Motive dispositions. In L. Wheeler (Ed.), *Review of personality and social psychology (Vol. 1)* (pp. 10-41). Beverly Hills, CA: Sage.
McClelland, D. C., Atkinson, J. W., Clark, R. A. & Lowell, E. L. (1953). *The achievement motive.* New York: Appleton-Century-Crofts.
McDougall, W. (1908). An introduction to social psychology. London: Menthuen.
Montgomery, K. C. (1955). The relation between fear induced by novel stimulation and exploratory behavior. *Journal of Comparative and Physiological Psychology, 48*, 254-260.
Moore, S. G. (1985). *Exploration in young children: Stability, variability and social influence.* Paper presented at the Symposium on Facets of Exploratory Activity

in Childhood, APA Conference, Los Angeles, CA.
Morgan, G. A., & Harmon, R. J. (1984). Developmental transformations in mastery motivation. In R. N. Emde & R. J. Harmon (Eds.), *Continuities and discontinuities in development* (pp. 263-291). New York: Plenum.
Purmann, E. (1988). *Neugier und Kompetenzstreben : Entwicklung eines Elternfragebogens* (Diplomarbeit). Bochum: Ruhr-Universität.
Schneider, K., & Flohr, B. (1989). *Explorationsverhalten bei Kindern im Vorschulalter.* Bericht Nr. 69, Ruhr-Universität Bochum, Fakultät für Psychologie.
Schneider, K., Moch, M., Sandfort, R., Auerswald, M., & Walther-Weckman, K. (1983). Exploring a novel object by preschool children: A sequential analysis of perceptual manipulating and verbal exploration. *International Journal of Behavioral Development, 6,* 477-496.
Schneider, K., & Schmalt, H.D. (1981). *Motivation.* Stuttgart: Kohlhammer.
Trudewind, C., & Husarek, B. (1979). Mutter-Kind-Interaktion bei der Hausaufgabenanfertigung und die Leistungsmotiventwicklung im Grundschulalter - Analyse einer ökologischen Schlüsselsituation. In H. Walter & R. Oerter (Eds.), *Ökologie und Entwicklung* (pp. 229-246). Donauwörth: Auer.
Voss, H. G. (1984). Curiosity, exploration and anxiety. In H. van der Ploeg, R. Schwarzer, & C. D. Spielberger (Eds.), *Advances in test anxiety research.* (Vol. 3) (pp. 121-146). Lisse: Swets.
Weisler, A., & McCall, R. B. (1976). Exploration and play. Résumé and redirection. *American Psychologist, 31,* 492-508.
White, R. W. (1959). Motivation reconsidered: The concept of competence. *Psychological Review, 66,* 297-333.
Yarrow, L. J., Morgan, G. A., Jennings, K. D., Harmon, R. J., & Gaiter, J. L. (1982). Infants' persistence at tasks: Relationships to cognitive functioning and early experience. *Infant Behavior and Development, 5,* 131-141.
Zuckerman, M., Kolin, E. A., Price, L., & Zoob, I. (1964). Development of sensation seeking scale. *Journal of Consulting Psychology, 28,* 477-482.

CHAPTER III.9

Preschoolers' Exploratory Behavior: The Influence of the Social and Physical Context

Klaus Schneider, Lothar Unzner

Children's Exploration in Natural Contexts

Children's exploration of the physical and social environment has been studied almost exclusively under restricted laboratory conditions. There are only a few studies demonstrating the power of curiosity-instigating incentives in preschoolers' under natural conditions. In addition, as Henderson (1984a) noted, most of children's exploration in natural settings happens in the presence of others: parents, siblings, peers, and teachers. This may change dramatically the power of incentives for exploration and for other competing behavioral systems and the relative importance of different instigated motivations in the situation.

Henderson, Charlesworth, and Gamradt (1982), for example, observed preschool and gradeschool children in a natural science museum, either in peer groups or when accompanied by parents. They found striking differences in the exploratory and general behavior of the children depending on the social context. Children in peer groups moved around more and explored the exhibited objects less thoroughly than children who were accompanied by their parents; they inspected and touched objects less often and also asked fewer questions. Thus, the same objects instigated more or less and qualitatively different exploration depending on the social interactions in the situation.

From a motivational perspective, we have to explain such situational differences in manifest behavior as functions of differently perceived incentives in these different social contexts - curiosity-instigating incentives in competition with incentives related to other motives,

---

We thank the parents for the cooperation in this study and Sabine Blöscher and Anne Petters for their assistance in the collection of data.

such as the tendency to engage in social interactions and play (see Ainsworth, Blehar, Waters, & Wall, 1978). In addition, curiosity-instigating novel objects also arouse fear or neophobia at the same time (see James, 1890; McDougall, 1908; Montgomery, 1955). Thus, manifest exploration is a compromise between both behavioral tendencies - curiosity, which can only be satisfied by approaching, inspecting and touching an object, and anxiety, which leads to withdrawal and behavioral inhibition. Individual differences at one age and developmental changes in manifest exploration can be caused by differences and changes in both behavioral systems (see Bischof, 1975).

Additionally, Henderson et al. (1982) suggested that parents presumably further their children's exploration by catching their attention and directing it to different facets of the environment, by pointing to features of the environment, by asking questions, answering questions and so on, or indirectly as behavioral models (see also Belsky, Goode, & Most, 1980; Endsley, Hutcherson, Garner, & Martin, 1979; Henderson, 1984; Johns & Endsley, 1977; Saxe & Stollak, 1971). However, they may simultaneously not allow specific kinds of exploration like touching, grasping, etc. This means that the motivation-activating power of curiosity-instigating object variables, Berlyne's (1960) "collative variables," is partly socially mediated. At the same time, some of the inhibiting factors in the situation, the negative incentives, are socially mediated as well. Depending on the physical and social context, we have to take into account a greater or lesser amount of social (parental) control of children's exploratory behavior. From a developmental perspective, we also have to assume that curiosity incentives as well as parents' control behaviors, change with increasing age, depending on the children's growing intellectual capabilities, the development of their interests and self-control and parents' perception of these changes (see Baldwin & Baldwin, 1977; Keller, 1987).

How can we explain age-related changes in the motivation-instigating potential of curiosity incentives? Following Berlyne (1960, 1978), we assume that the immediate cause of the investigatory response to features of the environment is the state of subjective uncertainty, which is instigated in man and animals alike by novel, complex, and ambiguous aspects of objects and situations (see Schneider, 1987; Schneider, Barthelmey, & Herrmann, 1981). From a developmental as well as from an individual difference perspective, it has been assumed that the intensity of that state differs due to the existence and/or the differentiation of the children's representations or schemata of the world's objects and the amount of incongruence between percepts and internal schemata (Dember & Earl, 1957; Hunt, 1965). When the discrepancy between the encountered object and cognitive schemata is extreme, anxiety or neophobia might be instigated in addition to or instead of curiosity (Hebb, 1955).

This hypothesis seems useful for understanding ontogenetic changes as well as individual differences in behavior as a function of individual differences in and changes of the cognitive structures of the children. Besides that, there might be individual differences and age-related changes in the hypothesized disposition to explore. Because both motive systems, curiosity and neophobia or more general anxiety, have their own adaptive value, we can assume a genetic basis for both systems (see Gray, 1982; Lorenz, 1969). Part of the individual variance might be determined by this.

Summarizing, we assume that children with more or less strong motives to explore and to avoid novel objects or situations and more or less elaborate cognitive schemata of objects in the world around them, either in an ontogenetic or in an individual difference perspective, manifest differences in their curiosity behavior in different environments due to the perceived positive and negative incentives (including anticipated punishment by authority figures) in these situations. Incentives for exploration are the "collative" qualitites (Berlyne, 1960) of objects. Under certain conditions these very same qualities will instigate anxiety and, thus, become effective as negative incentives (James, 1891). Others with whom children interact in such situations will make positive and negative incentives more or less salient. Partners in the social interaction may also activate additional behavioral systems, for example, the motivation to interact with others socially, or the motivation to please others or to avoid reproof of parents and other authority figures.

An Empirical Study

The first aim of the present study was to document that the amount, kinds (quality) and structure of children's exploratory behaviors depend on the physical environment: (a) directly via the activating and inhibiting power of object-related incentives in these environments and (b) indirectly via eliciting different supporting (positive) and controlling (negative) behaviors from the accompanying person. A second goal was to examine how the exploratory behaviors in different environments change during the second year of life. Changes observed in children's curiosity behavior over this age range could be either caused by a change in the force of these curiosity- and fear-instigating incentives related to the cognitive development of the children or caused by changes in their parents' tutoring and controlling behaviors.

Towards this goal, we observed toddlers at home and during a visit to a market with their mothers. We assumed that the mothers' controlling behavior, especially in the shopping market, would diminish considerably during this time. We also assumed that the incentives in the market for children's exploration and the fear-instigating power of the strange surrounding would change considerably. Children would habituate to the positive and negative incentives of this environment to which their mothers bring them normally once a week. However, because our study was an observational field study, it is not possible to separate the influence of these factors on children's exploration. At home we did not expect dramatic changes in children's curiosity and fear and in their mothers' control behaviors. Therefore, at home, no systematic changes in children's exploratory behavior should become manifest during the age range studied here.

Method

Due to the field-based character of our study there was generally little control of the situational factors. The families of the first sample lived in a small town with one market, the families of the second sample lived in a middle-sized university town and its rural neighborhood and visited different markets. However, all markets were similar in their arrangement of goods. Before the first observation, the observers visited the families once to familiarize themselves with the children and to discuss with the mothers the arrangement of the study. The mother was instructed to walk the child beside her in the market and not put them in the cart as they usually do, but besides that, mothers should behave as usually.

At home the observer brought, at each visit, a little present wrapped in paper for the child: 1. a Lego Duplo set; 2. a mini picture book (Ravensburger Mini-Bilderbuch) and 3. a little zoo in a box made of wood: animals, trees and fences. All children began by unwrapping the present and playing with it before they changed to other activities. All mothers watched their child first and then went ahead with their household activities. Most observations were done in the morning between 9 and 10 am; no siblings or other children were present during the observation at home. More details of the procedure will be presented in a journal article in preparation (Unzner, Blöcher, Petters, & Schneider, in prep.).

Subjects

Two samples of children were observed: Sample 1 (S1) consisted of 11 children (5 males and 6 females), 21 months old at the first observation. Sample 2 (S2) consisted of 13 children (6 males and 7 females), 27 months old at the first observation. Among the children there were first-, second- and third-born children. Children were of German origin and from middle-class backgrounds.

Procedure

The study was carried out as a short-term longitudinal study. The children of Sample 1 were observed three times; the children of Sample 2 were observed twice, in each case at 2-month intervals. At each time of observation, the children were observed in the two settings: (1) in the home during normal morning activities of mother and child when the mother was busy with routine activities (observation time: 30 minutes), and (2) during a visit to a shopping market where the mother let the child walk beside her (observation time about 20 minutes). At the beginning of each home observation, the child received the little gift (toy) from the observer.

Data collection and coding

The observer spoke a specimen record into a portable tape-recorder. Transcriptions of these protocols were the basis for further codings. We developed category systems to code the behaviors of the mother and the child on the basis of the results of a pilot study with 3 children and coded the first 10 minutes of the transcribed protocols according to these categories. Table 1 presents the coded behaviors of the children and mothers in both environments and the percentages of observer agreement for two coders (A. Petters and S. Blöscher). For this, the two coders coded the transcriptions of two observed children from Sample 2 in both situations on the basis of a mutual segmentation of the protocol. Except for question asking and children's negative reactions to mothers' interventions based on few codings, observer agreements are quite high, demonstrating that these described behaviors can be reliably coded.

Table 1. Coded children's and mothers' behaviors in the two environments

	Exploratory behavior of the child
Unorientated locomotion: (100%[a]; n = 7)	The child moves around without a visible goal.
Looking around: (100%; n = 7)	The child visually scans the surroundings (objects and persons).
Oriented locomotion: (93%; n = 31)	The child looks to a certain object and moves toward it.
Visual exploration: (91%; n = 37)	The child tries to attain information about the things around by looking to a certain object or inspecting visually the details of a certain object.
Tactile exploration: (84%; n = 49)	The child touches, fingers, or grasps an object.
Asking questions: (57%; n = 5)	The child asks object-related questions.
Simple manipulation: (92%; n = 13)	The child manipulates an object with simple unrelated movements.
Complex manipulation: (92%; n = 21)	The child manipulates one or more objects in a coordinated sequence; the action involves two or more separate behaviors which relate to one another (e.g., building up blocks, opening a box, or sorting objects). All behaviors in the market that relate to the chart are excluded.
Play: Pretend or symbolic play (not observed in these two children)	

Maternal control behavior

Supporting behavior: (77%; n = 102)	The mother gives an object to the child, explains and describes objects, comments her own behavior, neutrally or positively comments the behavior of the child, directs the attention of the child to certain features of the environment, asks questions, agrees with the child, asks the child to do something, supports or finishes actions of the child.
Coercive behavior: (80%; n = 5)	Mother interrupts the child's behavioral sequence (physical intervention), forbids the child in action, blames the child, comments the child's behavior negatively, denies child's request or postpones the child's reaction.
Mother's contact - seeking maintenance behavior (83%; n = 34)	Approacing, expressive behavior (facially and gestural), verbal contact, and close physical contact.
Children's reaction (79% and 50%; n = 20 and 6, respectively)	Positive reactions (agrees, gives in) and imitating of the mother as a model. Negative reactions (ignores, runs away, physical resistance, verbal protest, and crying).
Children's contact - seeking maintenance behavior (83%; n = 13)	Approaching, expressive contact and behavior, verbal contact behavior, and close physical contact.

[a] Number of segments in which a behavioral category was coded by at least one coder and coder agreement (in%) between two coders (S.B and A.P.)

Results and Discussion

Exploratory Behavior of the Children

Figure 1 presents the averaged frequencies of the different exploratory behaviors, looking around, goal-oriented locomotion, visual and tactile exploration, and simple and complex manipulation, for both samples and the three or two times of observation, respectively.

Figure 1. Averaged frequencies of children's exploratory behaviors at the market and at home for the 3 times of observation for Sample 1 (21 to 25 months) and the 2 times of observation for Sample 2 (27 to 29 months)

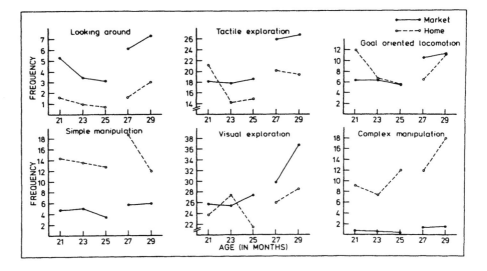

As assumed, the distal behaviors, looking around and visual exploration, and also the cautious tactile exploration, are on the average more often exhibited in the unfamiliar than in the familiar environment in both samples. However, the differences in visual and tactile exploration are significant only for the older children (Sample 2). Simple and complex manipulations, on the other hand are more often observed at home than in the market. These differences are highly significant in both samples (see Table 2). Table 2 presents a summary of the significant effects of the two factors Situation and Times of Observation for all variables of locomotion, exploration, and play as assessed in the two situations.

Table 2. Results (F-values) of statistical analyses (ANOVAs) of the differences between the two situations, the three respectively two times of observations and the interaction of these factors

	Situations (S)	Time of Observation (TO)	S x TO
Unoriented locomotion			
S1[a]	32.62***	-	-
S2[b]	-	-	-
Looking around			
S1	35.27***	4.02*	-
S2	31.65***	5.17*	-
Oriented locomotion			
S1	1.76	4.54*	2.67*
S2	-	-	-
Visual exploration			
S1	-	-	2.83*
S2	4.74*	6.90*	5.65*
Tactile exploration			
S1	-	2.83**	-
S2	10.22***	-	-
Asking questions			
S1	-	-	-
S2	-	-	-
Simple manipulation			
S1	49.83***	-	-
S2	43.29***	12.44**	9.32*
Complex manipulation			
S1	33.42***	-	-
S2	42.58***	6.64*	5.98*
Playing			
S1	8.08*	-	-
S2	3.64+	-	-

[a]df in Sample 1: S (1,9); TO (2,18); S x TO (2,18)
[b]df in Sample 2: S (1,11); TO (1,22); S x TO (1,22)
+p < .10; *p < .05; **p < .01; ***p < .001

In both samples, the number of questions asked do not differ between the home situation (M[S1] = 1.0 and M[S2] = 3.8, respectively) and the market (M = 0.5 and M = 3.5, respectively). The difference between the groups of children, however is significant (t[22] = 2.77, p < .01).

Playing with objects was not observed often, it appeared more at home (M = 1.6 and M = 0.8, respectively) than in the market (M = 0.1 and M = 0.0, respectively). The difference is significant in the younger sample (p < .05) and approaches significantly in the older one (p < .10). Unoriented locomotion appeared in the market more often than at home in Sample 1 only (p < .001).

Control Behaviors of the Mothers

The distance between child and mother was rated (observer agreement between two coders:91%) for all behavioral observations of the child: 1 = body contact; 2 = grasping distance; 3 = viewing distance; 4 = out of view. Children and mothers were further away from each other at home (M = 2.8 and M = 2.7, respectively) than in the market (M = 2.2 and M = 2.2, respectively). For both samples these differences are significant (S1: F[1,9] = 66.05, p < .001; S2: F[1,11] = 13.65, p < .01).

Mean values for mothers' supportive and coercive behaviors, children's positive and negative reactions to their mothers' controlling behavior, and the contact behavior of mothers and children are presented in Figure 2.

Figure 2: Averaged frequencies of mothers' tutoring behaviors (supportive and coercive), of children's positive and negative reactions to their mothers' tutoring, and of children's and mothers' contact-seeking behaviors at the market and at home for the 3 times of observation for Sample 1 (22 to 25 months) and the times of observation for Sample 2 (27 to 29 months)

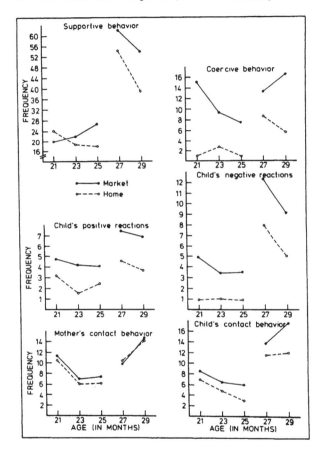

For the mothers' supporting behaviors we do not find any differences between the two situations and between the times of observations in both samples. However, mothers in Sample 2 demonstrated this supporting behavior generally more often (M = 52.8) than the mothers in Sample 1 (M = 22.0; t([2])= 5.30, p < .001).

As expected, mothers' control of their children's behavior was stronger in the market than at home (S1: F[1,9)]= 20.69, p < .01; S2: F[1,11] = 11.00, p < .05). In Sample 1 there is, in addition, an interaction effect, Situation x Times of Observation. In the market the coercive behavior of the mothers is high at the first observation and declines afterwards whereas at home it remains nearly on the same level (F[2,18] = 4.13, p < .05).

Children's positive and negative reactions to mothers' supportive and coercive behaviors are stronger in the market than at home (Sample 1, positive reaction: F(1,9) = 3.80, p < .10; negative reaction: F(1,9 = 31.10, p < .001; Sample 2, positive reaction: F(1,11) = 4.20, p < .10; negative reaction: F(1,11) = 6.37, p < .05).

Similarly, the contact seeking and the maintenance of contact of children is stronger in the strange environment than at home (S1: F(1,9) = 5.25, p < .05; S2: F(1,11) = 2.63, n.s.). In the sample of the younger children (S1) the frequency of this contact behavior decreases over the studied age range from 21 up to 25 months.

Children's Exploration and the Distance between Mother and Children

For children aged 1 to 3 years, exploratory behavior is presumably strongly influenced through the distance to the caregiver. On the one hand, a secure attachment relationship, in which the mother serves as a secure base (Bowlby, 1969), enhances exploratory behavior. On the other hand, the child might be ready to leave the mother and to explore the interesting things which are usually not within reach when the child clings to the mother (Ainsworth et al., 1978). Furthermore, at a greater distance, the mother certainly cannot be as restrictive as at a nearer distance.

Table 3 shows that there is a strong association between the occurrences of the different behaviors of children and mothers and the distance between mother and child at home as well as in the market. Children of both samples in both settings exhibited most kinds of exploratory behavior more often at a greater distance from the mother. There are a few exceptions. The children of Sample 1 showed more "simple manipulation" in the market and played more often at home the nearer they were to the mother. In Sample 2 children asked more questions nearby the mother at home and also demonstrated more tactile exploration in the market when they were near to the mother.

Table 3. Association between children's exploratory behaviors and maternal control behavior and the distance between mother and child. (Distances in which the observed frequency is higher than the expected frequency are given. Only significant relationships based on a chi^2 analysis (contingency coefficients in brackets) are presented.)

Samples	S1		S2	
Situations	Home	Market	Home	Market
Exploratory behaviors				
Unoriented locomotion	distant (.58)	distant (.81)	distant (.42)	distant (.79)
Looking around	-	-	distant (.38)	-
Oriented locomotion	distant (.45)	distant (.54)	distant (.30)	distant (.40)
Visual exploration	distant (.13)	distant (.14)	distant (.26)	distant (.11)
Tactile exploration	distant (.13)	near (.13)	distant (.18)	near (.17)
Asking questions	-	-	near (.37)	-
Simple manipulation	distant (.30)	near (.28)	-	distant (.41)
Complex manipulation	distant (.27)	-	distant (.29)	-
Playing	near (.57)	-	distant (.61)	-
Maternal control behavior				
Supportive behavior	near (.60)	near (.19)	near (.41)	near (.32)
Coercive behavior	near (.75)	near (.62)	near (.66)	near (.59)

Mothers, however, were much more supportive and coercive when the child was quite near. This relation is to some extent trivial, because some forms of maternal control behavior require a close physical contact. But it might also be a hint that, at least in our sample, positive as well as negative forms of control behavior hinder some exploratory behaviors of the child. The correlations between the frequency of maternal control behaviors and the exploratory behaviors of the child (summed over the whole observation period) which are partly negative support this suspicion.

Relationship between Behaviors of Mothers and Children

In respect to most exploratory behaviors of the children, maternal support correlates negatively with children's exploration, especially at home. Only tactile exploration correlates positively with supportive behavior of mothers in Sample 2.

In the market, more positive than negative relationships between mothers' tutoring and children's exploration are found. Children of more coercive mothers asked more questions and touched objects more often than children of non-coercive mothers. This might be a reaction of the child to the mothers' interventions or of the mother to children's exploration. Without a detailed sequential analysis of the mother child interaction, this issue cannot be solved.

In most instances, however, children did not show any distinct reaction to the maternal behavior. Children contingently reacted to about 17% (both samples) of maternal supportive interventions and to 14% (Sample 1) and 25% (Sample 2) of coercive interventions. If they showed any reaction, the proportions of positive and negative reactions differed in respect to the maternal behaviors. Moreover, these effects were different in both samples (logit-analyses: Maternal behavior: $chi2(1) = 26.67$, $p < .001$; Samples: $chi^2(1) = 18.41$, $p < 001$; interaction Maternal behavior x Sample: $chi^2(1) = 2.77$, $p < .10$). In Sample 1, 65% of the unequivocal reactions to supportive behavior were positive and 63% of the reactions to coercive behavior were negative. In Sample 2, for both types of maternal behavior negative reactions were observed more frequently than positive reactions (53% to mothers' supportive behavior, 67% to mothers' coercive behavior).

The Structure of the Exploratory Behavior

In order to analyze the structure of the exploratory behavior, we computed as a first step toward this goal first-order transition probabilities between aggregated categories of exploration. Being confronted with a novel object, children show a sequence of behaviors which starts with a general orientation, goes on with perceptual and tactual exploration and manipulation, and finally ends in playing with the object (Hutt, 1970; Nunnally & Lemond, 1973; Schneider et al., 1983). Therefore, we analyzed the sequences of the following combined categories for both situations and both samples: (a) unoriented locomotion and looking around (searching behaviors), (b) looking to an object (orienting), (c) visual and tactile inspections (perceptive exploration), and (d) manipulating. Because of the low frequency of playing, we excluded this category from the analysis. Figure 3 shows the significant (see Haberman, 1973) first-order transition probabilities between these combined categories.

Figure 3: Significant first-order transition probabilities for children's exploratory behaviors at home and at the market pooled over all children of Sample 1 and Sample 2

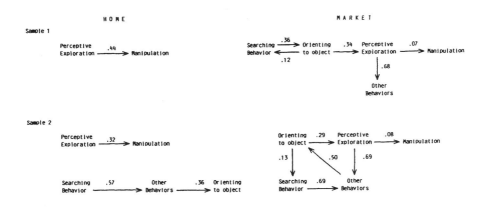

The observed sequential structures of exploratory behaviors mirror the proposed sequence more clearly in the sample of the older children and in both samples more clearly in the supermarket setting than at home. At home, only manipulation follows perceptive exploration in the proposed sequence in both samples. However, we could not find other predicted transitions in this situation.

In the market setting in Sample 2, more exploratory behaviors followed each other in the proposed sequence, although sometimes they were interrupted by other behaviors.

Summarizing, we conclude from these observations that toddlers' exploratory behaviors in natural environments are strongly influenced by the situations. Home and market settings differ in many aspects. The home is familiar and presumably offers a less complex array of objects, the market is rather unfamiliar and provides an overwhelming number and variety of objects to explore. In addition, mothers control their children's behavior differently in the two situations. We cannot decide on the basis of our observational data whether the incentives of the physical environment or mothers' controlling and tutoring behaviors exerts more influence on children's exploration and is responsible for the observed differences between the situations. However, there are some observations which indicate the importance of the curiosity-instigating incentives in the situations in comparison with the social influence of mothers.

1. Most exploratory behaviors appear more frequently than expected at a greater distance from the mother, whereas supportive and coercive control behaviors are observed more often when mother and child are at the nearest distance. This relationship holds for the market, where it might be trivial, as well as for the home.

2. In the market we observed more complex sequences of exploratory behaviors than at home. This might be explained by the fact that a market offers more powerful incentives for curiosity behaviors, which attract children longer.

3. In the market we also observed stronger relationships between children's exploratory behaviors than at home. Table 4 presents the averaged (over both samples) correlations between children's exploratory behaviors at home and in the market.

The highest association is between the two modes of manipulation. Additionally, children who moved directly to interesting objects explored these less often visually but more often by touch. Children who explored more often visually did less often grasp or touch the objects, but asked more questions and manipulated more often in a complex manner. At home all correlations are low and insignificant.

Table 4. Intercorrelations between children's exploratory behaviors at home and in the market

Home	V	T	AQ	SM	CM
Oriented Locomotion	-.16	.30	-.24	-.21	-.32
Visual Exploration (V)		-.31	.05	-.02	.12
Tactile Exploration (T)			-.20	-.26	-.13
Asking Questions (AQ)				.05	.08
Simple Manipulation (SM)					.20
Complex Manipulation (CM)					-

Market	V	T	AQ	SM	CM
Oriented Locomotion	-.45*	.70***	.10	-.07	-.04
Visual Exploration (V)		-.46*	.33	.26	.43*
Tactile Exploration (T)			-.11	.16	.02
Asking Questions (AQ)				.04	-.08
Simple Manipulation (SM)					.74***
Complex Manipulation (CM): pooled over both samples					-

The supermarket, therefore, seems to offer stronger incentives for toddlers' exploratory behavior than the home with a visiting, modestly familiar person bringing a little toy. The strong incentives in the market seem to organize the behavior of the children. Therefore, we find more goal-directed complete sequences of exploratory behaviors in the market than at home and also stronger relationships between different exploratory behaviors.

The behavior of the mothers, on the one hand, seems to be of minor importance compared with this situational influence. Children explore more when they are out of their mother's reach and do not react at all in more than 75% of all observations to the supporting or coercive behaviors of their mothers - they just went ahead with their activities.

Without denying the importance of a secure attachment between a child and her or his mother for the expression of curiosity (Ainsworth et al., 1978; Bowlby, 1969; Keller & Boigs, 1989, this volume) and the importance of parents' attention-guiding and modelling behavior (see Belsky, Goode, & Most, 1980), it seems to us that object-related qualities, the usual incentives for curiosity motivation, have for a while a greater importance than the mother in the two situations and in the age-group studied here. Children at this age demonstrate a high interest in the physical realities of their environment. To explore these seems to be an important task for which the mother, at least in the market, is more of a hindrance than a help. That curiosity motivation is of high priority in children's behavioral systems has also been demonstrated for preschoolers and grade school children by Henderson, Leone, & Loy (1985): Children of these age groups in a curiosity situation with "off limits" areas gave (a) conservative interpretations to what was allowed to explore and what not and (b) did not generalize this restricted prohibition to a second similar situation following immediately.

At home, mothers went ahead with their routine household work after the first minutes of the observation period; they seldom interfered, but also did not stimulate their children's exploratory behavior. Remember that they were not instructed to do so or to play with the child during the observation time. At the market, their primary goal was to do the shopping and due to the situational constraints they had to control their children's behavior. So the negative correlations between mothers' tutoring behaviors and children's exploration might be specific to situations where parents have no motivation to stimulate their children's exploration due to their own preoccupations or social fears and inhibitions.

Under playing or tutoring conditions and in the appropriate setting, this might be different. Several studies have demonstrated positive relationships between mothers' attention-guiding behavior and infants' exploration at home (Belsky et al., 1980), between preschoolers' exploration and mothers' modelling of exploratory behaviors as well as the importance of mothers' emotional support for their children's exploration (Endsley et al., 1979; Johns & Endsley, 1977; Saxe & Stollak, 1971). Thus, it seems to depend on the social context, the present occupations and goals of the parents, and the physical and social constraints of the situation whether exploration is enhanced, inhibited or not influenced by parents' behavior. Exploratory behavior seems to be of high priority in toddlers' and preschoolers' behavioral systems, but less so in parents' tutoring systems or educational philosophies - at least not in normal workday situations. In parents' behavior it might then be inhibited or at least postponed through other more important goals at the time. In addition, parents have stronger reactions to the constraints of the situation than their 2-year-olds. However, in a more educational context this might be different. Such different affordances of the situation might also explain the differences observed in children's exploration in the market situation studied here and in the science museum studied by Henderson et al. (1982), where parents might have gone with the intention to help their children explore the gadgets presented there. All these hypotheses are highly speculative and have to be tested by further field studies using microanalyses of parents' and children's behaviors and interactions in different situations and social contexts.

References

Ainsworth, M. D. S., Blehar, M. C., Waters, E., & Wall, S. (1978). *Patterns of attachment.* Hillsdale, NJ: Erlbaum.

Baldwin, J. D., & Baldwin, J. I. (1977). The role of learning phenomena in the ontogeny of exploration and play. In S. Chevalier-Skolnikoff & F. E. Poirier (Eds.), *Primate biosocial development: Biological, social, and ecological determinants* (pp. 343-xx). New York: Garland.

Belsky, J., Goode, M. K., & Most, R. K. (1980). Maternal stimulation and infant exploratory competence: Cross-sectional, correlational, and experimental analyses. *Child Development, 51,* 1168-1178.

Berlyne, D. E. (1960). *Conflict, arousal and curiosity.* New York: McGraw-Hill.

Berlyne, D. E. (1978). Curiosity and learning. *Motivation and Emotion, 2,* 97-175.

Bischof, N. (1975). A systems approach toward the functional connections of attachment and fear. *Child Development, 46,* 801-817.

Bowlby, J. (1969). *Attachment and loss: Vol. I. Attachment.* New York: Basic Books.

Dember, W. N., & Earl, R. W. (1957). Analysis of exploratory, manipulatory and curiosity behaviors. *Psychological Review, 64,* 90-96.

Endsley, R. C., Hutcherson, M. A., Garner, A. P., & Martin, M. J. (1979). Interrelationships among selected maternal behaviors, authoritarianism, and preschool children's verbal and nonverbal curiosity. *Child Development, 50*, 331-339.

Gray, J. A. (1982). *The neuropsychology of anxiety: an inquiry into functions of the septo-hippocampal system.* New York: Oxford University Press.

Haberman, S. J. (1973). The analysis of residuals in cross-classified tables. *Biometrics, 29*, 205-220.

Hebb, D. O. (1955). Drives and the C.N.S. (Conceptual nervous system). *Psychological Review, 62*, 243-254.

Henderson, B. B. (1984). The social context of exploratory play. In T. D. Yawkey & A. D. Pellegrini (Eds.), *Child's play: Developmental and applied* (pp. 171-201). Hillsdale, NJ: Erlbaum.

Henderson, B. B., Charlesworth, W. R., & Gamradt, J. (1982). Children's exploratory behavior in a novel field setting. *Ethology and Sociobiology, 3*, 93-99.

Henderson, B. B., Leone, C. T., & Loy, L. L. (1985). Children's exploratory behavior and the generalization of a prohibition. *The Journal of Genetic Psychology, 146*, 57-64.

Hunt, J. McV. (1965). Intrinsic motivation and its role in psychological development. In D. Levine (Ed.), *Nebraska symposium on motivation (Vol. 13)* (pp. 189-282). Lincoln: University of Nebraska Press.

Hutt, C. (1970). Specific and diversive exploration. In H. W. Reese & L. P. Lipsitt (Eds.), *Advances in Child Development and Behavior (Vol. 5)*. New York: Academic Press.

James, W. (1891). *The principles of psychology (Vol. II)*. London: Macmillan.

Johns, C., & Endsley, R. C. (1977). The effects of a maternal model on young children's tactual curiosity. *Journal of Genetic Psychology, 131*, 21-28.

Keller, H., & Boigs, R. (1989). Entwicklung des Explorationsverhaltens. In H. Keller (Ed.), *Handbuch der Kleinkindforschung* (pp. 443-464). Berlin: Springer.

Keller, J. A. (1987). Motivational aspects of exploratory behavior. In D. Görlitz & J. F. Wohlwill (Eds.), *Curiosity, imagination and play* (pp. 24-42). Hillsdale, NJ: Erlbaum.

Lorenz, K. (1969). Innate bases of learning. In K. H. Pribram (Ed.), *On the biology of learning* (pp. 13-93). New York: Harcourt.

McDougall, W. (1908). *An introduction to social psychology*. London: Methuen.

Montgomery, K. C. (1955). The relation between fear induced by novel stimulation and exploratory behavior. *Journal of Comparative and Physiological Psychology, 48*, 254-260.

Nunnally, J. C., & Lemond, L. C. (1973). Exploratory behavior and human development. In H. W. Reese (Ed.), *Advances in child development and behavior (Vol. 8)* (pp. 59-109). New York: Academic Press.

Saxe, R. M., & Stollak, G. (1971). Curiosity and the parent-child relationship. *Child Development, 42*, 373-384.

Schneider, K. (1987). Subjective uncertainty and exploratory behavior in preschool children. In D. Görlitz & J. F. Wohlwill (Eds.), *Curiosity, imagination, and play* (pp. 179-197). Hillsdale, NJ: Lawrence Erlbaum.

Schneider, K., Barthelmey, E., & Herrmann, P. (1981). Subjektive Unsicherheit und visuelle Exploration bei Kindern im Vorschulalter. *Zeitschrift für Entwicklungspsychologie und Pädagogische Psychologie, 2*, 106-115.

Schneider, K., Moch, M., Sandfort, R., Auerswald, M., & Walther-Weckman, K. (1983). Exploring a novel object by preschool children: A sequential analysis of perceptual manipulating and verbal exploration. *International Journal of Behavioral Development, 6*, 477-496.

CHAPTER **III.10**

A Developmental Analysis of Exploration Styles

Heidi Keller

Humans interact with their environment in different behavioral modes such as seeing, hearing, smelling, touching, approaching, manipulating, and talking. These interactions create information and belong to the behavioral system of exploration. Although they share the common purpose of information input, they differ significantly in structure. Shulamith and Hans Kreitler (1987, see also this volume; and Kreitler, Zigler, & Kreitler, 1974, 1975) who have done extensive research on the factor structure of exploration have repeatedly identified five factors in children as well as in adults. The most powerful factors with respect to the amount of explained variance in 6-to 8-year-old children were manipulative exploration (17.5%, focused on exploring by means of motor actions), followed by perceptual exploration (12.5%, focused on exploring by means of viewing, listening, or smelling) and conceptual exploration (9.3%, focused on exploring by checking meanings and their interrelations and by asking questions). The majority of scholars studying exploratory behavior in preschool and school age children have reported similar differentiations into manipulative or proximal exploration, visual or distal explorations and verbal or conceptual exploration (cf. Schneider et al., 1983; Voss & Meyer, 1987; Henderson & Moore, 1979; Henderson, 1984a; Keller et al., 1987). In a study with 3- to 40-year-old adults, Kreitler and Kreitler (1986) found the same factors, but with different prominence. Perceptual exploration (mostly visual) now accounted for 30% of the variance, conceptual exploration (mostly verbal forms) for 24% and manipulation for 20% of the explained variance. Although these results implicitly refer to developmental changes, the age comparisons are based on cross-sectional samples mainly.

From a developmental perspective, changes in the preferred modes of exploration have to be considered as being part of psychophysiological and neurological maturational processes which describe normal development. In a longitudinal study on the development of exploratory behavior in different modes over the first 9 years of life, we have demonstrated preferences for visual and manipulative behaviors due to different devlopmental courses for the two behavioral modes. We could also confirm that the visual and manipulatory exploration constitute two highly independent modalities (cf. Keller, 1991; Keller & Boigs, 1989, 1991).

Figure 1: Developmental course of visual exploratory behavior.

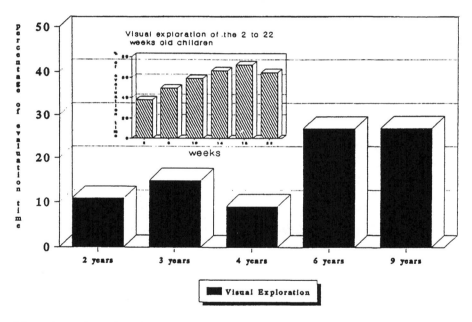

Visual exploration is the preferred mode of exploration during the first year of life. Accordingly, the visual exploration of the one year old relates substantially to the total exploration time at that time (r = .88[1], p = >.01). It decreases over the preschool period to recover again during school age. This change in meaning has been interpreted as a shift from pure visual information intake to more conceptual

---

[1] Correlation coefficients reported here are product moment correlations.

forms of information processing (cf. Keller & Boigs, 1989, 1991). Manipulatory behavior also follows an inverted U-shaped function, being especially prominent in total exploration time at that age, but also shows longer mean durations than any other behavioral mode (cf. Keller & Boigs, 1989, 1991).

Figure 2: Developmental course of manipulative exploratory behavior.

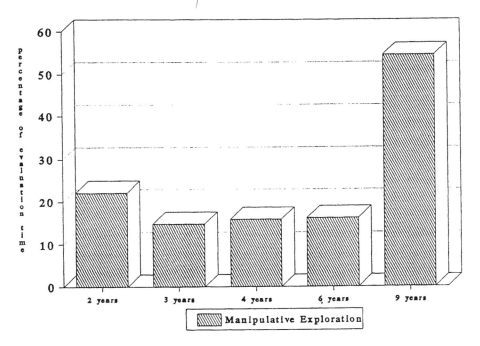

Total exploration time of the two year olds correlates significantly with the amount of time spent manipulating. The two year age span, thus, can be regarded as the focal time for the developmental task of manipulation covering high interindividual variability (cf. Keller et al., 1987; Keller, 1991; Keller & Boigs, 1989; 1991). With increasing age, manipulatory behavior becomes recontextualized. The mental capabilities of the children now allow to redefine the relationship between the disposition of fear and consequent avoidance of novel and unfamiliar situations as expressed in the presentation of a curiosity box and the readiness to approach those situations proximally (cf. Bischof, 1975, 1985, 1990; Gubler & Bischof, 1991).

Expressing different preferences for visual and manipulative exploration of children over the preschool and school period, thus, has to be expected from a normative developmental perspective in terms of adaptability of behavioral development.

With the different modes of exploration, different types of informations are necessarily associated (cf. Kreitler & Kreitler, see this volume; Keller & Boigs, 1989, 1991). Through visual inspection, informations concerning shape, colour, seize, texture and the like become available. By means of manipulation, functional properties and hidden qualities, as e.g. weight, might become apparent. It is obvious, that the visual channel is normally involved in any approach towards a novel object and situation, i.e. any manipulation is guided by vision after eye-hand-coordination has been developed. Manipulation, thus, includes visual exploration to a certain extent. Individuals, however, who restrict themselves to the visual mode for exploratory purposes, exclude significant informational input, thus, indicating impaired quantity and quality of exploration.

Since the modes of exploratory behavior refer to the h o w of the person-environment interaction, i.e. the stylistic components of behavior, they are conceived as exploratory styles.[2]

In explaining poor exploratory performance in children, insecure attachment relationships have been mainly quoted. Attachment theorists have pointed to the shortterm as well as the longterm consequences of the early quality of attachment (Bowlby, 1969; Ainsworth et al., 1978; Main, 1983; Arend, Gove & Sroufe, 1979; Hazen, 1982). In our own longitudinal study we have similarly presented evidence on the quality of early relationships between infants and their primary caregivers and the subsequent quality of manipulatory exploration (cf. Keller, 1991; Keller & Boigs, 1989, 1991). In a developmental analysis we have identified eye contact between mother (as primary caregiving person) and infants at the age of about 3 months as expressing an early but enduring form of relationship quality (cf. Keller, 1989; Keller et al., 1985; Keller & Gauda, 1987; Keller & Zach, 1991). This interactional parameter significantly correlates with manipulatory exploration at the childrens' age of two years (Keller, 1991; Keller & Boigs, 1989, 1991). Visual exploration as a distant form of interaction has not been expected to relate to the interactional quality, which could also be proven empirically.

---

[2] The definition of behavioral style is according to the stylistic approach in temperament research, as e.g. formulated by Thomas and Chess (1977).

Since the bulk of data, which have been reported, originate from first graders mainly and age comparisons are based on cross-sectional samples, the present chapter represents a longitudinal analysis of exploration styles and their developmental precursors in terms of earlier relationship experiences. We concentrate on the development of individual's preferences for manipulative and visual modes of exploration over the period of the first 9 years of life and relate these preferences to the quality of the early parent-child relationship.

An Empirical Study

The data presented here are part of a longitudinal study on the development of exploratory behavior and parent-child relationships. The total sample consists of 35 first born children and their parents (17 girls, 18 boys). The waves of data assessment are presented in Table 1.

Table 1: Observation waves and places of assessment

Number	Age	Home/Laboratory
0	3rd trimester of pregnancy	Home
1	2 weeks	Home
2	6 weeks	Home
3	10 weeks	Home
4	14 weeks	Home
5	18 weeks	Home
6	22 weeks	Home
7	8 1/2 months	Home
8	11 1/2 months	Home
9	2 years	Laboratory
10	3 years	Home
11	4 years	Home
12	5 years	Laboratory
13	6 years	Home
14	7 years	Home
15	9 years	Home

The selection on the volunteering families was based on German citizenship, normal pregnancies and deliveries, and medical low risk age span for reproduction (18 to 35 years for the mothers and 18 to 40 years for fathers at the time of birth) (for further sample information see Keller et al., 1982, 1985, 1987; Keller 1991). The data presented here refer to the early interactional quality and the exploratory quality of the preschool and school age period.

Interactional quality was assessed from videotaped free play interactions between infants and their mothers, as the primary caregiving persons. Infant looking into the mother's face and mother into the infant's face were coded by two trained observers. The observer were unfamiliar with other behaviors or data of the families and were not involved in the home visits. The observers had reached an interobserver agreement of at least 80 %, on a real time basis with a computer aided analysis technique (cf. Keller et al., 1987).

The overlap of these categories was defined as eye contact, which is reported in terms of percentage of interaction time as mean value during the 6, 10, and 14 weeks assessment (cf. Keller & Gauda, 1987; Keller et al., 1987, Keller, 1991).

On the basis of these data and behavioral evaluations of averion, two groups could be identified:

Gaze Averter: children whose visual engagement with the caregiving person at 6, 10, or 14 weeks is close to zero percent of the interaction time and who show signs of active averting by means of head and body movements and/or eye closing (cf. Keller & Gauda, 1987; Keller 1989);

Gazer: children with the expected amount of eye-contact behavior during the interaction who do not show any of the signs of aversion during interaction. As the lowest limit, 10 percent of looking behavior from the interactional time was determined. This datum is defined according to parental reports, indicating that parents only believe that their children look at them when this behavior comprises more than 10 percent of the interaction time.

The group of children, who gaze between 0 and 10 % of the interaction time at their parents, which was included in earlier analyses (cf. Keller & Gauda, 1987; Keller & Boigs, 1989, 1991), was excluded here, since they cannot clearly related to a theoretical assumption.

The *manipulatory* and *visual exploratory behavior* to be presented here, was assessed at childrens' ages of 2, 3, 4, 6 and 9 years. Assessment of exploratory behavior at age 1 and 5 is not comparable to these data. At the childrens' age of 7 years, exploratory behavior was not assessed. The presented analyses comprises data of 25 children, where at least three observations were completed.

The exploratory construct we are referring to with our procedure of assessment, is **specific exploration** according to the definition of Berlyne (1960). The children were confronted with a curiosity inducing novel box and generally instructed to do anything they would like to with the box. The different boxes are presented in Figure 2.

Figure 3: The curiosity inducing exploring boxes.

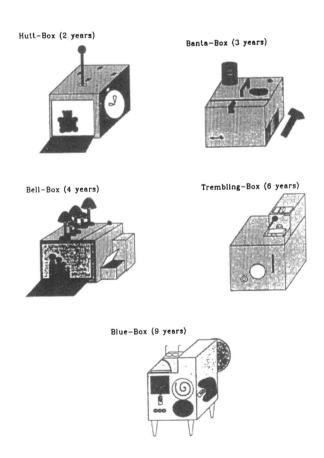

The observational situations took place in the childrens' homes. One or both parents were present, to especially account for the facilitating effect of familiar interactive partners present during exploration, thus, referring to the social support hypothesis as formulated by Henderson (1984 a,b). The parents were asked not to actively interfere into the child's activity. The behavioral sequences were videotaped by an experimenter who was also present in the same room. The first five minutes of exploration were analysed from the video-tapes by trained observers. Observer agreement was at least 80 %. The observers were unfamiliar with any other data or informations about the families and were not involved in the home visits.

Visual and manipulatory behavior was coded on a real time basis with a computer guided analysis technique (cf. Keller et al., 1987; Keller & Boigs, 1989, 1991; Keller et al., 1987; Schölmerich, 1990). The behavioral modes were defined as follows:

Visual exploration: child is regarding the objects or parts of the object without touching or manipulating it;

manipulative exploration: child touches and moves the object or parts of the object, elicits and tests functions while its attention is directed towards the object.

The amount of time children spent with visual and manipulatory exploration is computed as percentage of total exploration time. The median was then computed and each child was accordingly scored as low or high in visual and manipulative exploration. Through the combination of these scores, 4 styles can be identified (cf. Table 2):

Table 2: Styles of exploration

	behavioral characteristic	
style	visual	manipulative
1	-	-
2	+	-
3	-	+
4	+	+

Note: Style 1 indicates non-exploration, style 2 represents the distal mode, style 3 can be characterized as manipulative, and style 4 as a combination style of visual and manipulatory exploration.

Results

The distribution of the different styles of exploration at the different age levels is presented in Figure 4.

Figure 4: Exploration styles at different ages

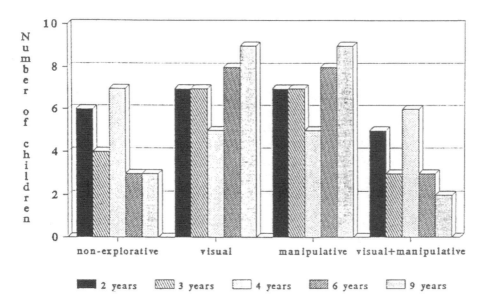

The data indicate similarities for the distribution of non-exploration and combined explorers over the years. The number of children in both styles are decreasing, except for the four-year-old ones. The four year assessment, however, seems to differ in many respects from the other assessments due to the more playful character of the curiosity box presented to that age group (cf. Keller, 1991). Visual and manipulatory exploration styles are equally distributed over the age span, generally increasing again, with exception of the four year olds, who use these two modalities less frequently than before and after that age span. Figure 4 indicates generally that the distribution of the exploratory styles differs from the general developmental course in these behavioral modalities (cf. Figure 1 and 2). It, thus, justifies the distinct consideration of the behavioral styles. Figure 4 gives information about the distribution of the exploratory styles per age level without referring to their individual stability. We, thus, do not know whether the same or different children compose the different style groups. All children who did not complete at least three assessments of exploratory behavior - and whose first year interactional data were not available - were excluded. From the remaining 20 children, 4 can be attributed as characterizing one style, which does not change over the entire time span from 2 to 9 years, two children representing the manipulative style and two children representing the visual style of exploration; 9 from the 16 changing children are unpredictable in their pattern of change; 3 children change from an earlier visual to a later manipulatory orientation and four children change from an early manipulatory to a later visual orientation.

In the next step of our analysis, we related the style of exploration to the early interactional quality. Table 3 presents the preferred mode of exploration of the children belonging to the two-eye-contact groups. Different total numbers of children in the different assessments are due to missing data, which constitutes one of the main problems of longitudinal research.

Table 3: Interactional quality and manipulatory exploration

years	averter manipulation +	averter manipulation -	high gazer manipulation [a] +	high gazer manipulation [a] -
2	2	4	5	2
3	2	5	3	3
4	2	4	3	2
6	1	5	3	2
9	3	2	2	5

[a] The manipulatory data are median splitted.

Table 3 indicates that at any age level, with exception of the 9-years-old children, formerly gaze averting children are more probably to be found among the low manipulators than among the high manipulators. Formerly high gazing children only differ with respect to their manipulatory behavior with two years of age. Earlier high gazers are more probably to the found among the high manipulatory children. The two year age period is, thus, further qualified for the analysis of the interrelationship between interactional quality and manipulatory exploration (cf. Keller & Boigs, 1989, 1991; Keller, 1991). Earlier high gazing children also show an unexpected behavioral differentiation of manipulatory exploration at the age of 9 years. This result equally underlines the interpretation of a developmental change in the meaning of manipulation in the context of general behavioral development (cf. Keller, 1991). As could be expected, these differentiations are not to be found in the visual mode (cf. Table 4).

Table 4: Interactional quality and visual exploration

years	averter "visual" +	-	high gazer +	"visual" [a] -
2	3	3	4	3
3	5	2	3	2
4	2	4	3	3
6	3	3	3	2
9	3	2	5	2

[a] The visual data are median splitted.

The general relationship between interactional quality and the extent of manipulatory exploration has also to be evaluated in terms of individual stability. The two children whose exploratory behavior has been classified as representing the manipulative style belong to the early gazer group, the two children representing the visual style belong one each to the gazer group and the gaze averter.

Discussion

The results underline the developmental function of exploratory behavior: increase of information by means of person-environment interactions in the behavioral modalities which are focal at specific developmental levels ranging from single behaviors (as e.g. looking) to complex and integrated behavioral responses (as e.g. epistemic activities). These behavioral changes are assumed to be part of the adaptive value of the exploration system. We, accordingly, find the majority of our children, 21 out of 25, changing between the behavioral modalities over time. The expression of behavioral stability in terms of preferring specific modalities over the preschool and school period can, thus, be proven as a-developmental. Since manipulatory exploration includes visual inspection, the lowest information increase can be expected by referring to the visual mode alone, thus excluding any information about the function and hidden qualities of the object. Only two children refer to visual information processing exclusively.

The exploratory quality which is obviously expressed in the modality preference can be related to the quality of early interactional exchanges between infants and their primary caregivers. Children with a more pronounced visual orientation in exploring novel and complex situations tend to have experienced less interactional quality as expressed in the early eye contact patterns as compared to children who are basically manipulatory oriented. The early interactional quality, thus, seems to relate to the mode of information seeking and processing over a significant period of childhood (cf. Keller, 1991; Keller & Boigs, 1989, 1991).

Poor interactional quality obviously reduces the readiness to physically interact in with unknown and unfamiliar situations, whereas exploration from distance seems to be unaffected. The prize for the withdrawal from direct interactional exchange with the physical environment is information loss. It has, however, to be stated, that the arguments presented here do not relate to the trait/state discussion of curiosity (cf. e.g. Boyle, 1989; Camp, Dietrich, & Olson, 1985). Manfred Koch has already stated 1959, that play, which at that time has not been differentiated from exploration in the German language literature, is the typical behavior of the human child. The mode of play, in this respect, typifies children and could serve important diagnostic functions. Here, as in many other respects, change can be identified as the essence of development.

References

Ainsworth, M.D.S., Blehar, M.C., Waters, E. & Wall, S. (1978). *Patterns of attachment. A psychological study of the strange situation.* Hillsdale, N.J.: Erlbaum.
Arend, R., Gove, F.L. & Sroufe, L.A. (1979). Continuity of individual adaptation from infancy to kindergarten: A predictive study of ego-resiliency and curiosity in preschoolers. *Child Development*, 50, 950-959.
Berlyne, D.E. (1960). *Conflict, arousal, and curiosity.* New York: McGraw-Hill.

Bischof, N. (1990). *Interconnections of exploration, attachment, and autonomy.* Paper read at the Conference "Curiosity and exploration" in Osnabrueck, March 19 - 22.

Bischof, N. (1985). *Das Rätsel Ödipus: Die biologischen Wurzeln des Urkonfliktes von Intimität und Autonomie.* München: Piper.

Bischof, N. (1975). A systems approach toward the functional connections of attachment and fear. *Child Development,* 46, 801-817.

Bowlby, J. (1969). *Attachment and loss, Vol. I: Attachment.* New York: Basic Books.

Boyle, G.J. (1989). Breath-depth or state-trait curiosity? A factor analysis of state-trait curiosity and state anxiety scales. *Person. Individ. Diff.,* 10 (2), 175-183.

Camp, C.J., Dietrich, M.S., & Olson, K.R. (1985). Curiosity and uncertainty in young, middle aged, and older adults. *Ecucational Gerontology,* 11, 401-412.

Gubler, H. & Bischof, N. (1991 in press). A systems' perspective on infant development. In M.E. Lamb & H. Keller (Eds.), *Infant development: perspectives from German-speaking countries.* Hillsdale, N.J.: Erlbaum.

Hazen, N.L. (1982). Spatial exploration and spatial knowledge: individual and developmental differences in very young children. *Child Development,* 53, 826-833.

Henderson, B.B. (1984a). Social support and exploration. *Child Development,* 55, 1246-1251.

Henderson, B.B. (1984b). Parents and exploration: The effect of context on individual differences in exploratory behavior. *Child Development,* 55, 1237-1245.

Henderson, B.B. & Moore, S.G. (1979). Measuring exploratory behavior in young children: A factor-analytic study. *Developmental Psychology,* 15, 113-119.

Keller, H. (1991 in press) The development cf exploratory behavior. In H. Keller & K. Schneider (Eds.), *Curiosity and exploration.* New York: Springer.

Keller, H. (1989). Kontinuität und Entwicklung. In H. Keller (Hrsg.), *Handbuch der Kleinkindforschung* (S. 163-180). Heidelberg: Springer.

Keller, H. & Boigs, R. (1991 in press). Distal and proximal exploration as behavioral styles. In H. Keller & K. Schneider (Eds.), *Curiosity and exploration.* New York: Springer.

Keller, H. & Boigs, R. (1989). Entwicklung des Explorationsverhaltens. In H. Keller (Hrsg.), *Handbuch der Kleinkindforschung* (S. 443-464). Heidelberg: Springer.

Keller, H. & Gauda, G. (1987). Eye contact in the first months of life and its developmental consequences. In H. Rauh & H.-Ch. Steinhausen (Eds.), *Psychobiology and early development* (pp. 129-143). Amsterdam, NL: Elsevier.

Keller, H. & Zach, U. (1991 in press). Developmental consequences of early eye-contact behavior. *Acta Paedopsychiatrica.*

Keller, H., Gauda, G., & Schölmerich, A. (1985). *Die Entwicklung des Explorationsverhaltens.* Osnabrück: Bericht für die Deutsche Forschungsgemeinschaft.

Keller, H., Gauda, G., Miranda, D., & Schölmerich, A. (1982). *Entwicklung des Explorationsverhaltens.* Darmstadt: Bericht für die Deutsche Forschungsgemeinschaft.

Keller, H., Schölmerich, A., Miranda, D., & Gauda, G. (1987). Exploratory behavior development in the first four years. In D. Görlitz & J.F. Wohlwill (Esd.), *Curiosity, imagination and play* (pp. 127-150). Hillsdale, N.J.: Erlbaum.

Koch, M. (1959). Wesensunterschiede menschlicher und tierischer Entwicklung. In H. Thomae (Ed.) *Entwicklungspsychologie,* 2. ed. (pp. 585-602). Göttingen: Verlag für Psychologie/Hogrefe.

Kreitler, S. & Kreitler, H. (1987). The motivational and cognitive determinants of individual planning. *Genetic, Social and General Psychology Monographs,* 113, 81-107.

Kreitler, S. & Kreitler, H. (1986). Types of curiosity behaviors and their cognitive determinants. *Archives of Psychology*, 138, 233-251.

Kreitler, S., Zigler, E., & Kreitler, H. (1975). The nature of curiosity in children. *Journal of School Psychology*, 13, 185-200.

Kreitler, S., Zigler, E., & Kreitler, H. (1974). The complexity of complexity. *Human Development*, 17, 54-73.

Main, M. (1983). Exploration, play, and cognitive functioning related to infant-mother attachment. *Infant Behavior and Development*, 6, 167-174.

Schneider, K., Moch, M., Sandfort, R., Auerswald, M., & Walther-Weckmann, K. (1983). Exploring a novel object by preschool children: A sequential analysis of perceptual, manipulating and verbal exploration. *International Journal of BehavioralDevelopment*, 6, 477-496.

Schölmerich, A. (1990). *Der Erwerb neuer Information im Verlauf des Explorationsnprozesses: eine sequentielle Analyse von Handlungsketten*. Inaugural-Dissertation, Universität Osnabrück: Fachbereich Psychologie.

Thomas, A. & Chess, S. (1977). *Temperament and development*. New York: Brunner/ Mazel.

Voss, H.G. & Meyer, H.J. (1987). *Entwicklung explorativen Verhaltens in der frühen Kindheit (2. - 4. Lebensjahr) und Genese des Neugiermotivs*. Darmstadt: Bericht für die Stiftung Volkswagenwerk.

CHAPTER **III.11**

Individual Differences in Experience-Producing Tendencies

Bruce Henderson

More than a decade ago, presenters at a conference on imagination, play and curiosity (e.g., Keller, 1987; Wohlwill, 1987) suggested that more attention needed to be paid to individual differences in curiosity and exploration. The presence of a section on individual differences in the present volume indicates that those earlier suggestions did not fall on deaf ears. In this chapter, I argue that individual differences in curiosity and exploration represent an important subcategory of behaviors that I will call experience-producing tendencies. I then review some preliminary research on the relation between individual differences in the tendency to produce experiences through exploration and individual differences in measured intelligence and achievement. My interpretation of the role of exploration in intellectual development is tentative and speculative in nature, but it is consistent with some recent views of how children create their own environments and thus influence their own intellectual abilities (e.g., Anastasi, 1985; Plomin, 1986; Scarr & McCartney, 1983).

Individual Differences in the Tendency to Explore

There is no argument about the existence of individual differences in the tendency to explore. As the other chapters in this book attest, the curiosity and exploratory behaviors of infants, children and adolescents have been assessed in myriad ways (see Maw & Maw, 1977, for a review). Measurement approaches have included observations of visual, verbal, and tactile exploration of novel objects, assessment of preferences for two-dimensional stimuli varying in novelty, complexity or incongruity, teachers, parents, peers and experimenters have rated children's curiosity, and children have rated their own levels of curiosity and exploration. Exploratory behaviors have been studied in a wide variety of settings, including homes, schools, laboratories, museums, and grocery stores. Curiosity and exploratory behaviors also have been conceptualized in many different ways. Distinctions have been made between exploration and play, between specific and diversive exploration, between intrinsically- and extrinsically-based exploration, and between affective and inspective exploration. Individual differences in exploration themselves have been characterized as differences in exploratory mode, visual, tactile, manipulative or verbal, or as differences in style, for example, breadth of exploration versus depth of exploration.

Regardless of the means used to assess or conceptualize curiosity and exploration, then, the reality of a wide range of individual differences in exploratory response is apparent to all who have studied the phenomenon. The variability in individual response in studies of exploration is usually greater than the observed differences due to gender, age, types of object, or social circumstance. What can we say about the nature and meaning of the patterns of differences in children's behavior in exploratory situations?

First, I believe that the considerable variability in the responses of children of the same age to similar stimuli makes a strictly situational or state interpretation of curiosity and exploratory behaviors untenable. Different children come to novel situations with different predispositions to explore.

Second, even if it is assumed that there are differences in some kind of predisposition to deal with novelty, it is clear that the tendency to explore is multi-dimensional. The child who asks many questions about a novel object may or may not locomote over a wide range, prefer complex over simple two-dimensional stimuli, play extensively with familiar objects, or show an interest in novel people. On the other hand, within a relatively broad domain, such as exploration of novel three-dimensional objects, there may be more consistency and stability of the tendency to explore than some of us have heretofore supposed (Henderson, 1988).

Finally, although individual differences are always apparent, there is more variability in curiosity and exploration in some situations and groups than in others. There appears to be, for example, more variability in verbal exploration (question asking) than in manipulative exploration, more variability in the exploration of younger children than that of older children, more variability in highly novel environments than in more familiar environments, and more variability when children explore independently or with peers than when they explore with the support of parents or other adults (Henderson, 1984a, 1988; Henderson & Moore, 1979). The latter finding, that social support decreases variability in exploration, also suggests there may be more variability in expressed exploratory behavior than in the potential underlying curiosity that a child may have difficulty in expressing under some circumstances.

The Category of Experience-Producing Tendencies

Given that individual differences in curiosity and exploration are apparent and are multi-dimensional in nature, is there anything more general that we can say about what different types of exploration have in common? At a micro-analytic level, different types of exploratory behavior are clearly distinguishable. However, I believe that at a broader level of analysis, most, if not all, of these behaviors serve a common adaptive function: the acquisition of information. Further, I believe that this common function is shared by other important classes of childhood behavior, both specific and diversive, including play and imaginative activity.

The term I will use to categorize exploration and related behaviors is adapted from a much-neglected article (see Anastasi, 1985; Scarr, 1981 for exceptions) published more than 30 years ago. In a paper entitled "Genes, Drives and Intellect," Keith Hayes (1962) proposed that individual differences in intelligence and aptitude, rather than being directly influenced by genetic factors, are mediated by genetically-influenced motivational biases. These hereditary, and presumably evolutionarily-selected, motivational factors, including curiosity and exploration, Hayes called "experience-producing drives," a term he thought was more descriptive than one like White's (1959) concept of "competence."

Because of the historical and conceptual baggage carried by the term "drive," I prefer the weaker label "experience-producing tendencies" (EPTs). They describe a category of behaviors, including curiosity and exploration, that function to produce new experiences, experiences that provide opportunities for learning. The individual who has relatively stronger tendencies to seek out and explore novelty in the environment will create more experiences, learn more, and is so doing, will become more knowledgeable about features of the environment that would afford opportunities for producing experiences through additional exploration, and so on.

The notion of a motivational mechanism like EPTs is consistent with recent reconceptualizations of the old nature-nurture issue and of ontogenetic adaptation. For example, Scarr (Scarr, 1981; Scarr & McCartney, 1983) suggests that intellectual competence results from learning histories largely determined by genetically-based motivational differences that interact with children's environments, which, in part, are correlated themselves with parent genotypes, including parental predispositions for producing experiences. Thus, a child's exploratory tendencies initiate a developmental feedback loop in which exploration leads to learning, which leads to intellectual competence, which leads to stronger tendencies to produce experiences, and on and on. In this view, EPTs provide a locus for the interaction of nature and nurture.

In a similar vein, Berg and Sternberg (1985) have hypothesized that responses to environmental novelty are a major source of developmental continuity in intelligence from infancy through adulthood. They support their argument with evidence suggesting that there are substantial correlations between individual differences in visual attention to novel versus familiar stimuli during infancy and individual differences in measured intelligence up to seven years later (for more recent reviews, see Bornstein & Krasnegor, 1989). Although this evidence has its own conceptual and methodological problems, it is suggestive of an important role for these behaviors that result from what I am calling EPTs.

EPTs and Intelligence

If, as I have argued so far, individual differences in exploration in particular, and in EPTs in general, lead to differential learning histories, then such differences should be related to differences in intelligence. In suggesting this, I assume that what is measured by intelligence tests as we know them (e.g., vocabulary, arithmetic skills, general knowledge) is learned. The argument here, again, is that intelligence test performance is a result of a learning history guided by EPTs. One empirical implication is that individual differences in intelligence test performance should be correlated with individual differences in exploratory tendencies.

What evidence is there of a correlation between measures of curiosity and exploration and of intelligence? I have briefly summarized the results of the relevant studies I have been able to identify in Tables 1 and 2. The results summarized in Table 1 include those from studies that used group or individualized intelligence tests and correlated them with behavioral, self-report, or ratings measures of curiosity and exploration. Table 2 summarizes studies in which groups defined by intelligence level are compared on a measure or measures of curiosity and exploratory behavior. An examination of the two reveals that, overall, the existing evidence for a relationship between individual differences in intelligence and individual differences in curiosity and exploratory behavior is not very impressive.

Table 1. Correlational studies of curiosity and intelligence

Reference	Subjects	Measures	Results
Coie (1974)	1st & 2nd graders	Ravens IQ 6 measures of object exploration	rs = -.08 to .30
Dollinger & Sellers (1988)	2nd-7th grade, clinic-referred	WISC-R Harter's intrinsic motivation scales	no significant correlations
Inagaki (1978)	kindergarden	Tanaka-Binet IQ toy manipulations exploration task	r = .14 r = -.32
Kreitler et al. (1975)	1st grade	Otis group IQ pref. for complexity pref. for unknown object exploration	no significant correlations
Langevin (1971)	6th grade	Ravens IQ Otis group IQ self-report of curiosity teacher rating	3 low, significant correlations, rs, .21 to .32
Maw & Maw (1975)	5th grade	PMA-Total self-rating	r = .38, boys r = .54, girls
Penney & McCann (1964)	4th-6th grade	Cal. Test of Mental Maturity self-report	rs, .03 - .24

Table 2. Group studies of curiosity and intelligence

Reference	Groups	Measures	Results
Harter & Zigler (1974)	retarded, CA & MA matched normals	pref. for unknown	normals higher than noninstitution retarded, institution retarded lowest
Henderson et al. (1982	5th-10th grade gifted & average	2 self-reports	gifted higher than average, curiosity and intelligence factors independent
Pielstick & Woodruff (1964)	2nd, 6th grade gifted & average	object exploration self-presentations of slides	no differences on object exploration; more frequent self-presentations by gifted
Vandenberg (1985)	retarded & normal CA & MA	pref. for unknown attention to complexity	normals higher than CA matches normals = MA matches

Why are the available correlations lower than might be expected on a theoretical basis? One obvious possibility is that the theory is wrong. Perhaps an individual's level of curiosity is simply unrelated to the acquisition and accumulation of information, at least the types of information measured by intelligence tests. Or, it is possible that some minimal level of EPTs (or other types of EPTs) are what are required for normal learning levels to be attained. Once that threshold level of motivation is reached, additional degrees of curiosity do not enhance the learning history of a child.

However, there are other explanations for why the obtained intelligence-curiosity relationships have been weak. For example, most of the studies have been of children of school age or older. Among school-age children, there may be an increasing domain-specificity in interests. Thus the curiosity of school-age children may be too narrow to be adequately assessed by general measures of exploration and to be reflected in the differential acquisition of general knowledge. Also, some of the studies have employed group-administered intelligence tests or measures of curiosity and exploration of questionable reliability and validity.

A Recent Study of Curiosity/Exploration Intelligence Relations

In order to further examine the relationship between individual differences in EPTs and intelligence, Susan Wilson and I conducted a study with preschool children (Henderson & Wilson, 1991). We studied preschool children because they are still at an age when they display curiosity about the environment in general and have not yet developed strong domain-specific interests. The assessment of intelligence was done individually with the Kaufman Assessment Battery for Children (K-ABC). The K-ABC (A. Kaufman & N. Kaufman, 1983) is appropriate for use with preschool children, yields a variety of indicators of intelligence, and is expressly marketed as being novel for children. The curiosity assessment was done with a battery of tasks that includes multiple indicators of curiosity and has been shown to be psychometrically sound for use with preschool children (Henderson & Moore, 1979).

Based on Berg and Sternberg's (1985) argument that response to novelty has two aspects that are integral components of intelligence, we predicted a positive correlation between individual differences in curiosity and exploration and in intelligence. The first aspect is what Berg and Sternberg call an "energizing" element that reflects a child's interest in, curiosity about, and preference for novel stimulation. Their hypothesis is that intelligence is related to tolerance for and attention to discrepancies between existing cognitive schemata and novelty in the environment. A second aspect linking curiosity and intelligence is the greater ability of more intelligent children to automatically extract information, thus freeing cognitive resources for dealing with novel tasks. It may not be possible to empirically disentangle the energizing and automaticity components, but the combination of the two aspects suggests the prediction of a substantial positive correlation between individual differences in curiosity and intelligence.

We studied 41 preschool children (mean age = 54.8 months). The K-ABC yielded four scores: measures of sequential and simultaneous processing which combine to form a mental processing index and an achievement score based on general informational knowledge. The curiosity battery included a measure of preference for two-dimensional figures varying in complexity, a measure of preference for unknown (preference for a hidden toy versus one that was in view) and observation of exploration of three novel objects: a Banta box, a box of 18 drawers each containing a novel object, and a curiosity board covered with manipulable objects. The questions asked about novel features of the three objects, comments about the novel features, different manipulations executed, and time exploring were summed across responses to the three objects.

The results of the intelligence assessments indicated that the children scored in a fairly wide range of the K-ABC scales (81-127 points on the Mental processing Composite scale and 80-141 points on the Achievement scale). The curiosity scores were similar to those obtained in other studies of preschool children (e.g., Henderson, 1988; Henderson & Moore, 1979). Canonical correlation analyses between sets of intelligence and curiosity measures were conducted on both raw scores and logarithmic transformations of the object exploration scores (because some of the distributions of the object exploration scores were skewed).

No significant canonical correlations were obtained. The correlations between intelligence scores and raw curiosity scores are presented in Table 3. Except for some low but significant correlations of intelligence scale scores with comments about novel objects, the correlations are quite low and statistically insignificant. The intelligence-comments correlations may be due to a common verbal ability factor. Thus, the results of our study are essentially consistent with those of previous studies indicating little relationship between individual differences in intelligence and individual differences in curiosity and exploratory tendencies.

Table 3. Correlations between K-ABC global scales and curiosity variables

	K-ABC Global Scale			
Curiosity Variable	Sequential	Simultaneous	MP-Composite	Achievement
Preference for Complexity	.08	.09	.16	.04
Preference for Unknown	.13	.05	.13	.09
Object Exploration				
Questions	.23	-.01	.03	.09
Comments	.19	.30*	.28	.35*
Manipulations	.05	.10	.10	.18
Time	.21	.17	.15	.10

*$p < .05$

The Need for Developmental Designs

Null results are difficult to interpret. Our study has low statistical power. The K-ABC has been criticized as having an inadequate and unsupported theoretical base (e.g., Sternberg, 1984), although the Achievement scale of the K-ABC has been shown to be highly correlated with other measures of general intelligence (Coffman, 1985). There are other possible reasons for the null results. However, upon reflection, we have decided that the main problem with our study and the others in Tables 1 and 2 is that the tests of the curiosity-intelligence relationship have been based on an inadequate design.

Hayes' original argument was that EPTs both influence and are influenced by a child's level of intelligence. Intelligent children have stronger tendencies to produce experiences and the production of experiences results in children becoming more intelligent. Thus, two children of equal concurrent intelligence may have arrived at that level of intelligence by different routes. If two children differ in their level of

EPTs, these differences will be predictive of changes in subsequent intellectual development, but will not be correlated with concurrent measures of intelligence. Of course, this perspective is consistent with the findings summarized by Berg and Sternberg (1985) indicating that infants' responses to novelty predict later levels of intelligence.

What are needed for older children are studies that parallel those done with infants in which individual differences in curiosity and exploratory behavior at Time 1 are correlated with measures of later intelligence, or perhaps even better, measures of change in intelligence or achievement between two times. One study of children older than infants that followed this tack comes from the Fels longitudinal study. Kagan, Sontag, Baker and Nelson (1958) examined the curiosity of two extreme groups of 70 children each. The children had been tested at 6 and 10 years of age on the Stanford-Binet (1937 forms L or M) and had been in the upper (ascenders) or lower (descenders) quartiles of IQ changes over the period. The measure of curiosity used was the Thematic Apperception Test (TAT) themes reflecting interest in the processes or phenomena of nature (only one TAT card elicited such responses). Significantly more of the ascenders (37.5%) than the descenders (9.7%) reported curiosity themes. The authors interpreted these results as indicating that curious children (as well as children with high achievement strivings) are more likely to show increases in IQ over time than are less curious children. Obviously, this single study, with its methodological shortcomings, provides minimal support for the EPT-intelligence hypothesis, but it is the only currently available direct test.

This approach addresses the EPT-knowledge acquisition problem at a very general level. The issue is being examined at a much more micro-analytical level by Keller and her colleagues (see Schölmerich's contribution in this volume). The issue also can be examined in a more fine-grained manner in terms of the content of exploration and interests (see the contributions of Krapp and Fink in this volume). Children, especially after the preschool years, will exercise their EPTs in areas of their particular interests, thus becoming more likely to acquire information in those areas and more likely to exercise further EPTs. With age, EPTs may be directed in increasingly domain-specific ways (though there also may be individual differences in the tendency to produce experiences in a relatively broad range of environments and stimuli). Studies of EPT-intelligence connections will provide only a gross indication of the knowledge acquisition-intelligence connection.

Moderating Influences on EPTs

One additional qualification on the overly simplistic description of the EPT-intelligence (or knowledge acquisition) relationship needs to be made. It is already clear that a number of factors moderate the effect of individual differences in EPTs on what experiences are produced and how they are produced in the natural environment. Just a few examples will illustrate how complex the picture is. Some of our own work (e.g., Henderson, 1984a, 1984b; Martin & Henderson, 1989) and that of others (e.g., Endsley, Hutcherson, Garner & Martin, 1979) has shown that parents and other adults can and do mediate the expression of EPTs in important ways. In addition, intraindividual influences of other behavioral systems may alter the course of exploration. These include attachment relationships and wariness (Caruso, 1989), cognitive attributions concerning ability and the perceived effects of effort (e.g., Dweck & Leggett, 1988) and the developmental course of specific interests, mentioned above. Many other examples could be given. The point is that an individual's tendency to produce experiences is susceptible to many influences, genetic, environmental, and internal, that are reflected in the developmental operation of curiosity and exploration.

Implications and Conclusion

I think that consideration of individual differences in curiosity and exploration as exemplars of the more general category of experience-producing tendencies has considerable potential heuristic value. If some form of the concept turns out to be valid, there are several important theoretical and practical implications. For example, the EPT construct provides a useful way to think about nature-nurture interaction. It may provide a start to answering Anastasi's old question about the "how" of nature-nurture interaction (see Anastasi's, 1985, own comment on this issue). Methodologically, it suggests that in real-life learning, EPT-type motivation may be more important than is apparent in laboratory situations in which external motivation is provided. Even more practically, the EPT view of individual differences in curiosity and exploratory behavior has implications for psychological assessment and intervention. In the area of assessment, if learning potential rather than current status is at issue, it may be as important to measure curiosity and exploratory tendencies (and other EPTs) as to measure intelligence or achievement. Likewise, if interventions are to have lasting effects, modifying EPTs may be more important than instruction in specific skills or knowledge.

All this may be premature. We need to find out more about the nature of EPTs, their development, and what influences them. We need to be more curious and exploratory about our tendencies to produce experiences.

References

Anastasi, A. (1985). Reciprocal relations between cognitive and affective development: With implications for sex differences. In T. B. Sonderegger (Ed.), *Psychology and gender* (Nebraska Symposium on Motivation, Vol. 32, pp. 1- 35). Lincoln, NE: University of Nebraska Press.

Berg, C., & Sternberg, R. J. (1985). Response to novelty: Continuity versus discontinuity in the developmental course of intelligence. In H. W. Reese (Ed.), *Advances in child development and behavior* (Vol. 19, pp. 1-47). New York: Academic Press.

Bornstein, M. H., & Krasnegor, N. A. (Eds.). (1989). *Stability and continuity in mental development: Behavioral and biological perspectives.* Hillsdale, NJ: Erlbaum.

Caruso, D. A. (1989). Attachment and exploration in infancy: Research and applied issues. *Early Childhood Research Quarterly, 4,* 117-132.

Coffman, W. (1985). Review of the Kaufman Assessment Battery for Children. In J. Mitchell, Jr. (Ed.), *The ninth mental measurements yearbook* (Vol. 1, pp. 771- 773). Lincoln, NE: The University of Nebraska Press.

Coie, J. D. (1974). An evaluation of the cross-sectional stability of children's curiosity. *Journal of Personality, 42,* 93-116.

Dollinger, S. J., & Seiters, J. A. (1988). Intrinsic motivation among clinic-referred children. *Bulletin of the Psychonomic Society, 26,* 449-451.

Dweck, C. S., & Leggett, E. L. (1988). A social-cognitive approach to motivation and personality. *Psychological Review, 95,* 256-273.

Endsley, R. C., Hutcherson, M. A., Garner, A. P., & Martin, M. J. (1979). Interrelationships among selected maternal behaviors, authoritarianism, and preschool children's verbal and nonverbal curiosity. *Child Development, 50,* 331-339.

Harter, S., & Zigler, E. (1974). The assessment of effectance motivation in normal and retarded children. *Developmental Psychology, 10,* 169-180.

Hayes, K. J. (1962). Genes, drives, and intellect. *Psychological Reports, 10,* 299-342.

Henderson, B. B. (1984a). Parents and exploration: The effect of context on individual differences in exploratory behavior. *Child Development, 55,* 1237-1245.

Henderson, B. B. (1984b). Social support and exploration. *Child Development, 55,* 1246-1251.

Henderson, B. B. (1988). Individual differences in exploration: A replication and extension. *Journal of Genetic Psychology, 149,* 555-557.

Henderson, B. B., Gold, S. R., & McCord, M. T. (1982). Daydreaming and curiosity in gifted and average children and adolescents. *Developmental Psychology, 18,* 576-582.

Henderson, B. B., & Moore, S. G. (1979). Measuring exploratory behavior in young children: A factor analytic study. *Developmental Psychology, 15,* 113-119.

Henderson, B. B., & Wilson, S. (1991). Intelligence and curiosity in preschool children. *Journal of School Psychology, 29,* 167-175.

Inagaki, K. (1978). Effects of object curiosity on exploratory learning in young children. *Psychological Reports, 42,* 899-908.

Kagan, J., Sontag, L. W., Baker, C. T., & Nelson, V. L. (1958). Personality and IQ change. *Journal of Abnormal and Social Psychology, 56,* 261-266.

Kaufman, A., & Kaufman, N. (1983). *Interpretive manual for the Kaufman Assessment Battery for Children.* Circle Pines, MN: American Guidance Service.

Keller, J. A. (1987). Motivational aspects of exploratory behavior. In D. Gorlitz, & J. F. Wohlwill (Eds.), *Curiosity, imagination, and play* (pp. 24-42). Hillsdale, N J : Erlbaum.

Kreitler, S., Zigler, E., & Kreitler, H. (1975). The nature of curiosity in children. *Journal of School Psychology, 13,* 185-200.

Langevin, R. (1971). Is curiosity a unitary construct? *Canadian Journal of Psychology, 25,* 360-374.

Martin, C. E., & Henderson, B. B. (1989). Adult support and the exploratory behavior of children with learning disabilities. *Journal of Learning Disabilities, 22,* 67-68.

Maw, W. H., & Maw, E. W. (1975). Note on curiosity and intelligence of school children. *Psychological Reports, 36,* 782.

Maw, W. H., & Maw, E. W. (1977). Nature and assessment of curiosity. In P. McReynolds (Ed.), *Advances in psychological assessment* (Vol. 4, pp. 526-571). San Francisco: Jossey-Bass.

Penney, R. K., & McCann, B. (1964). The Children's Reactive Curiosity Scale. *Psychological Reports, 15,* 323-334.

Pielstick, N. L., & Woodruff, A. B. (1964). Exploratory behavior and curiosity in two age and ability groups of children. *Psychological Reports, 14,* 831-838.

Plomin, R. (1986). *Development, genetics, and psychology.* Hillsdale, NJ: Erlbaum.

Scarr, S. (1981). Testing for children: Assessment and the many determinants of intellectual competence. *American Psychologist, 36,* 1159-1166.

Scarr, S., & McCartney, K. (1983). How people make their own environments: A theory of genotype-environment effects. *Child Development, 54,* 424-435.

Sternberg, R. J. (1984). The Kaufman Assessment Battery for Children: An information processing analysis and critique. *Journal of Special Education, 18,* 267-279.

Vandenberg, B. R. (1985). The effects of retardation on exploration. *Merrill-Palmer Quarterly, 31,* 397-409.

White, R. W. (1959). Motivation reconsidered: The concept of competence. *Psychological Review, 66,* 297-333.

Wohlwill, J. F. (1987). Varieties of exploratory activity in early childhood. In D. Gorlitz, & J. F. Wohlwill (Eds.), *Curiosity, imagination, and play* (pp. 59-77). Hillsdale, NJ: Erlbaum.

CHAPTER IV

The Interconnections of Exploratory Behavior and Other Behavioral Systems

CHAPTER **IV**.12

The Active Exploratory Nature of Perceiving: Some Developmental Implications

Ad W. Smitsman

Exploratory behavior refers to activities by which actors keep in touch with the environment in the search for information. Students of exploratory behavior (see Voss & Keller, 1983, for an elaborated overview and discussions) generally view personal states of uncertainty and conflict as being the ultimate cause of exploratory behavior.
However, recent theorizing and research within the field of perception and action from the perspective of J. J. Gibson's ecological approach to perceiving andacting (J. J. Gibson, 1966, 1979) and of dynamic actionsystems (Beek & Bingham, 1991; Kugler, Shaw, Vincente & Kinsella-Shaw, 1990; Reed, 1982; Turvey, 1977; Turvey, Carello & Kim, 1989) indicate that exploratory activities are not necessarily confined to particular states of an actor, but form an essential component of any action, he/she performs. Goal-directed behavior requires active, self-organizing perceptual systems that continuously search for information during the course of action. The exploratory activities need to be coordinated with the person's goals and tasks. The search for information forms an integral part of any perception-action cycle. Exploratory activity originates from the need to coordinate action properties to environmental properties rather than from the need to reduce uncertainty. Reduction of uncertainty may be an effect of exploratory activity but not necessarily the motivating cause.
Eleanor J. Gibson (1988) has recently applied the dynamic action systems perspective and J. J. Gibson's ecological approach to perceiving and acting to the study of exploratory behavior and the way it develops during the first years of life. The present paper discusses exploratory behavior from J. J. Gibson's (1979) perspective on information that has to be obtained in order to guide action. It will be

228   CHAPTER IV

argued that mutual constraints of perception and action (i.e., information obtained and activities employed) require task specific exploratory search activities. Some research findings will be reviewed that document the specificity of exploratory search in infants. The idea of amutual constraining relationship of perception and action will be used to highlight problems that face young children to coordinate and modify perceptual search to newly evolving task demands, when their action repertoire expands with age.

The Active Exploratory Nature of Perceiving

An old experiment of James J. Gibson on active touch (Reed, 1988) nicely illustrates the exploratory nature of perceiving. To investigate active touch, Gibson used a set of household cookie cutters. It is easy to get an impression of the outline of cookie cutters by pushing the sharp edges of these objects into the palm of a hand. Gibson asked observers to identify the cookie cutter they were feeling. In one condition observers were allowed to obtain information by actively touching the cutter for themselves. In other conditions stimulation or information was imposed.

The experimenter provided a sensation either by pressing a cookie cutter into a subject's palm or by rotating the cutter in the palm. When information was obtained by active touch, that is, when observers could explore the cutters, identification was nearly perfect. However, when stimulation was imposed, that is, when the object was only pressed into the palm, correct identification dropped dramatically. Identification was still far from perfect when the experimenter varied stimulation by rotating the object.

Gibson's experiment on active touch indicates that a passively registering sensory system is of little or no use to a person who needs to be aware of what is going on. The value of the system improves when the stimulation varies across time. However, to function optimal, the system needs to be active. Apparently stimulation has to covary very carefully with the exploratory activities that are employed by the system itself to make perceiving successful.

The active, exploratory nature of perceiving may be further clarified when it is considered in relation to its complementary entity, namely, the information that is searched for. The concept of information has been conceived frequently as a product of an organism's cognitive system, constructed on the basis of static impressions imposed on the organism's sensory registers. The structure that is supposed to be constructed ordinarily denotes a set of meaningful relationships between the organism and its environment. J. J. Gibson (1979),has made it clear that this constructive view ignores the fact that static impressions are rare occurrences for an active human being.

Even the slightest eye and head movements turn static impressions into a flow of stimulation, optic flow for the eye, acoustic flow for the ear, and a flow of pressure waves for the haptic system. J. J. Gibson (1979) has worked out several examples that indicate that the structure of such flowfields, or what Gibson has called an ambient array, covaries not only precisely with a person's actions, including his/her exploratory behavior, but also with environmental variation in, for example, surface layout and texture (the things that persist in the environment) and changes that occur in layout and texture (the events that take place). This covariation entails information to an actor for coordinating his/her actions to environmental properties that constrain those actions (such as animate and inanimate objects, surfaces that maybe looked at, touched and moved over, and events that take place with respect to those objects and surfaces). A simple example such as forward locomotion may be used to illustrate how the structure of ambient energy flows varies according to both actions and environmental properties. When a person starts to move forward, expansion patterns emerge at the center of the optic flowfield. The expansion pattern consists of outflow of texture elements from the center of this field and indicates for ward motion towards the place that corresponds to the center of the flow field. This might be a place at the horizon, when the terrain is open or an upright barrier nearby that obstructs free passage. When forward motion is changed into backward motion, outflow turns into inflow and backward motion is specified. Flow patterns also vary according to exploratory actions such as eye and head movements. When the person makes eye and head movements during locomotion, flow patterns caused by eye and head movements become superimposed on the flow patterns that emerged by locomotion and indicate places to which visual attention is directed. Flowfield properties also vary according to variation inenvironmental layout and changes that take place with respect to this layout. The flowfield for a cluttered environment differs from that for an open terrain.

Adequate tuning of perceptual systems to covariation that emerges in ambient energy flows means awareness of one's environment and the activities one employs. Covariation of flowfield properties with action properties and environmental properties means also that a person may control exploratory search and actions that are guided by exploratory search. To move forward, the person should employ activity that generates a flowfield that belongs to forward movement. To move forward to a particular place, the center of outflow should be located at that place. To focus attention on special parts of the environment the person should constrain flow caused by exploratory activity of eye and head to those parts. J. J. Gibson (1966; see also Reed, 1988) has raised the question of why making head and eye movements is not disruptive to seeing. The answer seems to be straightforward. It is not disruptive because there is information available to control head and eye movements such that they will be informative instead of disruptive. Because of the covariation of characteristics of the optical flowfield, head and eye movements, and environmental layout, a person may be aware of the exploratory activities of the visual system and control visual search and tuning to that layout.

The significance of simple activities such as head movements for perceiving has been demonstrated nicely by the recent research of Smets and her colleagues (Smets, Overbeeke, & Stratmann, 1987). Optic flow caused by head and eye movements provides information about depth when a spatial layout is observed that consists of surfaces that extend not only in the frontal plane but also in depth, such as solid objects located behind each other, or even a single solid object. By moving eyes and head, it is possible to look around an object. The slightest motion of head and eyes induces a shift of an object's edges that separates its visible from its invisible parts. Such a shift of occluding edges does not occur for eye and head movements when patches on a flat surface such as a picture or a motion picture on a video screen are observed. Smets and her colleagues designed a system by which head movements also induced a shift of occluding edges for displays that were observed on a video screen and recorded by a camera. By connecting the camera to an observer's head, the recording position of the camera was varied according to head movements made to explore visually the display that was shown on the screen. Judgements about depth in the screen image were nearly perfect when these were made for displays that were recorded by a camera that was connected to the observer's head, whereas such judgements were far from accurate when camera position did not vary according to an observer's head movements.

Perceptual Search and Information for Action

Not only do exploratory activities of the head and eye affect the optical flowfield of an observing person, butmotions of other body parts may do the same. Locomotion, grasping and manipulation of objects, to name a few activities, each involve particular disturbances of an optical flowfield that entails specific visual information about a person's relationship to the environment and how this relationship is maintained or changed by the particular action. The earlier example of forward locomotion can be elaborated to illustrate this tight relationship between environment information and action. When a person looks around and moves forward along some path, an expanding flowfield will emerge that differs for small obstacles that can be easily stepped over and larger obstacles that need to be circumvented. Information may be obtained about the moment of contact to these objects when a person continues the path (Lee & Reddish, 1981), and about how a pedestrian may avoid collision with obstacles. The optical flowfield specifies the environment to the acting person in a meaningful way, that is, in a specific relationship to his/her activities. J. J. Gibson has described environmental properties that are specified in this way with the term 'affordance'. An obstacle is an example of an affordance. Obstacles do not exist independently of an actor and the actions that are performed. Obstacles that may be stepped over by large pedestrians need to be circumvented by smaller pedestrians. However, obstacles need not to be invented by pedestrians. These properties are specified by flowfield characteristics that emerge when a pedestrian navigates through a cluttered environment. The close relationship between information and action has consequences for how we should consider exploratory activity. The coordination and control of action require perceptual systems that organize their search according to affordances of the environment.

This does not mean that other information is ignored. For example, when a pedestrian can walk easily along a sidewalk, he/she may engage in information search about, for example, the particular architecture of buildings along the street, or people who live there, in addition to how and where to go. However, when the terrain becomes uneven and full of obstacles, exploratory activity would be more narrowly confined to the pickup of information relevant to remaining upright. E. J. Gibson et al. (1987) showed that such is the case for exploratory activities even with infants. A flexible surface imposes specific demands on the locomotory system when locomotion consists of walking in comparison to crawling. E. J. Gibson and her colleagues required infants who could not yet walk and infants who were beginning to walk to cross a floor that was partly rigid and partly made up of some kind of waterbed. A sheet of plastic covered the floor so that a change in rigidity could not be perceived unless pressure was exerted on its surface. As soon as the infants reached the waterbed, the change in rigidity of the surface was noted and elicited increased exploratory activity in walkers but not in crawlers. Exploratory activity consisted, forexample, of exerting force on the waterbed by hand or feet, such that its deforming capacity could be observed.

Action constrains exploratory search. Action also provides possibilities for the pickup of information. The fact that the environment structures an optic flowfield in ways related specifically to actions, indicates that the information that may be obtained, for example, by visual search, depends to a large extent on the actions within which the visual search is embedded. In many cases relevant visual information can only be obtained when the actor does not confine his/her exploratory search to just head and eye movements. For example, the flexibility of a surface can only be observed when pressure is exerted on that surface. In some cases manipulation by hand may be sufficient. However, when the information needs to be scaled to a system of a larger mass, that system has to be used. For example, the resistance to pressure of an icy floor for walking can be observed adequately only when at least an amount of pressure is exerted on that floor that equals that of the mass of the whole body.

Developmental Implications

The framework discussed here has several important implications for the way in which children's exploratory activities may be conceived and studied (E. J. Gibson, 1988). From birth, children face the tremendous task of developing and controlling their body such that it enables them to orient successfully in their physical and social environment, and to deal successfully with the changing demands of this environment. Such a task can be performed only if exploratory activities are directed to what the environment affords to the developing action repertoire.

To obtain information about affordances, exploratory activity should be specific to the tasks at hand and to the information that is searched for, directed to affordances that arise when action skills develop, and make use of possibilities for further exploration provided by developing action skills.

The Task Specificity of Early Exploratory Activities

Although infant's exploratory activities are often supposed to be nonspecific, new studies of exploratory behavior in infancy indicate that these activities may be specific from birth (E. J. Gibson, 1988). The rigidity of substances is one of the properties for which exploratory activities of infants have been investigated. In the first months of life, infants are able to discriminate rigid from deforming surfaces, both visually and haptically by mouth. Intermodal recovery of looking time indicates that 2-month-old infants expect to see a non-rigid substance after they had just felt such a substance in their mouth (E. J. Gibson & Walker, 1984). Moreover, Rochat (1987) showed that exploratory activity of the tongue varied according to the rigidity of the substance that is put into the mouth of newborn babies.

Specificity of exploratory activity has also been found in infants' manipulatory activities as soon as these begin to develop (Palmer, 1989; Rochat, 1989). At about 4 to 6 months, infants' exploratory possibilities are extended by development of the manipulatory capacity of the arm-hand system (E. J. Gibson, 1988). Palmer (1989) showed that the manner of manipulation and fingering of objects of 6- to 12-month-old infants varied according to characteristics of the objects that were given to them. For example, stuffed objects were more often fingered, while spongy objects were more often sqeezed. Specificity of manipulation was also found in an unpublished study that was run with some of my colleagues. In this study, a complex object was given to 6- to 9-month-old infants. The object consisted of a small cube, a ball, and a cone, glued together via a tripod-shaped-wire. The compound object could be grasped by either of the three component object parts, each of which fit easily into the hand of a 6-month-old infant. Because the apex of the cone extended outwards, it provided the least stability for grasping. Even 6-month-old infants preferred the ball and cube side of a tripod for grasping.

However, when the size of the cube or ball was diminished, the larger cube or ball was grasped more often. The cone was the most preferred component for mouthing, because its shape fit easily into the enclosure of the mouth.

Specificity of exploratory activity in infancy exists not only for manipulatory activities and haptic search, but also for visual search. Results of studies on numerosity perception in infancy indicate that 5- to 6-month-old infants are able to perceive the number of objects that are available in small sets of up to four objects (e.g., van Loosbroek & Smitsman, 1990; Starkey, Gelman, & Spelke, 1985; Strauss & Curtis, 1981; Treiber & Wilcox, 1984). Van Loosbroek and Smitsman (1990) raised a question about the information that enables infants to perceive number. In their study, infants were habituated to video-displays of objects that moved continuously along independent trajectories. Independent motion of objects entail information about object unity to infants (Kellman & Spelke, 1983), while pattern characteristics vary continuously. The number of displayed objects varied from one to five objects.
Van Loosbroek and Smitsman found that infants are still able to perceive number when objects are in motion and follow independent trajectories that cross each other repeatedly. Moreover, number was also perceived when the moving objects were heterogeneous in shape and information to categorize objects was reduced to a minimum (van Loosbroek & Smitsman, 1993). This latter result provide additional support for van Loosbroek and Smitsman's (1990) hypothesis that infants perceive number by constrained sampling of individual elements. Information about number emerges when each object of a collection is sampled in a specific way, that is, only once, until the whole collection has been sampled. The development of counting that arises at the age of about 2 to 3 years (Gelman 1982) might be conceived of as an elaboration of this procedure, especially for collections larger than three or four objects. A specific property of this latter procedure is that the use of symbols might provide control on sampling of individual elements for collections of a size for which the initial procedure will fail. Perhaps, sampling of up to four objects can be visually controlled, whereas sampling of larger sizes may need additional support of tools such as symbolic media.

Exploration and the Development of Action Skills

The development of new action skills both facilitates and constrains the exploratory activities of children. New opportunities for perceiving arise when a child's action repertoire is enlarged. However, the development and control of new actions also imposes demands on the organization of perceptual systems and the information that has to be attended to.
An important action skill that infants develop during their first years of life is the ability to manipulate objects. Development of this skill, which may take several years (Bigelow, 1981), provides the child with a powerful system for haptic search. However, increased motor control of hand and finger movements also makes it possible for infants to perform manipulatory activities discriminately to different objects (Palmer, 1989). Manipulatory skill is of importance not only to haptic search. It affects the organization and activities of other perceptual systems as well. For example, visual activity may be facilitated by grasping a distant object and moving it closer to the eyes. Manipulation not only makes it possible to bring something into the visual field at particular places, but also to generate particular disturbances in parts of the visual field that environmental properties become more fully specified to the visual system. Such a disturbance is provided when an object is rotated in front of the eyes. It facilitates the pickup of information about shape. Rotation has been observed as one of the manipulatory categories infants use to explore objects. Object manipulation may also be used to affect the acoustic flowfield. Actions such as banging, throwing, and waving generate sound patterns that are specific to an object's substance, shape, and mass, and to the surface which the object strikes (Warren Jr., Kim, & Husney, 1987). Palmer (1989) showed that from 6 months on, infants begin to employ these actions with increasing discrimination as a function of the nature of an object as well as of the surface to which an object is made to interact. In general, manipulatory skill makes it possible to initiate and control events that reveal information about environmental properties. Such events may be simple at early age, such as rotating an object or banging it on a table, but exploratory search may become organized in increasingly complex ways as the child ages. When complexity increases, exploratory search may require the coordinated activity of several perceptual systems, pertaining to controlled interactions of several objects, that are planned in nested sequences of actions.

Typical examples of such highly complex search processes may be found in scientific research in which objects and substances are made to interact to obtain information about specific properties.

The coordination of different systems and actions that are nested in larger perception action cycles imposes special demands on exploratory search and planning. Planning and control of nested sequences of actions requires information about constraints or affordances on a larger time scale. Information needs to be obtained for the control of each action separately, but also for the proper order in which actions have to be performed. Moreover, actions that are nested in larger cycles mutually constrain each other. That is, actions that occur later in the sequence impose special demands on earlier actions and earlier actions constrain the degrees of freedom of later actions. Such nested cycles occur, for example, as soon as a child begins to use an object as a tool. The use of a spoon for eating imposes special demands on how it is grasped and manipulated. These demands do not exist when the spoon is grasped just for the sake of grasping. For the control of grasping alone, it is sufficient to have information that specifies whether the object is located within a reachable distance and whether its shape and size provide possibilities for grasping.

However, when the spoon is grasped for eating, a new set of constraints arise that concern loading to spoon and transport of food to the mouth, activities that occur after the spoon is grasped. This means that the spoon has to be grasped by its handle and not by its scoop, and that it should be held in particular ways to provide control of scooping and transport of food. Research by Connolly and Dalgleish (1989) indicates that it may take at least the whole second year of life until young children discover the mutual constraints of actions that eating with a spoon entails. Our research on tool use shows that information about mutual constraints of actions is not easily obtained by young children (van Leeuwen & Smitsman, 1991; van Leeuwen, Smitsman & van Leeuwen in press; Smitsman, van Leeuwen, & Peters, 1987). Children of 1 to 3 years of age were required to discover the function of a hook for moving an object to a target position. Both the hook and the object were placed on a table, with only the handle of the hook within reach of the child. To control object motion by means of a hook, the hook has to touch the object by the inside of its crook. In some conditions, this was arranged for the child. In other conditions, the spatial position of the hook to the object had to be changed by one or more transformations before the proper position was acquired.

The results of several studies using this type of task showed that discovery of the hook function became much more difficult for young children when it involved a longer sequence of nested actions to realize this function.

Actions that are nested into larger perception action cycles differ in the goals or subgoals at which they are aimed. These differences may concern performatory goals, such as the grasping, loading, and transporting of a spoon, but also exploratory purposes to obtain information for actions that have to be performed later. Many daily tasks involve cycles in which performatory phases are interspersed with exploratory phases. This does not mean that perceiving stops after an exploratory phase has been completed successfully, or that exploratory behavior does not include performatory aims that have to be satisfied. Perceiving and performing concur and mutually constrain each other for any action that takes place, irrespective of whether the action serves performatory or exploratory purposes. However, when separate actions are performed for exploratory purposes, the search for information becomes constrained by ongoing behavior as well as by performatory goals that have to be achieved later on. The search for and pickup of information, of course, continues when actions for which information has been obtained in an earlier phase of a cycle are finally executed. Continued search is needed for control of these actions, and because the information that has been obtained by earlier search may be insufficient to guide execution of these actions. It may be appropriate to plan the actions properly but, nevertheless, may be insufficient for action control. For example, when a carpenter starts to work on a piece of wood, he/she may perfectly diagnose the sharpness of his/her chisel by touching its cutting edge carefully with the fingertips. When the condition of sharpness is satisfied he/she will start chiseling. However, the information that is obtained by the manipulatory activities of touching the cutting edge does not specify precisely how the chisel has to be used for cutting wood. Chiseling and touching of a cutting edge by the fingertips are activities that involve different constraints. For touching, the chisel has to be grasped such that its sharp edge may be turned into the center of the visual field and may be touched by the fingertips to check its sharpness. To chisel, the handle has to be grasped such that the exchange of forces may be controlled when wood is cut by the chisel and information may be obtained about the pattern of forces that emerges at the interface of the wood and the chisel's edge, rather than at the interface of the fingertips and the sharp edge.

The coordination of exploratory search and actions in extended perception action cycles may be a difficult task to achieve for young children. The appropriate planning of action cycles seems to be a skill that gradually develops as children grow up (Gibson & Spelke, 1983; Pick, Frankel, & Hess, 1975). Parents and teachers have to instruct young children to look, touch, and listen first, before performing actions. Young children are often characterized as captured by what they see, hear, or feel and are unable to control and deploy their attention. An important reason for their lack of control and inattentiveness may be their inability to coordinate exploratory search to actions that have to be performed later on, in addition to insufficient insight and control of environmental events that reveal information about environmental properties. Coordination of exploratory search with subsequent actions requires the ability to constrain exploratory search by actions and goals that will be achieved later on. Although exploratory search is necessarily constrained by ongoing activity and in this sense task specific, it may be less task specific with respect to later actions. Task specificity of exploration for larger perception action cycles is a skill that only gradually develops in children.

References

Beek, P. J., & Bingham G. P. (1991). Task-specific dynamics and the study of perception and action: A reaction to von Hofsten (1989). *Ecological Psychology*, 3, 35-54.

Bigelow, A. E. (1981). Children's tactile identification of miniaturized common objects. *Developmental Psychology*, 17, 111-114.

Connolly, K. & Dagleish, M. (1989). The emergence of a tool-using skill in infancy. *Developmental Psychology*, 6, 894-912.

Gelman, R. (1982). Basic numerical abilities. In R. J. Sternberg (Ed.), *Advances in the psychology of human intelligence* (Vol. 1, pp. 181-205). Hillsdale, NJ: Erlbaum Associates, Inc.

Gibson, E. J. (1988). Exploratory behavior in the development of perceiving, acting, and the acquiring of knowledge. *Annual Review of Psychology*, 39, 1-41.

Gibson, E. J., Riccio, G., Schmuckler, M., Stoffregen, T., Rosenberg D., & Taormina, J. (1987). *Detection of the traversability of surfaces by crawling and walking infants.*, 13, 533-544.

Gibson, E. J., & Spelke, E. S. (1983). The development of perception. In J. H. Flavell & E. M. Markman (Eds.), *Handbook of child psychology*, (Vol.3), Cognitive development. New York: Wiley.

Gibson, E. J., & Walker, A. S. (1984). Development of knowledge of visual-tactual affordances of substance. *Child Development*, 55, 453-460.

Gibson, J. J. (1966). *The senses considered as perceptual systems*. Boston: Houghton-Mifflin.

Gibson, J. J. (1979). *The ecological approach to visual perception*. Boston: Houghton-Mifflin.

Kellman, P.J., & Spelke, E. S. (1983). Perception of partly occluded objects in infancy. *Cognitive Pschology*, 15, 483-524.

Kugler, P. N., Shaw, R. E., Vincente, K. J., & Kinsella-Shaw, J. (1990). Inquiry into intentional systems I: Issues in ecological physics. *Psychological Research*, 52, 98-121.

Lee, D. N., & Reddish, P. E. (1981). Plummetting gannets: A paradigm of ecological optics. *Nature*, 293, 293-294.

Palmer, C. F. (1989). The discriminating nature of infant's exploratory actions. *Developmental Psychology*, 25, 885-993.

Pick, A. D., Frankel, D.G., & Hess, V. L. (1975). Children's attention: The development of selectivity. In E. M. Hetherington (Ed.), *Review of child development research*, Vol. 5. Chicago: University of Chicago Press.

Reed, E. S. (1982). An outline of a theory of action systems. *Journal of Motor Behavior*, 14, 98-134.

Reed, E. S. (1988). *James J. Gibson and the psychology of perception*. New Haven: Yale University Press.

Rochat, P. (1987). Mouthing and grasping in neonates: Evidence for the early detection of what hard or soft substances afford to action. *Infant Behavior and Development*, 10, 435-449.

Rochat, P. (1989). Object manipulation and exploration in 2- to 5-month-old infants. *Developmental Psychology*, 25, 871-884.

Smets, G. J. F., Overbeeke, C. J., & Stratmann, M. H. (1987). Depth on a flat screen. *Perceptual and Motor Skills*, 64, 1023-1034.

Smitsman, A. W., van Leeuwen, L., & Peters, J. (1988, April 21-24th). *Object exploration and tool use at early age*. Paper presented at the Sixth International Conference on Infant Studies, Washington D. C.

Starkey, P., Gelman, R., & Spelke, E. S. (1985). Detection of number or numerousness by human infants. *Science*, 222, 179-181.

Strauss, M. S., & Curtis, L. E. (1981). Infant perception of numerosity. *Child Development*, 52, 1146-1152.

Treiber, F., & Wilcox, B. (1984). Discrimination of number by infants. *Infant Behavior and Development*, 7, 93-100.

Turvey, M. T. (1977). Preliminaries to a theory of action with reference to vision. In R. Shaw & J. Bransford (Eds.), *Perceiving, acting, and knowing: Toward an ecological psychology* (pp. 211-265). Hillsdale, NJ.: Erlbaum Associates, Inc.

Turvey, M. T., & Carello, C., & Kim, N. G. (1989). Links between active perception and the control of action. In H. Haken & M. Stadler (Eds.), *Synenergetics of cognition* (pp.269-295). Heidelberg, West Germany: Springer.

van Leeuwen, L., & Smitsman, A. W. (1991). Perzeption von Handlungsmöglichkeiten und die Entwicklung von Werkzeuggebrauch im frühen Kindesalter. In F. J. Mönks & G Lehwald (Hrsg.), *Neugier, Erkundung, Begabung bei Kleinkindern*. München: Reinhardt.

van Leeuwen, L., Smitsman, A. W. & van Leeuwen, C. (in press). Tool use in early childhood; perception of a higher-order relationship. *Journal of experimental psychology: human perception and performance.*

van Loosbroek, E., & Smitsman, A. W. (1990). Visual perception of numerosity in infancy. *Developmental Psychology,* 26, 916-922.

van Loosbroek, E., & Smitsman, A. W. (1993). *Heterogeneity of objects in numerosity perception.* Manuscript submitted for publication.

Voss, H.G., & Keller, H. (1983). *Curiosity and exloration: Theories and results.* New York: Academic Press.

Warren, W. H., Jr., Kim, E. E., & Husney, R. (1987). The way the ball bounces: Visual and auditory perception of elasticity and control of the bounce pass. *Perception,* 16, 309-336.

CHAPTER **IV**.13

The Process and Consequences of Manipulative Exploration

Axel Schölmerich

What is exploratory behavior? There are numerous definitions of exploration, but all agree on the result of exploratory behavior: exploration leads to the acquisition of information about unknown objects or situations. The concept of unknown as a feature of an object is quite difficult, and the different definitions of exploration simply assume that something unknown can become known to an exploring individual. Keller and Voss (1976) provide a definition of exploration:

There is an interaction between the stimuli and structures in memory (cognitive representations of earlier events) with the goal of decoding the stimuli (p. 144, translation by the author). It is not clear how Keller and Voss (1976) conceptualize the decoding of stimuli, but one can assume from their extensive discussion of discrepancy theories that they suggest a process of modifying cognitive structures by experiencing and increasing the match between object and structure. Lorenz (1973) writes:

Behavior is exploratory, if the organism acts to make experiences (p. 186, translation by the author). How this could be separated from other actions remains an unresolved question. Hassenstein (1980) thinks that the concept of play is adequate to comprise the four behaviors exploration, curiosity, play, and imitation which fulfil their biological function .... because they are directed towards individual gain of information (p. 64) which allows the individual animal to maximize experience with a minimum of endangerment, thereby increasing or maintaining a certain level of skill. The system is unspecific and general, which means that it can be activated by the novelty of objects, and does not require specific stimulus characteristics to be set in operation. From a psychological point of view, I cannot help noticing a mixture of behavioral and motivational concepts in Hassensteins argument, but the important point here is the explicit reference to the acquisition of information (see also Lorenz, 1969). In this chapter, I focus on the relationship between information acquisition and exploratory behavior in the presence of a novel object. I am aware of other ways to use the term exploration, for example, in the context of boredom or tedium-motivation, or diversive exploration. In fact, large parts of the literature deal with the distinction between specific types of exploratory behavior and other behaviors (cf. Keller & Boigs, 1989; Nunnally, 1981; Voss, 1987; Voss & Keller, 1985; Wohlwill, 1981, 1984).

The conceptual core of the cognitive view on exploration is the idea that the amount of attention an object receives somehow reflects the discrepancy between the observers internal schemata and the object. If this discrepancy is within a moderate range, the attention to that object will be maximal, too little discrepancy will not lead to any exploratory behavior, and a very strong discrepancy might lead to avoidance. Dember and Earl (1957) use a similar concept about the change of cognitive structures when one is exposed to stimuli that are optimally different from schemata. Hunt (1965) uses the term optimal incongruence between organismic and environmental variables. The driving force here is the need to reduce cognitive uncertainty. It is interesting to note that n er theories of cognitive development also explicitly refer to exploration as a generator of new experiences (see Case, 1985; Klahr & Wallace, 1976). The cognitive outcome of exploratory behavior is the enrichment of the individuals knowledge structures. This process is largely determined by existing knowledge and is guided by interests, which seem to be specific for each individual from an early age on (Renninger, 1989; see also Krapp, this volume).

The ethological view contributes a contextual perspective on the research on exploratory behavior. Bowlby (1969) outlined a control-systems model of attachment behavior which contains a component promoting proximity to the caregiver and a component that promotes acquisition of knowledge of the environment and adaptation to environmental variations (Ainsworth & Bell, 1974, p. 110). The exploration-attachment balance (Ainsworth & Wittig, 1969) expresses the interconnectedness of orientation towards the new and the infants need for a stable and familiar frame of reference, usually provided by the childs caregiver. Numerous environmental circumstances can tip the balance, towards exploration by introduction of appropriate objects and towards proximity-seeking by removal of the caregiver. The infants readiness to engage in exploratory behavior is not only a function of the cognitive system and features of the object, but it also reflects the specific condition of the caregiver-infant dyad. According to this view, exploratory behavior is unlikely as long as the infant engages in establishing or maintaining contact with the primary caregiver, which is predominant as long as the mother does not provide a secure base for the infants exploratory activities. Ainsworth and Bell (1974) noted:

Obviously, if attachment behavior were constantly activated at a high level, a childs development would be greatly hampered, for he would not be attracted away from his attachment figure to explore his world (p. 110). The ethological tradition emphasizes the general importance of exploration for development, but pays little attention to the process of exploration itself.

The long-term effects of different intensities or quantities of exploration during early childhood are unknown. Early speculations (see for example Carey, 1977 and other chapters in the volume edited by Grossmann, 1977) on the importance of exploration for cognitive development have been displaced by work focussing on emotional and motivational organization as a background condition for intellectual competence (Grossmann, 1989; Schildbach et al., 1990). The simple idea that the child accumulates knowledge by exploring the outer world, which implies that the more a child explores, the more it will know, has not been demonstrated by empirical studies. It appears possible that the differences in early interactional experiences do not lead to quantitative differences in knowledge about the external world, but to the use of different styles or strategies of exploration. The current chapter is an attempt to separate the behavioral appearance and the acquisition of information during exploration, thereby offering some insight into interindividual differences in the consequences of exploratory behavior.

The Process of Exploration

Different Models in the Exploration Literature

There are different models of the process of exploration in the literature. Each model falls into one of the three following categories: (a) chronological models, that imply order in the process in a one-way sequence of events, but without reference to physical time;
(b) structural models, that imply order between subprocesses, like manipulation and visual inspection, but without the one-way directionality

of chronological models. Conditioned probabilities are the most common form of such a model; (c) chronometrical models, that describe the occurrence of exploratory behavior or subprocesses along a physical-time dimension. Nunnally´s & Lemond´s (1973) well known model belongs to the first category, as does the more elaborated model proposed by Keller, Schölmerich, Gauda, and Miranda (1984). Schneider and his colleagues (Schneider & Flohr, 1989; Schneider, Moch, Sandfort, Auerswald, & Walter-Weckmann, 1983) also have contributed to the structural model. Wohlwill (1984) and Schneider (1985) have given chronometrical descriptions, both with special attention to the distinction between play and exploration. Play and exploration appear similar in their behavioral manifestations, but represent different psychological processes (Wohlwill, 1984). I see the main exploration-play difference to be the central role of acquisition of new information during exploration, relative to the assimilation of existing schemata to the outer world during play.

A Comparison with Habituation

Habituation can serve as a model for the process of exploration, because the paradigm deals with the process of becoming familiar with a previously unknown stimulus. Intake of information is operationalized as a decrease in looking time or in other measures of attention. The term decrease in looking time suggests a gradual decrease, but from Cohen´s and Menten´s (1981) work we know that the gradual decrease in the common graphical plots in habituation experiments might be an artifact produced by averaging across subjects. The individualized plots Cohen and Menten reported show little variation in the attention measure up to the interval preceeding the significant drop used as a criterion, during which fixation time sharply increases to its peak. This indicates a comparatively rapid intake and processing of information, with the decrease of attention following immediately. Whereas the idea of decreasing exploration and increasing play during a session (Belsky & Most, 1981; Wohlwill, 1984) resembles the gradual decrease in attention to a single novel stimulus, the decomposing of exploratory behavior into a sequence of single information intake units arranged one after the other would be compatible with the process suggested by Cohen and Menten (1981).

Problems of Obversation: From Quantities of Behavior to Quantities of Information

Exploratory behavior is observable, whereas information intake is a hypothetical process (see Nunnally & Lemond, 1973, p. 63), which cannot be observed. There is no direct relationship between amount of behavior and amount of information, because individual differences with respect to speed of information processing, accuracy of perception, stability of images in the visual system and the like have to be expected. All those factors will influence the amount of time an individual spends exploring the surroundings. Taking the possible exploration of inner spaces, the individuals own cognitions about the world, into consideration, the search for the relationship between the observable and the hypothetical becomes even more likely to fail. If it is possible to explore ones own feelings about a specific object, why should we count only visual inspection or manipulation of the object as exploration? Other (nonsensory) approaches can reveal something about a specific object as well.

Unfortunately, there are no definite indicators that qualify behavior as exploration. Exploration of objects can be distal or proximal (Keller, Schölmerich, Gauda, & Miranda, 1984), and specific behavior identifies each modality. But the same behavior can occur in nonexploratory contexts, too. Lorenz (1969) observed that a young corvide bird, confronted with an object it has never seen, runs through practically all the inventory of its behavior patterns, except social and sexual ones (p. 56). Specific behavior is necessary for exploration, but no single behavior is sufficient to qualify activity as exploration. Because all studies on the development of exploration use quantitative behavioral measures, it is necessary to investigate the relationship between the quantitative aspects of information acquisition and of observable behavior. Whereas there are straightforward ways to measure the quantity of observable behavior, acquisition of information is, by definition, an internal event which cannot be observed. Extensive testing of acquired knowledge would be one interesting procedure, but this is very difficult to do in a semi-naturalistic setting, and differential factors of response to testing could blur the picture. There the acquisition of information can be monitored only by taking advantage of specifically designed objects. The sequential organisation of small action units can reveal the acquisition of information, if the object consists of numerous details that disclose their interrelatedness only if they are explored in the proper sequence. Such a procedure, which I describe in the following paragraph, is restrictive insofar as it relies on observable units of actions, and therefore it will not count any instances of acquisition of information that deal with features of the object in a distal or associative fashion.

The Sequential Organization of Exploratory Behavior

Searching for information about an object is a process in time, which we can describe as a minimally-ordered sequence of single actions. According to the objects design, some of the sequences of such single actions reveal information, others do not. If a subject performs two actions involving details of the object which are of relevance for each other, then we can assume acquisition of information. Our system of observational categories operates on the level of single actions, like "open lid on left side," "look in opening," "push button X," or "move lever to inflate balloon." So, in order to pick up the information that pushing button X sets visual pattern Y in motion, which is covered by lid Z, the subject has to open lid Z, push button X, and look at visual pattern Y. The objects design determines the sequence. Opening lid Z without looking inside or while manipulating other details of the object, will not reveal relevant information about the relationship between pushbutton X and visual pattern Y. Only positive evidence reduces uncertainty, which is one definition of information (Brody, 1970). In designing objects, we are free to choose the number of internal relationships or the maximum amount of information that a subject can pick up from this object. The subjects sequence of single actions can be decomposed into those patterns that correspond to the features built into the objects (functionally relevant sequences), and those sequences that are irrelevant with respect to the object. Additionally, we assume that a single occurrence of the proper sequence of actions is sufficient for the subjects to pick up the information. This is acceptable for the relationship between a push button and a sound or a visual effect, and as long as there are no probabilistic effects built into the box. In our study, occasional malfunctions of the object simulated probabilistic effects, e.g., that an action no longer has the effect it has had before. The behavioral pattern around those malfunctions was markedly different from the regular flow of actions while exploring simpler effects. We did not study this systematically due to rare occurrence, but the regular effects were sufficiently simple to satisfy the assumption. Now we can specify the first occurrence of a functionally relevant sequence to be the point in time at which the subject acquires the corresponding information (IA). Repetitions of functionally relevant sequences after their first occurrence do not indicate information acquisition. This operationalization leaves much to be desired, because it excludes events of potential significance and it rests on assumptions that are not tested empirically. But both the importance of a restrictive definition and the available evidence from studies on the process of habituation seem to justify this definition, that is essential for the quantification of exploratory behavior and of acquisition of information. A more extensive discussion of the rationale behind this approach can be found in Schölmerich (1990).

An Empirical Study

Data Collection

The Object

We designed an exploratory object for the type of analysis described above. It consists of 32 details that have functional relationships with at least one other detail. For example, there is a series of switches that control speed, direction, and an on/off switch of a small electric motor rotating a black and white spiral-pattern. The details are attached to a box approximately 25 inches wide, high, and deep. The box is painted metallic-blue on the outer surfaces and is mounted on four legs about 12 inches high. The electrical effectors are battery powered. The subjects can inspect the inner wiring and construction by opening several flaps held by small bolts. The effects include experiences of motion, optical illusions, tactile experiences, visual effects, and an inflatable balloon.

Theoretically, we could have $32^2 = 1024$ two-element combinations of the 32 details. Only 62 (6.06 %) of those are functionally relevant sequences. The likelihood of performing a functionally relevant sequence by chance is therefore rather small.

The Sample

The sample consists of 23 children who were 9 years old at the time of the home visits. The children participated in a longitudinal study covering the development of exploratory and interactional behavior from birth to middle childhood. We saw the children and their families eight times during the first year and at least once a year from the second year on. Similar tasks were given each year. All children are healthy first-borns from middle-class families. During the observation, at least one of the parents was present. The parents were instructed not to interfere with the child´s actions. Some children have smaller siblings, and we tried to keep them away from the box for the first ten minutes, which were later analyzed from videotapes.

The Coding Procedure

In order to code the behavior from the videotapes, we used two sets of codes. The more comprehensive set of codes consisted of the modes of exploratory behavior (visual, tactile, manipulative, verbal). The finer set of codes consisted of the childrens actions directed towards specific details of the object (outer surface, the switch on the left side, lever to inflate balloon, etc.). The coders reached 90% agreement after training with material from the earlier years of the longitudinal study. The 10-min sequence was played back in normal speed with sound, and the observers had to press keys on a special keybord to indicate begin and duration of the behaviors according to the predefined codes. Two observers worked simultaneously on separate keyboards, and each coded up to five different codes in one run. The tape was replayed several times until all codes had been entered. This process resulted in a real-time transcript of the visible events on the videotapes, which was then analyzed with software calculating the frequency, duration, mean duration, percentage of criterion-time, and rate per minute for each of the codes. Disruptions or sequences where the childs activities were not visible on the tapes were excluded from the calculation.

Results

Amount of Exploratory Behavior

The object has many details affording manipulation. Therefore we obtained the high total amount of exploratory and manipulatory actions. The children focussed their attention on the object for ten minutes with some exceptions. Table 1 shows the means and standard deviations of total time coded and the percentages of different exploratory modalities.

Table 1: Percentages of Time Spent With Different Modalities of Exploration

	Time	SuPart	VisEx	TacEx	ManEx	TotEx
M	418.68	84.09	73.08	6.77	53.92	74.33
SD	160.68	29.81	15.08	8.50	19.75	14.94

Note. Time: seconds of coded sequence; SuPart: sum of all percentages of time where the childs activities were directed towards the coded details of the object; VisEx: percentage of total time spent exploring visually; TacEx: percentage of total time spent with tactile exploration; ManEx: percentage of total time spent with manipulative exploration; TotEx: total exploration (percentage of time during which the child explored visually, tactile, or manipulative).

The high level of exploratory and object-directed behavior is caused by the specific features of the object we used, and exceeds the quantities of exploratory behavior elicited in previous years with other objects (Schölmerich & Boigs, 1990). This is a precondition for analyzing the process of exploration in a quantitative manner.

In the following section, I use cumulative graphs for each subject to show amount of behavior or information over time. The advantage of cumulative graphs is that they indicate the result of previous behaviors and current action simultaneously. Current actions are shown as small increases in the graph, and the parts that run parallel to the time-axis indicate no current exploratory activity or acquisition of information. If such a graph approximates a straigth line, it indicates a process which operates with the same speed or intensity throughout the total episode. Alternatively, a graph can change steepness during the episode, thereby indicating changes in intensity or speed over time. We expect the acquisition of information to start with high steepness, and to level off to some maximum value during the process of exploration, because there is only limited information available in the object. For manipulatory behavior, there is little evidence of instability during the first ten minutes. The subjects appear to exhibit manipulatory exploration at a constant rate over time, which produces approximately straight lines in the cumulative graphs. Figure 1 shows the cumulated amount of manipulative exploration over time for all subjects in the sample. The differences in the overall steepness between the individual tracelines indicate individual differences in basic behavioral speed. The comprehensive category manipulatory exploration, with longer mean durations produces jumps in these tracelines. At the end of each manipulatory activity, the new cumulated amount is shown, so the line increases vertically. The length of the vertical lines indicates the duration of the preceeding manipulatory action. The finer analysis of the actions directed towards specific details of the object reveals a similar picture.

Figure 1. Cumulated manipulative exploration.

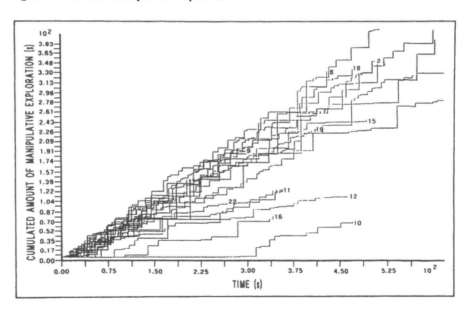

Note. X-axis indicates time in seconds, Y-axis cumulated amount of manipulative expolration in seconds.

Here we look at those sequences in the behavioral stream that indicate functionally relevant behavioral patterns. Figure 2 shows the cumulated graphs for the complete sample. The cumulated graphs of the functionally relevant sequences can be approximated by straight lines, but the overall appearance is much smoother, because the level of analysis is finer. The individual differences correspond to those described above.

Figure 2. Cumulated functionally relevant sequences.

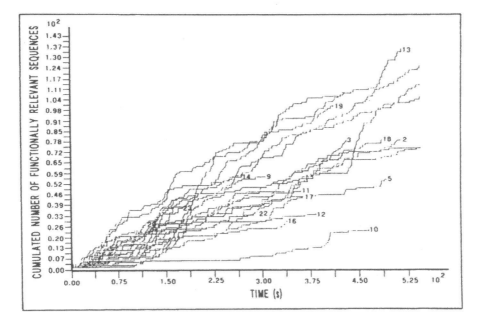

Note. X-axis indicates time in seconds, Y-axis cumulated number of functionally relevant sequences.

The correspondence of manipulatory exploration and functionally relevant sequences was expected, but the correlation is not as high as the graphs suggest. I discuss the correlations between the different aspects in the last section of the results. The functionally relevant sequences do not necessarily indicate information acquisition, which is defined by the first occurrence of a functionally relevant sequence.

Amount of Information

Information acquisition must necessarily be a subset of the functionally meaningful sequences. During the process of exploration it becomes increasingly less likely to discover anything new, because some aspects of the object are not novel any more. Figure 3 shows the actual acquisition of information over time as defined by our restrictive operationalization. Again, the individual curves follow relatively clear general lines.

Figure 3. Cumulated acquisition of information about the object.

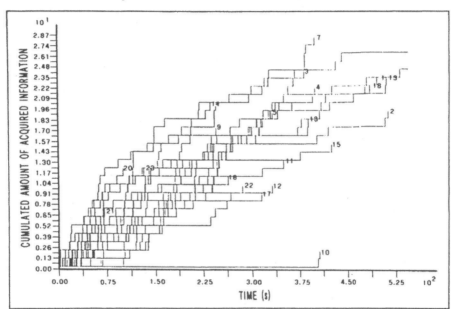

Note. X-axis indicates time in seconds, Y-axis cumulated amount of acquired information as measured by the first occurrances of functionally relevant sequences.

In the beginning, the sequences indicating acquisition of information are relatively frequent and between them are time segments without action corresponding to the model Keller and Boigs (1989) describe. But, in contrast to manipulative exploration and functionally meaningful sequences, there is a decrease in steepness of the curves indicating that with increasing time, less information is gained per time-interval. This finding corresponds to our expectations based on the analysis of the limited quantity of information that the subject can acquire from an object of a given structure. The subjects in our sample did not exhaust the potential of information in this object. On the average, a child performed 16.91 information acquisition sequences (27.3 % of the total 62 that the object afforded).

Interrelationships between Exploration and Information Acquisition

I have described three different aspects of the exploratory process so far: the aspect of exploratory behavior using different modalities, a pattern-recognition aspect, as defined by sequences of actions directed towards specific details of the object, and the actual acquisition of information, as defined by the first occurrence of functionally meaningful sequences. Table 2 shows the correlations between the different measures reported here. The modalities of exploration and the sequential patterns are coded in independent runs with independent sets of categories. Information acquisition is a subset of functionally meaningful sequences. The inspective handling of a detail would be coded as manipulative exploration, regardless of that behaviors position in the sequence of actions. Of course one would expect a highly manipulative child to show more functionally meaningful sequences as well, but this is a conceptual expectation, and not a numerical consequence. There is a one-direction dependency between functionally relevant sequences and information acquisition, the former being a necessary, but not sufficient precondition for the latter. It is, in theory, possible to perform many actions directed to details of the object without there being functionally meaningful sequences or information acquisition.

Table 2: Correlations Between Modalities of Exploration, Functionally Relevant Sequences, and Information Acquisition.

	VisEx	TacEx	ManEx	TotEx	FRS	IA
VisEx	1.00	0.17	0.11	0.90	0.15	0.21
TacEx		1.00	-0.30	0.30	-0.33	-0.32
ManEx			1.00	0.08	0.56	0.59
TotEX				1.00	0.13	0.18
FRS					1.00	0.92
IA						1.00

Note. VisEx: percentage of total time spent exploring visually; TacEx: percentage of total time spent with tactile exploration; ManEx: percentage of total time spent with manipulative exploration; TotEx: total exploration (percentage of time during which the child explored visually, tactile, or manipulative); FRS: functionally relevant sequence; IA: information acquisition.

Total exploration is computed as either visual, tactile, or manipulative exploration, including the overlaps between the modalities. The high correlation (r =.99) between total exploration and visual exploration reflects the nonexclusive definition of the exploratory modalities and the high percentage of total time during which visual exploration was coded. Visual exploration is an integral part of manipulative and other exploratory acts, but does not correlate with other aspects of the exploratory modalities or the sequential patterns. The three modalities are independent of each other. Visual exploration does not correlate with functionally relevant sequences or IA (r =.15 and r =.21), tactile exploration shows a mild negative correlation with functionally relevant sequences and information acquisition (r =-.33 and r =-.32), and manipulative exploration correlates positively with functionally relevant sequences and information acquisition (r =.56 and r =.59). This finding reflects, at least in part, the high importance of manipulatory acts for the acquisition of information with this specific object, where we put special emphasis on the observability of single actions. But the correlation is only in the moderate range, which emphasizes the partial independence of information acquisition and manipulatory actions. The differential patterns of the three exploratory modalities correspond to findings from other studies on the independence of the different modalities of exploration (Keller & Voss, 1976; Kreitler, Zigler, & Kreitler, 1975).

The Regulation of Exploratory Action and Interindividual Differences in Strategies of Exploration

The stability of manipulative exploration during the session and the similar stability of actions directed at the object is a surprising finding. Several studies have reported the decrease of exploratory behavior over time. The question is, what is the proper time frame to observe such a process? It appears to be likely that the children in this study would have eventually ceased to interact with the object, had the observer only waited long enough. For the acquisition of information, the time apparently was just right, because we could observe the expected decrease in the cumulative graphs. The difference in the process characteristics between manipulative exploration and information acquisition has implications for the internal dynamics of the exploratory system. Exploration seems to be initiated by presentation of a new object (something with information to be acquired), but the fine regulation of the behavior once the exploratory system is in operation appears to be independent of the amount of information still to be discovered. The (constant) manipulative exploration cannot be regulated by the (decreasing) amount of information in the object, at least not in the limited period of time that we studied.

The reported relationships between behavioral modalities, functionally relevant sequences, and information acquisition, both on the correlative level and on the level of differences in the process of such activities, point in the same direction. The relationship between observable behavior and information acquisition is by no means simple, and this should caution us against using quantifications of observable explorative behavior as a predictor of later (cognitive) outcome. The same caution is needed when the high amount of exploratory behavior an infant shows is interpreted as a result of the infants secure attachment. Individual differences, for example, in basic speed of action or in processing of information, can actually lead to more acquisition of information in shorter periods of time. The fast explorer, who gets low scores in observed e loratory behavior, but who acquires high amounts of information, will be misclassified by exclusive usage of time-on-task analysis. The temporal and the sequential organization of behavior are the determinants of productive interactions with the environment, and we have given the former much more attention than the latter. We can describe the sequential organization as a behavioral strategy, and the results indicate that there are strong and interesting individual differences in the quality of such strategies.

References

Ainsworth, M.D.S., & Bell, S.M. (1974). Mother-infant interaction and the development of competence. In K.J. Connolly & J.S. Bruner (Eds.), *The growth of competence*, (pp. 131-164). London, New York: Academic Press.

Ainsworth, M.D.S., & Wittig, B.A. (1969). Attachment and exploratory behavior of one-year-olds in a strange situation. In B.M. Foss (Ed.), *Determinants of infant behavior* (Vol 4., pp. 111-136). London: Methuen.

Belsky, J., & Most, R.K. (1981). From exploration to play: A cross-sectional study of infant free-play behavior. *Developmental Psychology, 17*, 630-639.

Bowlby, J. (1969). *Attachment and loss: Vol.1. Attachment.* New York: Basic.

Brody, M. (1970). Information theory, motivation, and personality. In H. M. Schroeder & P. Suedfeld (Eds.), *Personality theory and information processing*. New York: Ronald.

Case, R. (1985). *Intellectual development: A view from birth to adulthood.* Orlando, FL: Academic.
Cohen, L.B., & Menten, T.G. (1981). The rise and fall of infant habituation. *Infant Behavior and Development, 4,* 269-280.
Dember, W.N., & Earl, R.W. (1957). Analysis of exploratory, manipulatory, and curiosity behaviors. *Psychological Review, 64,* 91-96.
Grossmann, K.E. (Ed.). (1977). *Entwicklung der Lernfhigkeit in der sozialen Umwelt.* München: Kindler.
Grossmann, K.E. (1989, April). Die Entwicklung emotionaler Organisation und ihre Beziehung zum intelligenten Handeln. Vortrag auf dem *Symposion Kompetenzentwicklung im sozialen Kontext,* Gesellschaft fr Psychologie der DDR, Leipzig.
Hassenstein, B. (1980). *Instinkt, Lernen, Spiel, Einsicht. Einfhrung in die Verhaltensbiologie.* München: Piper.
Hunt, J. McV. (1965). Intrinsic motivation and its role in psychological development. In D. Levine (Ed.), *Nebraska symposium on motivation,* (pp.189-282). Lincoln: University of Nebraska Press.
Keller, H., & Boigs, R. (1989). Die Entwicklung des Explorationsverhaltens. In H. Keller (Ed.), *Handbuch der Kleinkindforschung,* (pp. 443-464). Heidelberg: Springer.
Keller, H., Schölmerich, A., Gauda, G., & Miranda, D. (1984). *Entwicklung des Explorationsverhaltens.* Bericht an die Deutsche Forschungsgemeinschaft.
Keller, H., & Voss, H.G. (1976). *Neugier und Exploration.* Weinheim: Beltz.
Klahr, D., & Wallace, J.G. (1976). *Cognitive development: An information-processing view.* Hillsdale, NJ: Erlbaum.
Kreitler, S., Zigler, E., & Kreitler, H. (1975). The nature of curiosity in children. *Journal of School Psychology, 13,* 185-200.
Lorenz, K. (1969). The innate basis f learning. In K.H. Pribham (Ed.), *On the biology of learning.* New York: Harcourt, Brace & World, Inc.
Lorenz, K. (1973). *Die Rückseite des Spiegels.* München: Piper.
Nunnally, J.C. (1981). Explorations of exploration. In H.I. Day (Ed.), *Advances in intrinsic motivation and aesthetics.* New York: Plenum.
Nunnally, J.C., & Lemond, C. (1973). Exploratory behavior and human development. In H. Reese (Ed.), *Advances in Child Development and Behavior,* (Vol. 8, pp. 60-109). New York: Academic Press.
Renninger, K.A. (1989). Individual patterns in childrens playinterests. In L.T. Winegar (Ed.), *Social Interaction and the development of social understanding,* (pp. 147-172). Norwood, NY: Ablex.
Schildbach, B., Loher, I., Volland, C., Köglmeier, E., Lacler, B., Winkler, S., & Gmeiner, R. (1990, September). *Entwicklung intellektueller Kompetenz im Kontext der Erfahrung emotionaler Unterstützung.* Poster presented at the 37th Kongreß der Deutschen Gesellschaft für Psychologie, Kiel.
Schneider, K. (1985). Subjektive Unsicherheit, Neugierverhalten und Spiel mit einem neuen Objekt. In W. Einsiedler (Ed.), *Aspekte des Kinderspiels,* (p. 127-143). Weinheim: Beltz.
Schneider, K., & Flohr, B. (1989). *Explorationsverhalten bei Kindern im Vorschulalter.* Unpublished manuscript, Ruhr-Universität Bochum.
Schneider, K., Moch, M., Sandfort, R., Auerswald, M., & Walter-Weckmann, K. (1983). Exploring a novel object by preschool children: A sequential analysis of perceptual manipulating and verbal exploration. *International Journal of Behavioral Development, 6,* 477-496.

Schölmerich, A. (1990). *Der Erwerb neuer Information im Verlauf des Explorationsprozesses: Eine sequentielle Analyse von Handlungsketten.* Unpublished doctoral dissertation, Universität Osnabrück.

Schölmerich, A., & Boigs, R. (1990, September). *Die Entwicklung des Explorationsverhaltens: Vorläufer strategischer Aspekte.* Poster presented at the 37th Congress of the Deutsche Gesellschaft für Psychologie, Kiel.

Voss, H.-G. (1987). Exploration and play: Research and perspectives in developmental psychology. In D. Görlitz & J. Wohlwill (Eds.), *Curiosity, imagination, and play* (pp. 43-59). Hillsdale, NJ: Erlbaum.

Voss, H.-G., & Keller, H. (1985, September). *Exploration und Spiel in der frühen Kindheit.* Paper presented at the Congress of the Fachgruppe Entwicklungspsychologie in der Deutschen Gesellschaft für Psychologie, Trier.

Wohlwill, J.F. (1981). A conceptual analysis of exploratory behavior. In H.I. Day (Ed.), *Advances in intrinsic motivation and aesthetics.* New York: Plenum.

Wohlwill, J.F. (1984). Relationships between exploration and play. In T.D. Yawkey & A.D. Pellegrini (Eds.), *Childs play: Developmental and applied,* (pp. 43-170). Hillsdale, NJ: Erlbaum.

CHAPTER **IV.14**

Motivational and Cognitive Determinants of Exploration

Shulamith Kreitler, Hans Kreitler

A large body of research showed that exploration plays an important role in cognitive development, problem solving, thinking and creativity (e.g., Cagle, 1985; Gibson, 1988; Kagan, 1978; Singer & Singer, 1990; Voss & Keller, 1983). Thus, it is a major factor in the construction and maintenance of a cognitive system that comprises the contents and processes necessary for coping intelligently with a complex and changing external and internal environment. Therefore it is of great theoretical and practical importance to clarify the motivational and cognitive determinants of exploration. This is the major purpose of the present chapter. It summarizes empirical findings and theoretical conclusions obtained by applying to the study of curiosity and exploration two theoretical frameworks: the theories of cognitive orientation and of meaning (Kreitler & Kreitler, 1976a, 1990a). To be sure, these theories and the methodologies they have generated have been applied also to a broad range of issues other than exploration and curiosity. Yet it may be of interest to note that the investigations of exploration have played a seminal role in the development and refinement of these theories and methods. Exploration, motivational determinants and cognitive determants are the three major terms in the chapter's title. We will deal with each in turn before we describe an empirically based theoretical attempt at their integration.

The Nature of Exploration: The Five Factors of Exploration

Studying the determinants of any phenomenon presupposes at least some clarity about its nature. It is evident that such clarity does not exist in regard to exploration and curiosity, or at least did not exist 15 years ago when we started our explorations in this domain. The great number of terms traditionally has been used to refer to the phenomenon of interest include constructs such as exploration, curiosity, play, manipulation, and stimulation seeking (Hutt, 1966; Keller, Schölmerich, Miranda, & Gauda, 1987; Voss & Keller, 1983), some of which denote a behavior (e.g., exploration), whereas others denote an assumed underlying drive or motive (e.g., stimulation seeking). We have opted to distinguish strictly between the overt behavior and its motivation, and have therefore selected the term exploration in order to emphasize the fact that we focus on behavior. However, even a cursory view of the literature reveals the great diversity of behaviors subsumed under the headings of exploration or curiosity. Visual exploration, alternation behavior, object manipulation, preferential attending to specific stimuli and asking questions are only some examples (Berlyne, 1966; Keller et al., 1987; Vandenberg, 1984). The questions that have motivated our first stage of inquiry were: Do the different tasks used in the study of exploration assess one basic type of exploration, and if not, how many different kinds and which kinds of exploration do they assess?

Our first empirical step focused on a task based on responses to Berlyne's collative variables, because it has become one of the commonest and theoretically the best founded method for assessing curiosity (Kreitler, Zigler, & Kreitler, 1974). Our subjects were 84 first graders (42 boys, 42 girls), 6.7 to 8.4 years old. They were shown pairs of stimuli—one more complex and one more simple—exemplifying the five collative dimensions: heterogeneity of elements, irregularity of arrangement, amout of material, irregularity of shape, and incongruities. Each dimension was represented by four stimuli pairs. In one task the children were asked to observe the stimuli of each pair, in another to state their preferences for one or the other member of each pair. The response variables were duration of observation of each stimulus and stated preferences. Contrary to common expectations, the findings showed generally low interrelations within and between the collative variables, as well as between duration of viewing time and preferences. These results suggested that collative variables are more complex than is usually assumed, and that curiosity is probably a differentiated rather than a homogeneous phenomenon.

Therefore, in the subsequent study (Kreitler, Zigler, & Kreitler, 1975), we extended our search of the literature and came up with a list of over 60 tasks used for assessing curiosity. Three expert judges were asked to decide which of the tasks are highly similar and hence redundant. Thus, we ended up with six basic tasks that generated 19 variables plus the variable of teacher's evaluations. The tasks, administered individually to 84 first graders (42 girls, 42 boys), were:

Observation of simple and complex stimuli (e.g., Berlyne, 1966): 10 pairs of simple and complex stimuli exemplifying the five collative dimensions, were presented to the child for viewing in a box, so that only one stimulus could be viewed at a time. The variables were the mean time spent viewing the complex stimuli, mean difference between these two viewing times, number of times the child switched glances between the stimuli, and number of pairs whose complex stimulus was viewed longer.

Preferences for simple or complex stimuli: The child was shown in the sequence 10 pairs of stimuli (comparable to those used in Task a) and was asked to state his or her preference for one of the stimuli. The variables were the number of complex stimuli preferred, and the number of times the child switched preferences from a complex stimulus to a simple one or vice versa.

Preferences for the known and the unkown (devised by Professor Susan Harter): The stimuli were nine folded cardboards, each with one flap covering a picture identical to the one depicted on the outside and another blank flap covering a hidden picture. The variables were the number of flaps covering hidden pictures that the child opened, and the number of times the child switched from opening one kind of flap to opening the other kind.

Extent and structure of meaning of ordinary objects: The child was asked to communicate to another imaginary child, verbally or nonverbally, the meaning of four familiar objects presented as toys (e.g., an iron, a car). The variables were the number of different meaning dimensions (namely, content categories defined in terms of the system of meaning, Kreitler & Kreitler, 1990a, e.g., the object's material, its size, or location; see Table 4 for the list of meaning dimensions).

Choice of known or new toys: The child was asked to choose for playing one of the two sets of toys, a 'known' set (used in the immediately preceding Task 4) or a 'novel' set, both sets including toys comparable in attractiveness, stimulation power and familiarity. The variables were the child's choice to play with the known or new toys, and the reaction time of the choice.

Object manipulations: The child was left alone to play with the set of toys he or she chose, while two observers recorded through a one-way-mirror the kind, number and duration of performed manipulations, as well as the questions the child asked about the toys when the experimenter returned. The variables were the total number of inspecting manipulations (e.g., observing the toy while rotating it), the total number of manipulations using the object in its customary manner (e.g.,'phoning' on the toy phone), the total number of exploratory manipulations (e.g., taking the toy apart), total number of different manipulations, total duration of play, and a weighted index of the spontaneous and encouraged questions the child asked.

In addition, *teachers' ratings* (Maw & Maw, 1965) provided the variable of the mean rating of the child's curiosity by two teachers on a 5-point scale.

Table 1 shows that the results yielded five differentiated factors that may be identified (in a decreasing order of prominence) as manipulatory exploration (focused on exploring by means of motor actions, including handling, etc.), perceptual exploration (focused on exploring by means of viewing, listening, or smelling), conceptual exploration (focused on exploring by checking meanings and their interrelations, and by asking questions), exploration of the complex or ambiguous (focused on exploring especially the complex aspects), and adjustive-compliant exloration (focused on exploring in line with obvious demand characteristics of the situation, and especially when expected or stimulated to do so).

The same findings were replicated in a sample of adults (n=90, 45 men and 45 women, 30 to 40 years old, M=13.5 years of education) with somewhat different tasks adapted for adults (Kreitler & Kreitler, 1986). Again, we obtained similar factors but in a different order of prominence. The most prominent factor was perceptual exploration, primarily visual (accounting for 39% of the variance), followed by conceptual exploration, primarily verbal (24%), manipulatory exploration (20%), exploration of the complex (12%), and adjustive-compliant exploration (5%).

Identifying the five factors in adults seems to contradict findings which suggest the existence of only one (Olson & Camp, 1984) or two (Ainly, 1987) curiosity factors. The difference may, however, be due to the fact that we tested actual manifestations of exploration in a variety

Table 1. Factor loadings (above 0.25 or below -0.25) of the Exploration Variables

Task	Variables	Factor I	II	III	IV	V
1.	Observ. time of simple stimuli		.97			
	Observ. time of complex stimuli		.91	.35		
	Time dif. bet. complexs & simple				.96	
	No. of switchings		.63		.26	
	No. of pairs with complex stimulus viewed longer				.48	
2.	No. of complex stimuli preferred				.25	
	No. of switchings in preferences		.43			.25
3.	No. of choices of the unknown					.36
	No. of switching in choices					
4.	No. of response units			.85		
	No. of meaning dimensions			.84		
5.	Reaction time of choice	-.32		-.25		
6.	No. of `inspecting` manip.	.79				
	No. of `customary use` manip.			.36		.87
	No. of exploratory manip..	.93				
	Total no. of manip.	.97				
	Total time of manip.	.33				
	Weighted index of questions			.36		
	Teachers' ratings					.30
	Eigenvalues	3.33	2.37	1.77	1.42	1.04
	Percent of variance	17.5%	12.5%	9.33%	7.46%	5.48%
	Suggested title	Manip.	Perceptual	Conceptual	About complex	Adjustive compliance

Note. For the description of tasks and variables see the text. The table is based on Kreitler, Zigler and Kreitler (1974) and Kreitler, Zigler and Kreitler (1975). Manip. = Manipulations. Manipul. = Manipulatory. No. = Number. Bet. = Between. Dif. = Difference. Observ. = Observation.

of tasks, relying also on nonverbal responses, whereas the latter studies are based on the verbal reports of the subjects about their own exploration, which are both more restricted and less reliable.

The successful replication of the factorial findings serves to reinforce the status of the five factors as basic exploratory modes in both children and adults. Notably, each appears in one or more studies of curiosity. Thus, Keller et al. (1987) identified in children up to 4 years old visual, auditory and tactile exploration (corresponding to the factor "perceptual exploration"), manipulatory exploration (corresponding to the factor "manipulatory exploration") and verbal exploration (corresponding largely to the factor "conceptual exploration"). Similarly, Henderson (1984) classified children as belonging to the "watcher", "toucher" or "questioner" types, Dash and Dash (1983) distinguished between verbal and perceptual exploration, and Schneider , Moch, Sandfort, Auerswald and Walther-Weckman (1983) assessed manipulatory, verbal and perceptual curiosity. Complexity has been identified as differentiating between gifted and nongifted elementary school youngsters (Harty & Beall, 1985). Thus, our identification of the five exploratory factors serves to integrate a large body of research.

The five exploratory modes seem to appear with differential intensity in different individuals. Gender differences in children have been observed only in regard to manipulatory exploration, on which girls score lower (Kreitler, Zigler, & Kreitler, 1975, 1984). A replication in a new sample of first graders (n=75, 38 boys, 37 girls) showed that when three factors (manipulatory, conceptual, and complexity exploration) are assessed each by the variables (3 to 5) most highly saturated on it, the factor scores are not interrelated significantly (Kreitler, Zigler, & Kreitler, 1984). This finding implies that in assessing curiosity, for example in children, it is advisable to assess all five exploratory modes before a child is diagnosed as "curious" or "noncurious", and if assessment is more restricted to specify which modes were involved.

Yet, though they are all manifestations of exploration, the five modes differ in their psychological functions. This was demonstrated in regard to responses in a probably learning setup in which the subject was asked to press,100 times, one of three knobs, only one of which provided random reinforcement 33% of the cases. In first graders, the maximizing response (pressing only the reinforcement knob) was related negatively to the factor scores of manipulatory, conceptual and complexity exploration; variable pattern responses were related positively to all three exploratory factors; systematic variation responses (e.g., pressing right-middle-left) were related positively to complexity exploration; and perseverative responses (e.g., pressing the same knob) were related negatively to manipulatory and complexity exploration (Kreitler, Zigler, & Kreitler, 1984).

The five factors seem to respect major aspects of exploration from the age of 4 onward that persist into adulthood. Their salience, however, changes with age and personality, as has been shown already in regard to the more elementary exploratory modes up to the age of 4 years (Keller et al., 1987). This conclusion has implications, for example, in regard to education. Thus, it can be argued that if curiosity is to be mobillized for the enhancement of motivation in the study of the sciences, different exploratory modes are to be elicited at different ages, such as manipulatory exploration through the performance of experiments at younger ages or exploration of the complex and ambiguous through the presentation of intriguing issues at older ages (Kreitler & Kreitler, 1974)

The Motivational Determinants of Exploration: The Cognitive Orientation of Curiosity

Most of the work about the motivation for exploration has focused on showing that curiosity is an intrinsic motive (e.g., Tzuriel & Klein, 1983), related to cognitive needs (Olson, Camp, Fuller, 1984) and affected by emotions, such as depression and fear that may inhibit it (e.g., Caruso, 1989; Rodrigue, Olson, & Markley, 1987). In order to deepen understanding of the motivational determinants of exploration we studied it in the framework of the cognitive orientation (CO) theory because it is a comprehensive theory of behavior, which has been used successfully for predicting and changing behaviors, and has generated a systematic methodology for identifiying motivationally relevant factors (Kreitler & Kreitler, 1976a, 1982).

Briefly, the main tenet of the CO theory is that cognitive contents — namely, meanings, beliefs, attitudes — guide human behavior.The processes intervening between input and output may be grouped into four stages, each characterized in terms of metaphoric questions and answers. The first is initiated by an external or internal input and is focused on the question "What is it?". It leads to input identification in the form of initial meaning which enables unconditioned or conditioned responses. The second stage is focused on the question "What does it mean to me and for me?". It is initiated either by a meaning signalling the need to consider molar action or by feedback indicating failure of the previous responses to cope with the input. By means of enriched meaning generation it leads to a specification of whether action is required. A positive answer initiates the third stage which is focused on the question "What will I do?".

The answer is sought by means of relevant beliefs of four types:

(a) beliefs about goals, denoting actions or states desired or not by the individual (e.g., "I want to know about music"); (b) beliefs about rules and norms, denoting ethical, social, aesthetic and other standards (e.g., "A girl should not ask too many questions"); (c) beliefs about self, expressing information about oneself, such as one's habits, feelings etc. (e.g., "New things make me nervous"); and (d) general beliefs, expressing information about others and the environment (e.g., "Grownups do not like a child who knows too much").

If all four belief types point in the direction of the same behavior, a CO cluster is formed, generating a behavioral intent supporting the performance of that action. The fourth stage is focused on the question "How will I do it". The answer is in the form of a behavioral program guiding the actual performance of the action. Some programs are innate (e.g., reflexes), some are partly learned and partly innate (e.g., reproductive behaviors), and most are learned.

CO is a cognitive model of motivation but contrary to other models (e.g., Ajzen & Fishbein, 1980) it does not confound cognition with rationality and voluntary contral: rather than assuming that behavior is the product of rational decision or carefully reasoned weighting of benefits and losses, it specifies the underlying cognitive dynamics and shows how behavior proceeds from clustered beliefs and their directionalities.

The CO theory has been applied successfully to the prediction of over forty behaviors, such as being on time, achievement, or quitting smoking (Kreitler & Kreitler, 1976a, 1982). The predictions are based on information about the subject's CO cluster and behavioral programs in the examined domain of behavior. When it can be assumed that the program is available, an individual will show the expected behavior if there are enough relevant beliefs orienting toward that behavior in all four belief types, or at least three if the fourth does not point in a contrary direction. Beliefs are identified as relevant for, say, exploration or curiosity if they have occurred in the process of a systematic analysis of the meaning of exploration and related cues (for procedure see Kreitler & Kreitler, 1982).

In order to study the motivation for exploration, a CO Questionnaire of Curiosity has been constructed, designed to enable predicting the overt exploration of children (Kreitler & Kreitler, 1981; Kreitler, Zigler, & Kreitler, 1974). It includes questions of the four belief types, referring to 10 themes identified in pretests as basic to the meaning of curiosity and exploration in children, such as the value of knowledge, changing the given order of things, fear evoked by the unfamiliar, or intrusion of privacy. Each theme was presented in the form of a brief story and the questions referred to the story. Each subject got four scores, one for each belief type, that are used as continuous variables or as dichotomous ones (above the group's mean = 1, below it = 0) summed into a CO score (ranging from 0 to 4).

Table 2. Comparisons of mean responses along exploration variables in Groups defined by Cognitive Orientation Scores

Exploration variables	CO Score 4	CO Scores 3&2	CO Scores 1&0	F
No. of switchings (Task 1)	27.49 (9.33)	23.59 (5.95)	23.56 (3.89)	6.57**
No of choices of the unknown (Task 3)	6.11 (2.25)	5.94 (1.37)	4.74 (1.29)	3.29*
No. of "inspecting" manipulations (Task 6)	5.11 (3.45)	2.28 (2.63)	2.30 (1.56)	8.18**
No. of exploratory manipulations (Task 6)	13.63 (1.38)	12.33 (2.59)	9.93 (3.02)	12.88*
Total number of manipulations (Task 6)	22.79 (9.25)	16.94 (8.33)	16.52 (6.91)	3.85*

Note. The table presents only examples demonstrating the relations of Cognitive Orientation (CO) scores with exploration variables. It is based on Kreitler, Zigler and Kreitler (1974). The numbers in parentheses are standard deviations. For the description of the variables and the tasks see the text. The number of subjects in the groups of CO 4, CO 3 & 2, and CO 1 & 0 is 19, 13 and 27, respectively.
* $p < .05$.   ** $p < .01$.

Interrelating the belief measures with the measures of exploratory behavior (see examples in Table 2) showed that the belief scores were related significantly and positively to 14 of the 20 original measures, representing each of the five exploration factors. The unpredicted measures were mostly unrelated to exploration, according to the factor analysis. These findings support the conclusion that the higher the support of the four belief types for exploration the higher the overt manifestations of exploration. Notably, the beliefs predicted a broad range of ecploratory behaviors, some of which were not even interrelated. This indicates that though the manifestations of exploration are independent of each other they all share the same core of underlying motivation, which may be called curiosity.

Further, because the same CO of Curiosity predicts unrelated exploratory manifestations, it may be used for diagnostic purposes, to determine whether a certain behavior is motivated by curiosity or not. This was demonstrated in regard to behaviors characteristic of the probability learning setup (see p.9). In a sample of 64 first graders (32boys, 32 girls), we found that, as expected, the CO scores of curiosity were related negatively to the responses of maximizing (pressing the reinforced knob) and minimizing (pressing a nonreinforced knob) but positively to the responses of variable systematic switchings between the knobs (Kreitler & Kreitler, 1976 b). Thus, the results showed that the controversial response of perseverative alternation is a form of exploration rather than primitive stereotypy.

The demonstration that specific beliefs of four belief types predict exploratory behaviors conforms to the theoretical claim that the belief matrix called CO cluster guides behavior, but it does not prove the claim because the demonstration is based on a correlation. In order to determine the causal directionality it is necessary to show that exploration can be modified by changing the relevant beliefs. This demonstration was provided by a study with 59 preschoolers, whose mean age was 5.4 years (Kreitler, Kreitler, & Boas 1976). First, children were administered the CO Questionnaire of Curiosity and two curiosity tasks (a modified version of Task 1 and Task 6, see above), designed to provide five measures of the three exploratory factors-manipulatory, perceptual and complexity exploration (see Tables 1 & 3). Then they were divided randomly into three groups, that did not differ in mean age, CO scores, exploratory scores, and gender composition. Two of the groups served as experimental groups, and one as a control group. The former got a systematic training (5 half-hour sessions) designed to raise their CO scores in support of curiosity. The control group met for 5 playing sessions. Then all groups were retested on the CO Questionnaire and parallel forms of the exploratory tasks, one experimental group and the controls 1 week after termination of the training, and the other experimental group 2 months later.

There were no significant changes in CO scores in the three groups, and therefore we concentrated only on those subjects in the experimental groups who showed any rises in CO scores (63.3% in the first experimental group and 52.6% in the second). The changes in these subgroups affected all belief types similarly, were higher in children with low initial CO scores than in those with higher scores and ranged (in terms of mean standards scores) from 2.76 to 3.46, as compared with changes from -1.7 to -2.04 in the controls. This indicates that changes in CO scores are possible even in children, differ in extent across children and are sufficiently durable to remain constant even after 2 months.

Comparisons of the behavioral changes in the subgroups showed that rises in exploration were significantly higher in the experimental children whose CO scores of curiosity increased than in those in whom they did not increase or the controls (see Table 3). This was true in regard to all five behavioral measures in the first testing period and three measures in the second testing period.

These findings show that increases in the CO of curiosity are accompanied by increases in exploratory behaviors of the different modes. The changes are in the expected direction and persist at least for 2 months. Hence we may conclude that CO scores of curiosity represent the motivational matrix of exploration. This means that the motivation for exploration is a cognitively based vector orienting toward exploratory behavior. The vector—called behavioral intent—is generated by a set of specific beliefs relevant for exploration, along each of four belief types that support exploration. The beliefs can be assessed (by the CO Questionnaire of Curiosity, that proved valid in both American and Israeli samples) and can be modified by systematic training, even in children: Beyond their theoretical importance, the findings also have practical importance for they show how exploration may be increased in children so that it both persists and is amenable to generalization.

Table 3. Comparisons of experimental groups and control group with respect to changes in exploratory variables

Exploratory variables	Time of retesting After:	Experim. Rise in CO	Experim. No rise in CO	Control	Sig. if differenc
Duration of play	1 Week	278.08	142.71	69.52	a>b',
	2 Months	161.09	55.25	69.52	a>c**
No. of manipulations	1 Week	7.83	3.86	1.00	a>b',
	2 Months	4.91	1.50	1.00	a>c**
					a>c*
Duration of observ. simple stimuli	1 Week	23.71	2.21	-11.45	a>c*
	2 Months	34.38	7.14	-11.45	a>c*
Duration of observ. complex stimuli	1 Week	45.54	.93	-5.76	a>b', a>c*
	2 Months	42.81	13.64	-5.76	
Difference of observ. complex and simple	1 Week	17.38	-.71	-1.57	a>b', a>c*
	2 Months	8.94	6.50	-1.57	

Note. For the description of the variables and groups see the text. The number of subjects in the control group was 21, and in the Experimental "Rise" and "No rise" it was 12 and 7, respectively, after 1 week, and 10 and 9, respectively, after 2 months. The significance of differences was tested by the q_r statistic following Duncan's (1955) procedure. The numbers represent mean changes after training. The table is based on Kreitler, Kreitler and Boas (1976). Observ. = Observing.
' $p < .10$. * $p < .05$. ** $p < .01$.

Cognitive Determinants of Exploration: Patterns of Meaning Variables

According to the CO theory the occurrence of exploration depends on the two following major conditions: (a) a suffiently high number of beliefs relevant for exploration, along four belief types that support exploration and hence allow for the formation of a behavioral intent that orients toward exploration and (b) a behavioral program implementing the behavioral intent. The last section dealt with the first condition. The present section focuses on the second condition.

We have seen that there are five exploratory modes: manipulatory, perceptual, conceptual, about the complex, and reactive-compliant. Each is defined by specific eliciting stimuli (e.g., manipulatory exploration by objects that can be manipulated motorically) and mode of operation (e.g., for conceptual exploration thinking, asking questions) that determines in general terms the kind of information obtainable through that exploratory mode (e.g., manipulatory exploration may be expected to provide information about what the object can do, what can be done with it, how it operates, its weight, etc.) As we have seen, CO scores determine to what extent the individual would evidence exploratory behavior but does not determine which exploratory mode the individual would evidence. Prediction by means of CO scores is general. In a sense, this is its strength, for it makes for breadth of prediction and generality of posttraining effects, but it also leaves open the question: What determines the selection of a specific exploratory mode?

Our hypothesis was that the individual's preferences for specific meaning assignment tendencies (i.e., tendencies to process inputs in terms of specific meanings) would be involved in determining the exploratory mode. Two considerations undergird this hypothesis. First, because exploration provides information, it is likely that specific exploratory modes correspond to preferences for specific types of meaning processing. Second, because exploratory modes were found to be characteristic for individuals (Keller et al., 1987; Kreitler, Zigler, & Kreitler, 1984), they may bear the earmaks of personality traits, which in turn were shown to correspond to patterns of specific meaning-assignment tendencies (Kreitler & Kreitler, 1990a).

On the basis of former studies, we define meaning as a referent-centered pattern of meaning values, whereby referent denotes the representation of the input to which meaning is assigned (e.g., a word, an image, an object) and meaning values denote particular cognitive contents expressed verbally or nonverbally (e.g., red, dangerous, made of wood). The analysis of meaning is carried out by means of four sets of variables: (a) *Meaning Dimensions,* that characterize the contents of the meaning values in terms of general kinds of information they provide about the referent, for example, Sensory Qualities, Material, or Structure; (b) *Types of Relation,* that characterize the relations of the meaning value to the referent in terms of its immediacy or directness and include the four major classes of the attributive, comparative, exemplifying-illustrative and metaphoric-symbolic types of relation; (c) *Forms of Relation,* that characterize the relation of the meaning values to the referent in formal-logical terms, such as, positive, negative, conjunctive or disjunctive; (d) *Referent Shifts,* that characterize the sequential shifts in the referent in the course of meaning assignment, such as mentioning a modified referent (e.g., a part of the former one) or an association. Table 4 presents a list of the meaning variables.

Table 4. Major variables of the Meaning System

I. MEANING DIMENSIONS

Dim. 1. Contextual Allocation
Dim. 2. Range of inclusion (2a: Subclasses of referent; 2b: Parts of referent
Dim. 3. Function, Purpose and Role
Dim. 4. Actions and Potentialitis for Actions (4a: by referent; 4b: to/with referent)
Dim. 5. Manner of Occurence or Operation
Dim. 6. Antecedents and Causes
Dim. 7. Consequences and Results
Dim. 8. Domain of Application (8a: Referent as Subject; 8b: Referent as Object)
Dim. 9. Material
Dim. 10. Structure
Dim. 11. State and possible Changes in State
Dim. 12. Weight and Mass
Dim. 13. Size and Dimensionality
Dim. 14. Quantity and Number
Dim. 15. Locational Qualities
Dim. 16. Temporal Qualities
Dim. 17. Possessions (17a) and Belongingness (17b)
Dim. 18. Development
Dim. 19. Sensory Qualities (19a: of referent; 19b: by referent)
Dim. 20. Feelings and Emotions (20a: evoked by referent; 20b: felt by referent)
Dim. 21. Judgements and Evaluations (21a: about referent; 20b: by referent)
Dim. 22. Cognitive Qualities (22a: evoked by referent; 22b: of referent)

II. TYPES OF RELATION

TR 1. Attributive
　　1a. Qualities to substance
　　1b. Actions to agent
TR 2. Comparative
　　2a. Similarity
　　2b. Difference
　　2c. Complementariness
　　2d. Relationality
TR 3. Exemplifying-Illustrative
　　3a. Exemplifying instance
　　3b. Exemplifying situation
　　3c. Exemplifying scene
TR 4. Metaphoric-Symbolic
　　4a. Interpretation
　　4b. Conventinal metaphor
　　4c. Original metaphor
　　4d. Symbol

MODES OF MEANING

Lexical Mode   TR 1 + TR 2
Personal Mode TR 3 + TR 4

III. FORMS OF RELATIONS

FR 1. Positive
FR 2. Negative
FR 3. Mixed positive and negative
FR 4. Conjunctive
FR 5. Disjunctive
FR 6. Combined positive and negative
FR 7. Double negative
FR 8. Obligatory
FR 9. Question

IV. SHIFTS OF REFERENT

SR 1. Identical
SR 2. Opposite
SR 3. Partial
SR 4. Modified by adding a meaning value
SR 5. Previous meaning value
SR 6. Associative
SR 7. Unrelated
SR 8. Grammatical variation
SR 9. Linguistic label
SR 10. Several previous meaning values combined
SR 11. Higher level referent

Our concept of meaning is two-facted in the sense that, on the one hand, it serves for coding, characterizing, quantifying, and evaluating cognitive contents as well as the results of cognitive performance (namely, the static application of the meaning system) and, on the other hand, it can be applied as a set of interrelated strategies that explain major aspects of cognitive functioning.

The assessment of meaning is done by coding, in terms of the meaning variables, units of contents in the cognitive product, for example, a dialogue, a text, a solution to a problem, jokes, questions or the communications of meanings of words, phrases and other stimuli. The subject's meaning-assignment tendencies are assessed by the Meaning Test. This test requires the subjects to communicate to an imaginary other person the meaning (general and/or personal) of a standard set of stimuli. Coding the responses consists of assigning to each response unit four scores, one of each type of meaning variables, that is, one meaning dimension, one type of relation, and so forth. Thus, the response "blue" to the stimulus "ocean" is coded in terms of the meaning dimension Sensory Qualities, the attributive type of relation, the positive form of relation and no referent shift. Summing these scores across all response units yields the subject's meaning profile, namely, the distribution of the subject's frequencies of responses in each meaning variable. These frequencies were found to be characteristic of the individual and hence we call them meaning assignment tendencies are related to different cognitive activities, such as planning, memory, analogical thinking, or conceptualizations (Kreitler & Kreitler, 1987, 1988, 1990b). These findings demonstrate the cognitive function of meaning assignment tendencies. They also have a function in personality, because it was shown that each of over 100 common personality traits corresponds to a pattern of meaning assignment tendencies (Kreitler & Kreitler, 1990a).

In order to explore the relations of the five exploratory modes to meaning assignment tendencies, we administered to a sample of 80 first graders (40 girls, 40 boys) a brief version of the Meaning Test and five of the basic exploration tasks (all tasks except Task 4, see above), which yielded 18 variables of exploration, along the five exploratory modes (Kreitler & Kreitler, 1986). Correlations were computed between the subjects meaning profiles and their scores on the exploratory variables. In order to characterize the exploratory modes in terms of meaning variables, we focused only on those meaning variables that were correlated in the same direction with at least 33.3% of the exploration variables saturated on the exploration factor. The findings, summarized in Table 5, demonstrate first, that each exploratory mode is characterized by a unique set of meaning variables, which indicates that a child manifests preferentially that exploratory mode that matches most closely his or her characteristic meaning assignment tendencies; and second, that the set of meaning variables corresponding to the exploratory mode represents the kind of information which the exploratory mode is presumably geared to provide.

In the following, the meaning variables correlated with at least two exploration variables with loadings on the different exploratory modes are explained.

Manipulatory exploration. *Meaning dimensions:* Referent's constituent parts; Range of inclusion (referent's parts and subclasses); Referent's function, purpose and role (neg.); Referent's actions (neg.); Actions that can be done with/to the referent; Referent's manner of occurence and operation; Referent's results and consequences; Domain of application (with referent as subject or object) (neg.); Referent's state and possible changes in state; Referent's weight and mass; Referent's quantity and number; Referent's temporal qualities (neg.); Referent's sensory qualities (especially tactility, kinaesthesia, temperature, vision, acoustic); Referent's sensations; Referent's cognitive qualities (neg.); *Types of relations*: Attributive (qualities to substance); Comparative: similarity; Exemplifying-illustrative: instancre. situation (neg.); *Forms of relations*: mixed positive and negative: *Referent shifts*: grammatical variation of referent; defining the referent by combining several meaning values.

Perceptual exploration. Meaning dimensions: Referent's contextual allocation; Referent's constituent parts; Referent's actions (neg.); Actions that can be done with/to referent; Referent's antecedents and causes; Domain of application (with referent as subject); Referent's material; Referent's strucure; Referent's size and dimensionality; Referent's locational qualities; Referent's sensory qualities and sensations; Referent's evaluation (neg.); *Types of relation*: comparative: difference; *Forms of relations*: positive; number of different forms of relations; *Referent shifts*: no shift; shift to a grammatical variationof referent; shift to an associated referent.

Conceptual exploration. Meaning dimensions: Referent's constituent parts; Referent's subclasses; Referent's function ; purpose and role (neg.); Referent's manner of occurence and operation; Referent's antecedents and causes; Referent's consequences and results; Domain of application (with referent as object) (neg.); Domain of application (with referent as subject or object) (neg.); Referent's quantity and number; Referent's possessions (neg.); Referent's sensations (especially vision, acoustic and smell); total number of different meaning dimensions; *Types of relation*: total number of different types of relation; *Forms of relation*: mixed positive and negative; Conjunctive; Disjunctive; Number of different forms of relation; *Referent shifts*: shift to a previous meaning value; Shift to a referent which is a combination of previous meaning values; Shift to a modified referent; Total number of different shifts.

Exploration of the complex or ambiguous. Meaning dimensions: referent's constituent parts; Range of inclusion (referent's parts and subclasses); Domain of application (with referent as subject); Referent's senso-

---

The word 'negative' (= neg.) in paratheses denotes that the meaning variable is correlated negatively, when nothing is written after the meaning variable, the correlation is positive. In some cases it may seem that there is a repetition of the same meaning dimension: this occurs when both more specific dimension on the more general dimension were correlated significantly with the exploration variables. Only variables correlated significantly are presented (cf. Kreitler & Kreitler, 1986).

ry qualities; Referent's sensations; Referent's sensory qualities and sensations (especially internal sensations and vision); Feelings and emotions evoked by referent; Feelings and emotions experienced by referent; Referent's evalutaion; Referent's cognitive qualities; *Types of relation*: comparative: difference; Comparative: complementariness; Exemplifiying-illustrative: instance (neg.); Total number of types of relation; *Forms of relation*: mixed positive and negative; Total number of different forms of relation; *Referent shifts*: Grammatical variation of referent; Shift to an associated referent; Shift to a modified referent; Shifting to the linguistic label as referent.

Adjustive-compliant exploration. Meaning Dimensions: referent's contextual allocation; Referent's constituent parts; Referent's action (neg.), Referent's actions and actions that can be done with/to it (neg.); Domain of application (with referent as subject); Domain of application (with referent as object); Referent's state (neg.); Referent's locational qualities; Referent's belongingness; Referent's sensory qualities (especially visual); Referent's sensations; Referent's sensory qualities and sensations; Referent's evaluation (neg.); *Types of relation*: attributive (qualities to substance) (neg.); Comparative: difference (neg.); Comparative (neg.); *Forms of relation*: positive; *Referent shifts*: no shift; Grammatical variation of referent; Shift to a previous meaning value (neg.); Shift to a referent which is a combination of previous meaning values; Total number of responses (neg.).

Let us illustrate the latter conclusion. For example, high scorers on *manipulatory exploration* are concerned especially with how objects function, what can be done with them, their weight, quantity, sensory qualities but not their cognitive qualities. Further, they focus on concrete examples, dwell on the similarities between referents and tend to qualify their statements (does object manipulation demonstrate to them the futility of absolute categories?!). In contrast, high scores on *perceptual exploration* have meaning-assignment tendencies which show interest in those aspects of the referent that are likely to be revealed by means of perception, for example, sensory qualities, mainly visual and acoustic, and locational qualities. They also tend to focus on differences between referents and to stick closely to the presented referents, only occasionally deviating from them associatively. The meaning assignment tendencies of high scorers on *conceptual exploration* reveal concern with the conceptual classes the referent includes, its constituent parts, the manner in which it operates or occurs, its causes and results and other aspects, which indicate tendencies for analytical, logical, and consistent thinking, tolerance for ambiguities, extrapolations, and rich variation in contents and processes. The meaning-assignment tendencies of the high scorers on *complexity exploration* reveal concern with internal sensations, evaluations, feelings and emotions, coupled with avoidance of concrete examples, tendencies to focus on comparing referents and redefining presented inputs. These and other tendencies indicate focusing on the covert aspects of reality, including thinking and feeling, which perhaps produce more opportunities for encoutering complexities than the world of overt appearances, amenable to manipulation and observation. Finally, high scorers on *adjustive-compliant exploration* reveal concern with practical aspects of reality, for example, belongingness of things, as well as with the evaluations of others, and other tendencies indicating rigidity, low tolerance of ambiguities and conformity.

These examples suggest the possibility of interactions between the individuals' meaning assignment tendencies and their preferred exploratory mode. Thus, individuals who, say, tend to be concerned with how things function, will tend to prefer the manipulatory or conceptual exploratory modes, which are likely to satisfy their interests and in turn to further reinforce them.

As noted above, personality traits were found to correspond to patterns of meaning assignment tendencies (Kreitler & Kreitler, 1990a). Are we justified in concluding that the exploratory modes are personality traits? In order to answer this question it is necessary to examine if the patterns corresponding to the exploratory modes resemble those corresponding to personality traits in terms of five major empirically based criteria (see Table 5). At present we may give only a tentative answer to the question, because the Meaning Test administered to the children was a brief version, and the criteria for testing the resemblance of the patterns are based on data from adult subjects. Keeping these reservations in mind, we performed the test. Table 5 shows that none of the patterns corresponding to the five exploratory modes deviates from the patterns for personality traits in more than two of the five criteria. This degree of deviation is considered acceptable. Thus, the patterns of meaning variables corresponding to the five exploratory modes fall within the boundaries of variation of personality traits. The conclusion is that the five exploratory modes are personality traits. Our finding needs to be rechecked in a study using the full version of the Meaning Test and values of the five criteria based of a children's sample. Yet, even as a tentative finding, it is of importance in view of the fairly common tendency to consider curiosity as a personality trait (e.g., Ainley, 1987; Ben-Zur & Zeidner, 1988; Boyle, 1989; Camp, Dietrich, & Olson, 1985; Henderson, 1988; Joachimthaler & Lastovicka, 1984; Stoner & Spencer, 1986).

Table 5. Comparing the Patterns of Meaning Variables Corresponding to the Exploratory Modes with the Patterns Cooresponding to Personality traits

Criteria of comparison	Manip. Explor.	Perc. Explor.	Conc. Explor.	Compl. Explor.	Adj.-Com. Explor.
No. of different kinds of meaning var. in pattern [3 or 4]	3 [no dev.]	4 [no dev.]	4 [no dev.]	4 [no dev.]	4 [no dev.]
Proportions of dif. kinds of meaning var. in pattern: [meaning dim. 54.75% types of rel. 25.75% forms of rel. 5.90% ref. shifts 12.57%] Chi Square of dif.	69.56% 17.39% 4.35% 8.69% 8.32* [dev]	70.00% 5.00% 10.00% 15.00% 24.29** [dev.]	60.87% 4.35% 17.39% 17.39% 42.69** [dev.]	54.54% 18.18% 9.09% 18.18% 6.45 [no dev.]	63.64% 3.64% 4.54% 18.18% 9.96* [dev.]
Proportion of negative cor. [.38]	.35 z=.20 [no dev.]	.20 z=1.06 [no dev.]	.22 z=1.29 [no dev.]	.04 z=2.27* [dev.]	.41 z=.18 [no dev.]
No. of variables in pattern [range 7-20]	23 [no dev.]	20 [no dev.]	23 [dev.]	22 [dev.]	22 [dev]
Prop. of general to specific meaning var. in pattern [.44]	.27 z=1.00 [no dev.]	.15 z=1.60 [no dev.]	.25 z=.98 [no dev.]	.23 z=1.13 [no dev.]	.21 z=1.27 [no dev.]
Total of deviations	2	1	2	2	2

Note. The standard values of the criteria are presented in square parentheses. For a more complete presentation and illustration of the procedure for checking the similarity to patterns of personality traits see Kreitler and Kreitler (1990a, pp. 303-310). Dev. = Deviation from the pattern of personality traits. Rel. = Relation. Ref. = Reference. Cor. = Correlation. Prop. = Proportion. Var. = Variables. Dif. = Different. No. = Number. Dim. = Dimensions.
* $p < .05$. *** $p < .001$.

Cognitive Motivation and Cognitive Dynamics of Exploratory Modes

In view of the evidence that CO scores predict the direction and intensity of exploration (see *The Motivational Determinants of Exploration: The Cognitive Orientation of Curiosity*) whereas meaning variables predict the kind of exploratory mode the individual manifests (see *Cognitive Determinants of Exploration: Patterns of Meaning Variables*), it is likely that combining CO scores and meaning variables would provide for a prediction of both the level of exploration and its manner of manifestation.

Accordingly, in the last stage of our empirical work in the domain of exloration to date, we assessed, in a group of 100 (50 girls, 50 boys) first graders, the behavioral manifestations as well as the motivational and cognitive determinants of exploration. The behavioral manifestations were assessed with three exploration tasks (Tasks 1, 2, and 6, see above) that provided 10 measures (see Table 1), which were turned into factor scores assessing the three major exploratory modes—manipulatory, perceptual, and conceptual. The motivational determinants were assessed by the CO Questionnaire of Curiosity, and the cognitive ones by the Meaning Test, that provided scores for meaning variables. The latters were combined into meaning indices, one for each of the following exploratory modes—manipulatory, perceptual, and conceptual. The index represented the sum of the meaning variables in the subject's test that matched those in the pattern characteristic for the exploratory mode, whereby matching was defined as a frequency of a variable above the group's mean when the variable in the pattern has a positive sign, and a frequency below the group's mean if the variable has a negative sign. Thus, each child got seven scores: factor scores and meaning indices for manipulatory, perceptual, and conceptual exploration, and a CO score of curiosity.

Stepwise regression analyses were used for examining the relations of the predictors—CO scores and the meaning indices—with the behavioral manifestations of exploration. Table 6 shows that CO scores are invariably the first predictor entered into the prediction equation and they alone account for 46.24% to 53.29% of the variance in exploration. When the meaning indices specific for each exploratory mode are added, the percent of variance accounted for rises (by 28.8% to 36.96%) and ranges from 75.69% (for perceptual exploration) to 90.25% (for conceptual exploration).

Table 6 also presents the results of prediction when we add to the CO scorers a meaning index that is nonspecific for a given exploratory mode. The results indicate (a) that the meaning index for manipulatory exploration does not raise much the prediction of perceptual and conceptual exploration (merely by 2.76% to 4.47%), (b) that the meaning index for perceptual exploration does not raise much the prediction of conceptual exploration (merely by 2.96%) but does raise (by 12%) the prediction of manipulatory exploration, and (c) that the meaning index for conceptual exploration raises the prediction of manipulatory (by 21.84%) and perceptual exploration (by 11.52%). Do these findings suggest that the meaning indices are not specific for the exploratory modes? To our mind this is not a justified conclusion for two reasons. First, because adding the specific meaning provided always indices for the largest improvement in predicting the exploratory mode; and second, because the improvements attained by adding nonspecific meaning indices may be due to the overlap of the meaning indices (the conceptual exploration index includes 47.83% of the meaning variables in the manipulatory exploration index and 20% of those in the perceptual exploration index, whereas the perceptual exploration index includes 10.87% of those in the manipulatory exploration index and 13.04 of those in the conceptual exploration index).

Table 6. Multiple correlation coefficients based on multiple stepwise regression analyses with CO scores and meaning indices as predictors and factor scores on three exploratory modes as dependent variables.

	Manip. exploration	Percep. exploration	Concep. exploration
CO scores	.71***	.68***	.73***
CO scores + MI of manip. explorations	.89***	.70***	.76***
CO scores + MI of percep. exploration	.79***	.87***	.75***
CO scores + MI of concep. exploration	.85***	.76***	.95***

Note. CO = Cognitive orientation. MI = Meaning index representing the meaning variables correlated with the specific exploratory mode (see text and Table 5). Manip. = Manipulatory. Percept. = Perceptual. Concep. = Conceptual.
*** $p < .001$.

Main Findings, Major Conclusions, and Some Afterthoughts

1. On theoretical and empirical grounds we distinguish between the behavioral manifestations of exploration and the determinants of these behaviors in the spheres of motivation, cognition, and personality.
2. The behavioral manifestations of exploration lend themselves to a grouping in terms of five exploratory modes that are identifiable from at least the age of 5 years and persist into adulthood: manipulatory exploration, perceptual exploration, conceptual exploration, exploration of the complex and ambiguous, and adjustive-compliant exploration. Each has presumably specific eliciting stimuli, a specific mode of operation and provides for specific kinds of information. They differ in salience with age and personality.
3. The motivational determinants were examined in the framework of the CO theory and were found to be specific beliefs (about readiness to violate privacy, not being afraid of the unfamiliar, etc.), along four belief types (beliefs about goals, about norms, about the self and general beliefs). If the individual has a sufficiently high number of such beliefs of the four types, it may be concluded that he or she have a behavioral intent for exloration. The existence and strength of the intent may be assessed by the CO Questionnaire of Curiosity. Accordingly, CO scores account for the *why* of exploration.
4. The occurrence of exploration in the form of overt behavior requires implementing the behavioral intent through behavioral programs. These account for the *how* of behavior. It is likely that a specific behavioral program underlies each of the five exploratory modes.
5. The determinants for selecting one behavioral program rather than another were studied in the framework of the meaning theory. We identified specific patterns of meaning-assignment tendencies corresponding to each exploratory mode. Thus, and individual motivated for exploration (i.e., an individual with a behavioral intent for exploration) expresses this motivational tendency through the exploratory mode that corresponds most closely to the set of meaning-assignment tendencies characteristic for him or her. This set may be assessed by the Meaning Test. The preference for an exploratory mode is probably akin to a personality trait.
6. Together the motivational and cognitive determinants enable predicting to what extent and in which exploratory mode an individual would

explore. The predictors account for 76% to 90% of the variance in the exploratory behaviors.

7. Our findings may be used for diagnosing the reasons why a certain individual scores low on exploration (e.g., low motivation and in which respect, i.e., in which group or type of beliefs, or low correspondence between his/her meaning variables and any of the exploratory modes).

8. Our findings indicate the possibility of raising the level of exploration of an individual or group. This can be attained by training, whenever indicated, the CO or the meaning variables or both. CO scores have already been raised by training in the domain of exploration as well as other domains (Kreitler & Kreitler, 1982). Meaning variables have been trained successfully in other domains (e.g., Arnon & Kreitler, 1984) but not yet in the domain of exploration. It is, however, likely that the training will succeed in this domain, too. The training could proceed by strengthening the individual's most salient exploratory mode or by developing a broader range of modes.

9. It seems as if we have a solution to the problem of increasing curiosity and exploration of individuals: if we increase their motivation (CO) and set of cognitive tendencies for one or more of the exploratory modes, they will explore more. As conclusions in science go, this conclusion, too, is subject to constraints. One of the most evident ones is imposed by the environment. Exploration is conducted at least partly in the external environment and requires an exploration-friendly environment, at least in the first stages, before an individual can control the environment so as to shape it into an exploration-friendly one. Thus, in order to become curious and exploring individuals, children need an environment that is not too dangerous for exploration and can supply the answers to their questions. Such an environment is neither obvious nor highly common. But if we promote exploration in enough individuals there is hope that they may help to create an exploration-friendly environment for themselves and others.

References

Ainley, M . D. (1987). The factor structure of curiosity measures: Breadth and depth of interest curiosity styles. *Australian Journal of Psychology, 39,* 53-59.

Ajzen, I., & Fishbein, M. (1980). *Understanding attitudes and predicting social behavior.* Englewood Cliffs, NJ: Prentice Hall.

Arnon, R., & Kreitler, S. (1984). Effects of meaning training on overcoming functional fixedness. *Current Psychologicl Research and Review,* 3, 11-24.

Ben-Zur, H., & Zeidner, M. (1988). Sex differences in anxiety, curiosity, and anger: A cross-cultural study. *Sex Roles, 19*, 335-347.

Berlyne, D. E. (1966). Curiosity and exploration. *Science, 1953*, 25-33.

Boyle, G. J, (1989). Breadth-depth or state-trait curiosity? A factor analysis of state-trait curiosity and state anxiety scales. *Personality and Individual Differences, 10*, 175-183.

Cagle, M. (1985). A general abstract-concrete model of creative thinking. *Journal of Creative Behavior, 19*, 104-109.

Camp, C. J., Dietrich, M. S., & Olson, K. R. (1985). Curiosity and uncertainity in young, middle-aged and older adults. *Educational Gerontology, 11*, 401-412.

Caruso, D. A. (1989). Attachment and exploration in infancy: Research and applied issues. *Early Childhood Research Quarterly, 4*, 117-132.

Dash, J., & Dash,. S. (1982). Verbal and perceptual measures of children's curiosity. *Psychp-Lingua, 12*, 55-59.

Gibson, E. J. (1988). Exploratory behavior in the development of perceiving, acting, and the acquiring of knowledge. *Annual Review of Psychology, 39*, 1-41.

Harty, H., & Beall, D. (1985). Reactive curiosity of gifted and nongifted elementary school youngsters. *Roeper Review, 7*, 214-217

Henderson, B. B. (1984). Social support and exploration. *Child Development, 55*, 1246-1251.

Henderson, B. B. (1988). Individual differences in exploration: A replication and extension. *Journal of Genetic Psychology, 149*, 555-557.

Hutt, C. (1966). Exploration and play in children. *Symposia of the Zoological Society, 18*, 61-81.

Joachimthaler, E. A., & Lastovicka, J. L. (1984). Optimal stimulation level: Exploratory behavior models. *Journal of Consumer Research, 11*, 830-835.

Kagan, J. (1978). *Infancy: Its place in human development.* Cambridge, MA: Harvard University Press.

Keller, H., Schölmerich, A., Miranda, D., & Gauda G. (1987). Exploratory behavior development in the first four years. In D. Goerlitz & J. F. Wohlwill (Eds.), *Curiosity, imagination, and play* p. 127-150). Hillsdale, NJ: Erlbaum.

Keller, H., & Voss, H.-G. (1976). *Neugier und Exploration: Theorien und Ergebnisse.* Stuttgart: Kohlhammer.

Kreitler, S., & Kreitler, H. (1974).The role of the experiment in science education. *Instructional Science, 3*, 75-88.

Kreitler, H., & Kreitler, S. (1976a). *Cognitive orientation and behavior.* New York: Springer Publishing.

Kreitler, H., & Kreitler, S. (1976b). CO of Curiosity and probability learning. In H. Kreitler & S. Kreitler, *Cognitive orientation and behavior* (pp.228-235). New York: Springer Publishing.

Kreitler, H., & Kreitler, S. (1981). Die kognitive Determinante des Neugierverhaltens. In H.-G. Voss & H. Keller (Eds.), *Neugierforschung: Grundlagen, Theorien, Anwendungen* (pp. 144-174). Weinheim & Basel: Beltz Verlag.

Kreitler, H., & Kreitler, S. (1982). The theory of cognitive orientation:Widening the scope of behavior prediction. In B. Maher & W.B. Maher(Eds.), *Progress in experimental personality research* (Vol. 11, pp. 101-169). New York: Academic Press.

Kreitler, S., & Kreitler, H. (1986). Types of curiosity behaviors and their cognitive determinants. *Archives of Psychology, 138*, 233-251.

Kreitler, S., & Kreitler, H. (1987). The motivational and cognitive determinants of in-

dividual planning. *Genetic, Social and General Psychology Monographs, 113,* 81-107.
Kreitler, S. & Kreitler, H. (1988). Horizontal decalage: A problem and its resolution. *Cognitive Development, 4,* 89-119.
Kreitler, S. & Kreitler, H. (1990a). *The cognitive foundations of personality traits.* New York: Plenum.
Kreitler, H. & Kreitler, S. (1990b). Psychosemantic foundations of creativity. In K. J. Gilhooly, M. Keane, R. Logie, & G.Erdos (Eds.), *Lines of thought: Reflections on the psychology of thinking* (pp. Vol. 1, pp.15-28). Chichester, England: Wiley.
Kreitler, S., Kreitler, H., & Boas, T. (1976). Changing curiosity. In H. Kreitler & S. Kreitler, *Cognitive orientation and behavior* (pp. 238-253). New York: Springer Publishing.
Kreitler, S., Zigler, E., & Kreitler, H. (1974). The complexity of complexity. *Human Development,17,* 54-73.
Kreitler, S., Zigler, E., & Kreitler, H. (1975). The nature of curiosity in children. *Journal of School Psychology, 13,* 185-200.
Kreitler, S., Zigler, E., & Kreitler, H. (1984). Curiosity and demographic factors as determinants of children's probability learning strategies. *Journal of Genetic Psychology, 145,* 61-75.
Maw, W. H. & Maw, E. W. (1965). Personal and social variables differentiating children with high and low curiosity. (CRP 511) Washington, D.C.: United States Office of Education.
Moch, M. (1987). Asking questions: An expression of epistemological curiosity in children. In D. Goerlity & J. F. Wohlwill (Eds.), *Curiosity, imagination and play* (pp. 199-211). Hillsdale, NJ: Erlbaum.
Olson, K. R., & Camp, C. J. (1984). Factor analysis of curiosity measures in adults. *Psychological Reports, 54,* 491-497.
Olson, K. R., Camp, C. J., & Fuller, D. (1984). Curiosity and the need for cognition. *Psychological Reports, 54,* 71-74.
Rodrigue, J. R., Olson, K., R., & Markley, R. P. (1987). Induced mood and curiosity. *Cognitive Therapy and Research, 11,* 101-106.
Schneider, K., Moch, M., Sandfort, R., Auerswald, M. & Walther-Weckman, K. (1983). Exploring a novel object by preschool children: A sequential analysis of perceptual, manipulatory and verbal exploration. *International Journal of Behavioral Development, 6,* 477-496.
Singer, D. G., & Singer, J. L. (1990). *The house of make-believe: Play and the developing imagination.* Cambridge, MA: Harvard University Press.
Stoner, S. B., & Spencer, W. B. (1986). Age and sex differences on the state-trait personality inventory. *Psychological Reports, 59,* 1315-1319.
Tzuriel, D., & Klein, P S. (1983). Learning skills and types of temperament as discriminants between intrinsically and extrinsically motivated children. *Psychological Reports, 53,* 59-69.
Vandenberg, B. (1984). Developmental features of exploration. *Developmental Psychology, 20,* 38.
Voss, H.-G., & Keller, H. (1983). *Curiosity and exploration: Theories and results.* New York: Academic Press.

CHAPTER V

Applied Perspectives

CHAPTER **V.15**

Computer Systems as Exploratory Environments

Siegfried Greif

Exploratory behavior and curiosity are old subjects of experimental psychology (cf. Nunnally, 1981). Scientists designed artificial labyrinths for animals and studied incidental learning and the functions of exploratory behavior of "curious beings" (Lorenz, 1977). The modern labyrinths for humans are computer and software systems. Like traditional labyrinths, these systems are artificially designed environments. They elicit the curiosity of many people. When we enter these labyrinths, we often find ourselves at crossways where we do not know where to go and sometimes feel helpless and unable to find the way out of a problem situation. Because the redesign of software systems is easy and we can use the system for an automatic and precise protocol of all movements, these labyrinths are also ideal tools for research on exploratory behavior. Therefore, theory on human exploratory behavior and curiosity may profit from research and practical experience in this field.

There are also many practical reasons for undertaking research on exploratory behavior in computerized environments. Behavior observation and transcripts of learners thinking aloud while trying to use software systems have revealed seemingly chaotic trial and error exploratory processes (Carroll & Mack, 1983). For some researchers in the field of human-computer interaction (HCI), it seemed to be a surprising finding that professional experts also show an error-prone exploratory style of learning when utilizing software systems. The application of

---

I gratefully acknowledge the editorial help of Michael Grunwald, Miss Boschatzky and Bruce Henderson on an earlier version of this paper.

exploratory learning and the design of exploratory environments (Carroll & Mack, 1983), exploratory learning by errors (Greif, 1989, 1991) and derived concepts such as error-management training (Frese et al., 1991) have become popular approaches in HCI. Greif and Keller (1990) show how these approaches relate to general theory and research on human exploration and curiosity (Berlyne, 1960; Schölmerich, 1990; Voss & Keller, 1983, 1986). The chapter gives an overview of developments in this rather new field. This contribution concentrates on theories of complexity and on exploratory activities when coping with computer systems, especially in learning situations typically encountered with office software systems. Research perspectives on exploratory activities in everyday error situations with modern office software systems will be discussed in the last part of this chapter.

Simplicity in Human-Computer Interaction

The design of simple and easy-to-use computer and software systems is a basic goal which has been seldom questioned. In the introduction to their influential book on the psychology of HCI, Card, Moran, and Newell (1983, p. 1) state that "a scientific psychology should help us in arranging this interface so it is easy, efficient, error-free - even enjoyable." According to their simulation models of human learning and task performance, high complexity results in longer learning and performance time, and increases the risk of errors. Therefore, the basic design rule is to reduce complexity.

Roberts and Moran (1983) derive the assumption from their empirical data that there is a high positive correlation between ease of learning and performance time, and ease of learning and errors. The development of formal models and research on the complexity problem seem to support this perspective in HCI (Chapanis, 1991; Kieras & Polson, 1985). Research on knowledge acquisition and learning curves (cf. the review of Bösser, 1987) also favors simplicity.

Chapanis (1991), in his reanalysis of multiple easiness scores, has found very high correlations between ease of learning and performance. He admits that such correlations (where the same subjects are tested on different systems and tasks) may result from positive transfer of training effects (i.e., subjects with higher knowledge can perform better with easy systems). But his practical conclusion does not consider these methodological problems. After evaluation of several studies correlating multiple easiness scores with performance, he states that the "accumulated weight of it is compelling enough to allow me to conclude confidently that, in general, computers, computer programs, and manuals that are well designed will be easier for people to learn to use, allow users to do more work with fewer errors, and will be better liked than those that are not well designed" (p. 367).

Exploratory Environments and Minimalist Design

Carroll and Mack (1983) prompted novices to think aloud as they tried to learn to use a word processor. Video and sound recordings were made throughout the whole session of 4 half-days learning. As the transcripts show, the seemingly simple task of finding out how to delete a wrong letter may be a difficult problem for a computer novice. It is solved only after a long and straining zig-zag trial and error path through the computer labyrinth. The learners " in this situation see many things going on, but they do not know which of these occurrences are relevant to their immediate concerns. Indeed, they do not know if their current concerns are the appropiate concerns for them to have. The learner reads something in the manual, sees something on the display unit and must try to connect the two - to integrate, to interpret. ... Whatever initial confusions get into such a process, it is easy to see that they are at the mercy of an at least partially negative feedback loop: Things quite often get worse before they get better." (Carroll & Mack, 1983, p. 262)

Carroll and Mack (1983) conclude from their vivid observations that learning environments for computer novices should be "exploratory environments" supporting active exploratory learning processes. Basic properties of exploratory environments in their opinion are system simplicity, clarity, and reduced risks of errors. In addition, the learning setting should provide the learners with a sequence of manageable learning tasks. Carroll (1985) proposes a *Minimalist Design* solution to reduce the complexity of the system environment and the learning handbook:

1. Instead of thick handbooks and learning manuals, he favors *Minimal Manuals*. These manuals should be as short as possible (not more than about 45 pages), task- or action-oriented (describing the necessary performance steps for typical practical tasks), give error recovery information (how to manage typical error situations), and give the learner the initiative to actively fill in the missing information.

2. Lower system complexity is accomplished by *Training Wheels* where complex system functions are temporarily blocked (especially those which are risky for beginners). The learner is exposed to the complete menu, but if she uses a blocked function, she gets a message like "this function is not available in the Training Wheels System." Menu choices can be tried but the novice is "protected" from the risks of errors in the exploratory learning process.

The empirical evaluation of training novices following Minimalist Design seems to support this approach in comparison to traditional learning manuals and complex software systems. Carroll (1985) concludes that in comparison to other manuals Minimal Manuals are quicker and more effective for learners. Using the Training Wheels system, learners get started faster, spend less time on errors, and perform better in tests on system concepts.

The "Simple is always Best" Hypothesis

The mainstream of computer systems design and learning in HCI favors a "simple is always best" hypothesis. According to this hypothesis, simple systems are always the best with respect to all relevant performance criteria. They are universal (for all systems and user groups), easier and quicker to learn, efficient in their use and reduce the risk of errors (simplicity, easiness of learning, and good design are treated as near synonyms). Even the design of exploratory computer environments for supporting exploratory learning processes is based on a simplification concept (Minimalist Design). In the following sections I will try to modify the "simple is always best" hypothesis to a more refined "simple is sometimes good" hypothesis. Theories on exploration and research outside HCI can be applied to develop a more sophisticated hypothesis on the role of simplicity and complexity in computer environments.

Complexity and Exploration Theory

Cognitve exploration theories treat the problem of environmental simplicity or complexity in the "complexity hypothesis" (cf. Hunt, 1965) and in the "discrepancy hypothesis" as formulated by Kagan (1972). According to the complexity hypothesis, infants will attend longest to stimuli of moderate complexity. Attention is shorter for stimuli which are too complex or too simple. As they grow older, children look at complex stimuli longer. According to the discrepancy hypothesis, stimuli that are moderately discrepant from schemata that the subject has already encountered are likely to attract and maintain attention. Higher discrepancy elicits anxiety and avoidance behavior. Lower discrepancy results in feelings of boredom.

If we relate these hypotheses to knowledge acquisition processes in simple and complex computer environments and tasks, we would generally assume that moderately novel and complex situations stimulate active learning, as Carroll and Mack (1983) call it. According to this theory, simple computer environments or Minimalist Design may result in feelings of boredom and monotony. According to the complexity and discrepancy hypothesis, a higher but moderate level of complexity and novelty would be preferable.

Schölmerich (1990) summarizes the literature on the exploratory process of information intake. Exploratory information intake is a time-related process. Through exploration, an inner representation of the exploration object is developed. Berlyne (1960) distinguishes between diversive exploration and specific exploration. Curious playing around with the computer system without a clear task would be classified as diversive exploration. Exploratory strategies coping with problem situations aimed at acquiring new knowledge in problem situations which cannot be integrated into existing schemata without further information belong to the so called specific exploratory activities. HCI research is concerned with specific exploration and knowledge acquisition processes involved in operating new computer systems and performing new tasks with problem solving activities. Our chapter concentrates on such specific exploration processes. (We will not treat here successful task performance in computer environments when the performance goals are given. This is a problem which is explained by achievement motivation theories, cf. Atkinson & Birch, 1978; Heckhausen, 1980.)

There are individual differences in curiosity and exploratory behavior and the degree of novelty and complexity which stimulate specific exploratory knowledge acquisition (cf. Keller & Voss, 1976). If *Maximalist Design* (Greif, 1991) means that the individual learner is confronted with an environment of maximum *manageable* novelty and complexity, we should favor Maximalist to a Minimalist Design because it expresses the essential theoretical assumption of the complexity and discrepancy hypothesis. But if we assume that Carroll and Mack (1983) and Carroll (1985) are describing environments which, following from their vivid observations, obviously are much too complex for novices, a strategy lowering complexity for this situation may in fact be essentially the same as a Maximalist Design solution. It is, therefore, more the theoretical focus and interpretation that differs. A closer look at regulatory processes involved in coping with configurations of low discrepancy and simple or new and complex computer environments may still be necessary to get a better understanding of the dynamics of exploratory behavior in computer environments.

Bischof (1975; 1989) extended Berlyne's model of an optimal level of complexity (s. Berlyne, 1967) and assumes personal set points with dynamically changing levels, depending also on the phase of ontogenetic development of the individual. We may try to apply the basic model of an individual level of complexity and discuss the different reactions to simple and complex computer tasks, the resulting coping strategies and motivational situations.

If a simple computer task and environment elicits an inner excitation which remains below a person's competence and set point we would assume that the subject will engage in curiosity-motivated behavior (Berlyne's diversive exploration) in order to raise its excitation to the prefered level. Since the relationship between task complexity and arousal follows a curvilinear u-function and should be low in a balance situation and high in either an arousal deficit or in an overflow situation, inventive activities may observed. They can be interpreted as coping strategies in order to reach a balance where the resulting arousal equals the individual's set point. Therefore, we would assume that an individual confronted with a computer problem, which is not exciting enough, will try to invent new and challenging experiences, raising the novelty or complexity regulating the arousal level. But in a situation (e.g., a complex error situation) where it is confronted with a severe problem which stimulates an arousal level beyond its optimal set point it will engage in inventive activities to reduce the imbalance (e.g., asking an expert for help).

If we follow Berlyne's and Bischof's assumptions and apply them to the design of exploratory computer environments, the relationships between novelty and complexity and exploratory activities follow complex and dynamic functions, and elicit changing, individually different coping strategies. Depending on a person's knowledge and individual set point, and the problem situation, the resulting exploratory activity may serve the function of trying to either raise or lower the novelty and complexity level. Following Bishof (1989), we may assume that especially adolescents, who in the course of their ontogenetic development show a high spirit of enterprize, prefer novel and complex computer problems. (This may explain the development of so called computer kids with an extremely high motivation and competence.) An optimal exploratory environment should, therefore, allow the individual to choose between simplification *and* complication or newness to compensate for novelty and complexity levels which are either too low or too high in a given state or phase of development.

The practical design of optimal exploratory environments is a complex and dynamic problem in itself. Below, I shall describe a prototype software tool which might provide a possible solution to the problem of dynamic individual adaptations of complexity. I shall also give an overview of our empirical results as related to the assumptions of exploration theory. Before doing this, however, we should try to define what is meant by complexity.

Definition of the Complexity of a Task Situation

The complexity of the computer labyrinth results from complex hardware and software configurations interacting with the task demands and input behavicr of the user. The paths through the labyrinth may be very simple, predictable, and safe for routine tasks as long as we perform with precise, error-free input behavior and use our well-known, standard configurations. But if the task is new, if the system configuration is changed only a little bit (by a colleague using our computer), or if we vary our input behavior (pressing one wrong key at the wrong moment), the complexity of the whole environment may increase dramatically. The *overall complexity of the task situation* therefore is composed of the following four *complexity components*:

1. task,
2. hard, and software configuration,
3. input, and
4. interactions.

By performing a task efficiently (without detours and errors), we reduce the task complexity. If the task is done, there is no complexity left. But a minor input error or change of any other of the complexity components may increase the complexity and novelty of the whole task situation substantially. Because unpredicted increases of complexity and novelty are always possible in computer task performance, we should not treat complexity like a stable level of difficulty in psychological test theory. A situation which is simple and routine for an expert may be much too complex for novices. But even professional users can be confronted with new and complex situations if the configuration of systems has been changed without their noticing.

How can we define the state task complexity independent of expert knowledge and in a way which is appropriate enough to mirror the dynamic changes of complexity levels? In HCI research we can assess the state of the configuration and the complete individual input protocol (by logfiles). Therefore, after a task has been performed, we are able to repeat and simulate the whole input process. When we confront the subject with the simulation of her or his own input behavior or step-by-step input performance of other users, the subject - depending on the individual expert knowledge - is able to continue the task at any step of the simulation where we have interrupted the process. In the instruction, we simply demonstrate the preliminary input process, ask the subject to perform the remainder of the task from this point and find a solution with a minimum number of steps.

We may apply the simulation and interruption method of task analysis to an assessment of the task complexity for any interruption state. The state complexity of a task for an individual in the field of HCI now may be *defined* by the *number of steps or commands the individual needs to solve the task, beginning at the state where we interrupt the simulation.*

The individual number of steps necessary to complete the task seems to be a sensitive operationalization of the construct of individually different task state complexity. It is possible to assess the complexity at any state in the process and to compare different paths of action (with changing input behavior). For an individual who is not able to solve the problem within a reasonable number of steps (or gives up), the complexity is extremely high. If we confront a computer expert with the same state, she/he will be able to solve the task with a lower (maybe the lowest possible) number of steps. We also may change the configuration or simulate a difficult error situation (without telling the user) and ask the person to solve the resulting problem. In doing so, we may simulate and assess individual task complexity changes in special situations. (For differences between complexity and complicatedness see Frese, 1987.)

Design of Systems with adaptable Complexity

After our discussion of the dynamic relationship between complexity and exploration in the light of Berlyne's (1967) and Bischof's (1989) cybernetic model and our definition of state task complexity we are in a position to describe the implications for the design of exploratory computer environments more precisely. We have to give up the "simple is always best" hypothesis and the idea of easy design solutions which can be generalized for any person (irrespective of her or his knowledge) or problem situation. Depending on her/his knowledge, individual set point, and the problem situation, the individual should be able to either raise or lower the state complexity level when learning to manipulate the overall state complexity while performing tasks with computer and software systems.

The individual who is allowed to decide on the complexity and novelty of the learning tasks is given a simple but efficient way of controlling the overall learning task complexity. Because task complexity is merely one component of the overall complexity (see above), control is not complete. Another complexity component, which is alterable by means of special system design and which can be adapted by the individual, is the complexity of the software tool. If the individual controls the complexity of both the task and the tool, the resulting overall control of complexity may be very powerful. However, control always remains incomplete because unpredicted interactions between the complexity components can increase the novelty and complexity of the problem situation. Therefore, problem solvers and learners should be able to ask experts for concrete support.

The individual novelty of a problem situation may result from high complexity and low knowledge. The solution to a new problem may, however, sometimes be simple even for novices. Therefore, we should distinguish carefully between the novelty and the complexity of a problem, even if, according to theory, subjects show similar reactions to both variables. The measurement of the information intake of novel information is a difficult problem (cf. Schölmerich, 1990, for a methodologically interesting approach), a discussion of which we have excluded here. But we allow learners to determine by themselves which learning tasks they want to try and to perform at their own rate. This can be considered as an indirect design solution of individual control of the degree of novelty (cf. Greif & Keller, 1990). In a joint research and development project[1] Gediga (1989) has programmed a prototype office-software tool (known as the "individual System") and has been able to show that it is possible to design a system which allows the user to adapt the complexity level of the system deliberately.

The "Individual System"

The research prototype "individual System (iS)" is a multifunctional system for typical word processor, data-base, spreadsheet and business-graphic tasks, and a modifiable help-information system. It is programmed in C (with professional quality control). It is compatible with all standard personal computers and DOS systems. The special structural feature of the system (Gediga, 1989) is a menu manager controlling the different menus and individual adaptations. For laboratory experiments we use a version with modifiable structures and a logfile program for task analysis and simulation purposes.

At the lowest level of complexity, the features of the word processor system resemble those of an electronic typewriter, where the files are saved automatically (with serial numbers for the day). Knowledge of how to create file names, how to delete files or change names, is not necessary at this low level. For elementary file-managing tasks Level 2 competence has to be acquired and used. Level 3 gives a complete menu system (with task-orientated pulldown menus) without altering the file managing menu. This is differentiated at Level 4, including safe procedures for formating disks, creating backup files, copies of files, and delete functions. Level 5 is the complete professional system with sets of shortcut commands, modifiable Alt-key macros, and the refined use of installation versions for special adaptations. (We presented this version at the Orgatech, Köln, October 1988. We exhibited the nine-level multifunctional version of the prototype at the CeBIT, Hannover, March 1989.)

At each level of complexity it is possible to perform complete tasks such as writing, correcting, saving, and printing of files. For each level we have developed Minimal Manuals (cf. Carroll, 1985, see above) and

---

[1]The project "Multifunctional Office Software and Qualification" began in 1987 and ended in 1990. It was funded by the German Minister of Research and Technology, ("Work and Technology" program).

training curricula, applying a concept of exploratory learning by errors (Greif, 1989). The user can always decide to change the complexity level of the system using a menu option in the start system. Because the prototype software design embodies the idea of a stepwise and "natural" development from concrete thinking and direct manipulation toward actions which also integrate higher-order abstract thinking levels and flexible operations (cf. Piaget, 1969), we also call it a "genetically growing system" (Greif, 1989).

Natural human exploratory learning should be supported by a gradual beginning with the mastering of simple (but complete) learning tasks. The efficient performance of complex tasks often demands complex systems with abbreviated commands or special menus for special tasks or higher levels of thinking. Software systems with a simple structure and a small number of functions or too few ways of combining them often force the professional users to apply long path detours to perform the task. In other words, the state complexity of the overall tasks as defined above for experts is higher in this case than in systems with function-key- or other command-inputs which may be risky for novices but abbreviate the input for experts (cf. the concept of "tool functionality", Greif & Hamborg, 1991; or Wohlwill's concept of "affordance" in exploration research, 1984).

Empirical Research

According to exploration theory, computer novices who can deliberately choose the adequate level of system and task complexity should be able to balance their activation and inner arousal level (Bischof, 1989) to an individually preferred level. As a consequence, we would assume that in doing so they can acquire knowledge on complex or whole tasks more efficiently.

We evaluated this hypothesis by means of four experimental studies conducted in training courses for novices learning to use different word processor software systems and multifunctional office packages. All studies were conducted with typists (obtained from local newspaper advertisement) who had no previous experience with computer systems. In the last three studies, we employed professional trainers to allow a comparison of learning results. We applied experimental research designs (two to four conditions) and chance or matched-pair selection of subjects. Because the costs of trainers, staff, and technical equipment were high, we were forced to use small experimental groups (6-8 subjects). This somewhat risky design meant that only substantial differences had a chance of producing significant (p<.05) differences. In the following summary only the major (significant) results are mentioned.

Müller (1989) was the first to test the word processor subsystem of our prototype genetically growing system. One group learned to use the word processor with a stepwise introduction of the five complexity levels. The second group had the same learning tasks, but had the complete, highest-level version of the word processor from the beginning. Differing from our other studies mentioned below, Müller tested our first genetically growing system version with a mixed group of novices and students with prior computer knowledge. The subjects were not allowed to learn at their own pace nor to choose between complexity levels, as is the standard in our later experiments. Therefore, they were not permitted to adapt their activation level and learning speed to their individual preferences. In the group with the stepwise enlargement of complexity, the students with prior computer knowledge had to wait until the slower students performed the learning tasks.

Applying Berlyne's (1960, 1967) and Bischof's (1989) theoretical assumptions, we would assume, that in this setting, the students with higher background knowledge will not profit from the lower complexity levels and that they will react with coping strategies to increase their arousal level. As predicted, we observed several subjects in this group who spontaneously complained of boredom at the simple levels of the word processor. As expected, the learning results of this group were also lower than those of the group who had been allowed to learn with the complex system from the beginning. Subjects who had been classified as comparatively anxious on the basis of questionnaire ratings profited significantly in their performance from the stepwise growth in complexity.

A comparison of the test performance of the stepwise-growth-in-complexity group (novices and experts) with the high complexity group only shows performance gains for a divergent exploration task. Here, the subjects were instructed to explore a new function not included in the training, and had to find as many different ways as possible of performing the task. A subgroup analysis shows that, as predicted by exploration theory, subjects with previous computer knowledge are especially stimulated by the exploratory learning-by-errors training and make more errors. If they use the complex tool they explore and learn even more in the learning phase. These results are consistent with Berlyne's and

Bischof's theoretical assumptions. Depending on the relevant background knowledge, the same software system may be too complex for some learners and optimal for other subjects.

In our second experiment, Gediga, Lohmann, Monecke, and Greif (1989) tested an individually adaptable system for the first time with real novices (mostly typists from the region) and our prototype genetically growing system, Minimal Manuals and a five-level word processor subsystem of our multifunctional system. Our comparison group was trained by a highly recommended professional trainer, who also used a word processor system in a complicated multifunctional system (OPEN ACCESS, according to a field study, the multifunctional system which was applied by most regional firms at that time).

The results were very impressive. Neither group showed differences in elementary subtasks (e.g., correcting letters in a text) but did show remarkable differences in complex standard task performance like loading, correcting, and printing a text. In the group with the professional trainer, only 29% were able to do this task without external help at the end of a 2 1/2 day course. The subjects complained about the complexity of the system. In comparison, most of the subjects in the self-organized learning condition, who were able to set the complexity level of the system individually, were able to do this task after the first day, 88% after the second and 100% after the third day.

In a follow-up comparison, after 6 weeks, these differences became even more pronounced (a critical drop out in the small groups does not allow the application of statistical tests). A very impressive result was that the genetically growing systems group, confronted with the difficult task of learning to use the complex word processor subsystem of OPEN ACCESS was able to do this on its own in only two hours with better results in the file correction and printing task than the professional trainer group after 2 1/2 days of training. We were impressed how actively they explored the system and that the complexity obviously did not bother them.

In our third study, Gediga, Janikowski, Lemm, Monecke and Pezalla (1989) and Sauvageod (1990) designed and evaluated the complete multifunctional individual System (the nine-level version, Gediga, 1989). We developed self-organized training concepts (Minimal Manuals for all levels) both for our system and for OPEN ACCESS (the standard system with a similar exploratory-learning setting and a stepwise growth in complexity over the learning sessions). Once again we used a professional trainer for a comparison of our training effectiveness. The subjects were real novices from the region and the courses lasted one week. The results show that once again the genetically growing systems group nearly reached a 100% success rate in our standard text correcting and printing tasks and also showed better performance on some other tasks, e.g., a database task. The professional-trainer group had significantly lower values in these tasks but not in elementary word processing and complex calculation tasks. The results of the OPEN ACCESS group trained using our concept lay between both groups or near the professional trainer group.

The fourth study (Lohmann, in prep.) was designed to repeat the previous study and to test a better design of our training program for the somewhat abstract calculation module. The results show the same stable test performance values as recorded in the previous study in the genetically growing systems group and the OPEN ACCESS group trained under our concept. As intended, performance on the calculation tasks in the genetically growing group were much better.

The results for the genetically growing group are of particular interest as the training program was badly disrupted on the second day when one of the participants had a severe epileptic fit, which came as a shock to all present. The program was interrupted in order to take care of him and bring him to the local hospital and also to allow the other participants to discuss the incident and calm down. Because his physician recommended the participation and the client wanted it, he rejoined the group on the following day and managed to complete the course successfully. However, the effect of the incident on the whole group can be seen clearly in the learning curves, high decline steeply directly after the incident. It was not until the fourth day that the group's performance reached the high levels expected.

The performance results for the group under the professional trainer also differed from our typical results of groups of external trainers. The professional trainer reported that she concentrated more heavily on the training of the complex standard tasks (e.g., the correction and printing tasks). Consequently, the results showed performance values in the same range as the genetically growing group. But at the same time, the subjects performed significantly lower in several other test tasks. The best overall performance values in this experiment were reached in the OPEN ACCESS group with the self-organized training concept.

As the results of the four studies show, we have to be cautious with premature generalizations. We should not assume that our approach is the single best way to reach high learning performance on whatever task we may apply for testing. However, the different approaches result in differences which seem to be theoretically plausible. Even the performance response of the subjects following the observed fit of epilepsy is consistent with our theoretical assumptions. The research results show dynamically changing individual reactions to the complexity and novelty of the tasks. Self-determination of the learning task and tool complexity seems to be a powerful approach optimizing individual control over the complexity of computer tasks. The most important result from the perspective of psychology may be that the results show that the "simple is always best" hypothesis cannot explain the differences in the results. Reducing complexity may be disadvantageous for some people and some situations. In contrast to professionals, novices with low computer knowledge seem to profit from exploratory-learning task and tool environments with individual control of the task and tool complexity.

Colleagues have applied our approach to other standard software-training courses and also have used modern computer-based training technology. Lately, together with practioners, we developed and applied training courses with a genetically growing WINWORD system (using the macro-facilities of the system) and Dietlind Ehlers, a design student from Bielefeld, produced a manual for it. In comparison with standard programmed tutorials in these courses, the subjects are not guided by immediate feedback but have to perform a set of exploratory learning tasks. They are instructed to use heuristic problem solving rules (Skell, 1980) and to check the resulting solution themselves.

Frese et al. (1991) developed a similar error training concept, also trying to reduce stress in the error situations and training error-coping heuristics. They evaluated an error-training in comparison with an error-avoiding strategy (cf. also Lohmann and Mangel, 1988). Their results also support the hypothesis that, in comparison to avoiding errors, an error-training approach in the field of HCI results in greater knowledge. But they are not trying to interpret these results in terms of exploration theory and their subjects had no control over the complexity of the tasks and tool during the learning process.

As a growing body of research research in this application field indicates, exploratory or discovery learning and an adaptable level of tool novelty and complexity may be relevant for the design of exploratory learning environments supporting successful acquisition of new knowledge of complex tools and tasks. But there are also several other design criteria which subsequent research on exploration and discovery learning must take into consideration. Relevant criteria for the design of

exploratory environments which should not be overlooked in future research and development are (see Greif & Keller, 1990):
1. aesthetically pleasing environments,
2. activation of different channels of information input (sensory, auditory, tactile, etc.),
3. sufficient time, and
4. self-determination of the exploration process.

Research Perspectives: Exploratory Activities in Error Situations

It seems impossible to design computer and software systems which are completely reliable and reduce the risk of errors to zero. New releases of programming tools or application software systems (especially in the so-called DOS-family) have often increased the risk of errors instead of reducing it. Professionals often mistrust the marketing propaganda of software producers selling simple, easy-to-use, reliable systems. It may seem paradoxical, but the goal of designing simple and reliable software tools for multiple tasks and diverse user groups has resulted in the development of extremely complex programs. By following the "simple is always best" hypothesis, the risk of complex error situations with these systems has become higher than before. No software producer can test all possible hardware and software configuration states and possible inputs to their complex systems.

A problem for intermittent users or people with little background knowledge in computer and error management is that the complexity of the configuration state of the hardware and software is hidden from the user. In an error situation the user is confronted with an unpredicted problem and a sudden increase of the complexity of the problem (in terms of necessary problem solving steps) without the necessary and comprehensible information about the complex state and risky procedures required to solve the problem situation.

The modern system, with its hidden complexity, may work perfectly if the user applies standard procedures and performs as precisely as a robot. But humans make errors. Unobserved minor action slips (like pressing a neighboring function key) may result in completely new problem situations, which even for professionals are difficult to diagnose. Users who do not have the the sufficient diagnostic, tool-specific procedural knowledge required for handling complex error problems react completely helpless, and need expert help to find a safe way out of the computer labyrinth.

New types of horror stories depicting the destruction of important files or whole directories and disks after system or user errors (and insufficient safety routines) circulate in our modern computerized world and a new type of fear of data destruction seems to be growing even among professional users. Psychological error research (cf. Hacker, 1986; Hoyos, 1980; Reason, 1990; Wehner & Stadler, 1989; Zimolong, 1990) has, therefore, become a subject of general theoretical and practical interest. But again, the major strategy is to design reliable systems which hide the complexity from the user and try to reduce the risks by programmed procedures which compensate for human errors, without helping the user to understand what has happened, giving him low control over the error management process (Frese, Irmer, & Prümper, 1991).

In the field of office software, Zapf, Lang, and Wittman (1991) have shown that typical users of WORD 4.0 in industrial organizations have to cope with about three errors per hour. The study shows the limitations of the system identifying user errors. Only in 12% of the situations did the system detect the errors. Two-thirds of the errors had to be found by the users themselves. The most frequent types of errors observed were lack of knowledge, wrong reasoning, wrong movements (slips), or habitual errors. The mean coping times for the first two types of errors were 225 and 81 seconds. In comparison it is a rather quick process to correct wrong movements (19 sec) or habitual errors (24 sec). From their results, they conclude that errors are unavoidable. It is not so much the avoidance of errors through system design which should be supported, but rather the provision of resources and the development of competences for error management (Frese & Brodbeck, 1989). Wehner and Stadler (1989) demand the design of "error friendly" systems or work environments, where the error does not result in irreversible negative consequences and allows us to learn from our errors. They revert to the gestalt theory of Köhler (1917) and Duncker (1935) and the position that there are "good errors" which help us to develop the necessary insight into problem solving.

Chapanis (1991) has reanalyzed several easiness measures for text editors and showed that the average error handling time correlates very highly (.72 to .78) with the ease of learning or the necessary learning time for novices. As cited above, Chapanis concluded that it is possible to design systems which are easy to learn and to use, and reduce error handling time at the same time. But we should always be careful not to draw premature causal inferences from correlations. In a recent study (Krieger, 1993) frequency of errors, looking at the help key, or asking an expert for help on different tasks together are correlated with expert knowledge. As our results show, these correlations may be interpreted as a trivial result, due to a correlation of background knowledge and the quality of task performance, and the transfer of knowledge between different systems and procedures. The typical correlations found between error handling times and easiness scores, therefore, may be simply explained by the relation between knowledge and resulting performance. Correlations between easiness scores and errors do not prove the validity of the "simple is always best" hypothesis. Experimental research is necessary if we want to discover the complex and dynamic causal interdependencies. In several research groups, experimental research is being conducted which, in the future, may help to understand the psychological processes involved in coping with computer systems as exploratory environments.

References

Atkinson, J. W., & Birch, D. (1978). *Introduction to motivation* (2nd ed.). New York: Van Nostrand.
Berlyne, D. E. (1960). *Conflict, arousal, and curiosity.* New York: McGraw Hill.
Berlyne, D. E. (1967). Arousal and reinforcement. In D. E. Levine (Ed.), *Nebraska symposium on motivation* (pp. 1-110). Lincoln, NB: University of Nebraska Press.
Bischof, N. (1975). A systems approach towards the functional connections of attachment and fear. *Child Development*, 46, 801-817.
Bischof, N. (1989). *Das Rätsel Ödipus.* München: Piper.
Bösser, T. (1987). *Learning in man-computer interaction.* New York: Springer.
Card, S.K., Moran, T.P., & Newell, A. (1983). *The psychology of human-computer interaction.* Hillsdale, N.J.: Erlbaum.
Carroll, J. M. (1985). Minimals design for the active user. In B. Shackel (Ed.), *Human-computer interaction, INTERACT '84* (pp. 39-44). Amsterdam: North-Holland.
Carroll, J. M., & Mack, R. L. (1983). Actively learning to use a word processor. In W. E. Cooper (Ed.), *Cognitive aspects of skilled typewriting* (pp. 259-282). Berlin: Springer.
Chapanis, A. (1991). Evaluating usability. In B. Shackel & S. Richardson (Eds.), *Human factors for informatics usability* (pp. 359-395). Cambridge: Cambridge University Press.
Duncker, K. (1935). *Zur Psychologie des produktiven Denkens.* Berlin: Springer.
Frese, M. (1987). A theory of control and complexity: Implications for software design and integration of computer systems into the work place. In M. Frese, E. Ulich, & W. Dzida (Eds.), *Psychological issues of human-computer interaction in the work*

place (pp. 313-338). Amsterdam: North-Holland.
Frese, M. & Brodbeck, F. C. (1989). *Computer in Büro und Verwaltung.* Berlin: Springer.
Frese, M., Brodbeck, F., Heinbokel, T., Mooser, C., Schleiffenbaum, E., & Thiemann, P. (1991). Errors in training computer skills: On the positive function of errors. *Human-Computer Interaction, 6,* 77-93.
Frese, M., Irmer, C. & Prümper, J. (1991). Das Konzept Fehlermanagement: Eine Strategie des Umgangs mit Handlungsfehlern in der Mensch-Computer Interaktion. In M. Frese, Chr. Kasten, C. Skarpelis, & B. Zang-Scheucher (Hrsg.), *Software für die Arbeit von morgen* (S. 241-251). Berlin: Springer.
Gediga, G. (1989). Das Funktionshandbuch zum System iS (5.05). *Schriftenreihe "Ergebnisse des Projekts Multifunktionale Bürosoftware und Qualifizierung",* Heft Nr. 15, Osnabrück.
Gediga, G., Janikowski, A., Lemm, H.-D., Monecke, U., & Pezalla, C. (1989). Ein Seminar- und Testkonzept zum Bereich integrierte Software-Systeme. *Schriftenreihe "Ergebnisse des Projekts Multifunktionale Bürosoftware und Qualifizierung",* Heft Nr. 13, Osnabrück.
Gediga, G., Lohmann, D., Monecke, U., & Greif, S. (1989). Exploratorisches Lernen in einer Seminarumgebung: Erste empirische Ergebnisse. *Schriftenreihe "Ergebnisse des Projekts Multifunktionale Bürosoftware und Qualifizierung",* Heft Nr. 14, Osnabrück.
Greif, S. (1989). Exploratorisches Lernen durch Fehler und qualifikationsorientiertes Software-Design. In S. Maaß & H. Oberquelle (Hrsg.), *Software-Ergonomie '89. Aufgabenorientierte Systemgestaltung und Funktionalität* (S. 204-212). Gemeinsame Fachtagung des German Chapter der ACM und der Gesellschaft für Information in Hamburg. Stuttgart: Teubner.
Greif, S. (1991). The role of German work psychology in the design of artifacts. In J. M. Carroll (Ed.), *Designing interaction. Psychology at the human-omputer interface* (pp. 203-226). New York: Cambridge University Press.
Greif, S., & Hamborg, K.-C. (1991). Aufgabenorientierte Softwaregestaltung und Funktionalität. In M. Frese, Chr. Kasten, C. Skarpelis & B. Zang-Scheucher (Hrsg.), *Software für die Arbeit von morgen* (S. 107-122). Berlin: Springer.
Greif, S., & Keller, H. (1990). Innovation and the design of work an learning environments: The concept of exploration in human-computer interaction. In M. West & J. Farr (Eds.), *Innovation and creativity at work: Psychological approaches* (pp. 231-249). N.Y.: Wiley.
Hacker, W. (1986). *Arbeitspsychologie.* Bern: Huber.
Heckhausen, H. (1980). *Motivation und Handeln.* Berlin: Springer.
Hoyos, C. Graf (1980). *Psychologische Unfall- und Sicherheitsforschung.* Stuttgart: Kohlhammer.
Hunt (1965). Intrinsic motivation and its role in psychological development. In D. Levine (Ed.), *Nebraska symposium on motivation* (pp. 189-282). Lincoln, NB: Univ. of Nebraska Press.
Kagan, J. (1972). The determinants of attention in the infant. In J.O. Whittaker (Ed.), *Recent discoveries in psychology* (p. 237-245). Philadelphia: Saunders.
Keller, H., & Voss, H.-G. (1976). *Neugier und Exploration.* Beltz: Weinheim.
Kieras, D., & Polson, P.G. (1985). An approach to the formal analysis of user complexity. *International Journal of Man-Machine Studies, 20,* 201-213.
Köhler, W. (1917). *Intelligenzprüfungen an Menschenaffen.* Berlin: Springer.

Krieger, R. (1993). *Zusammenhänge zwischen Selbstbeschreibungen und Leistungsunterschieden bei Textverarbeitungsaufgaben.* Unpublished thesis (Diplomarbeit), University of Osnabrück.

Lohmann, D. (in prep.). *Vergleich unterschiedlicher Trainingskonzepte zur integrierten Software.* University of Osnabrück (dissertation manuscript).

Lohmann, D., & Mangel, I. (1988). *Alternative Trainingsmethoden für ein Textverarbeitungsprogramm.* Unpublished thesis (Diplomarbeit), University of Osnabrück.

Lorenz, K. (1977). *Die Rückseite des Spiegels.* München: Piper.

Müller, M. (1989) *Exploratorisches Lernen mit einem gestuften System.* Unpublished thesis (Diplomarbeit), University of Osnabrück.

Nunnally, J. C. (1981). Explorations of exploration. In H. I. Day (Ed.), *Advances in instrinsic motivation and aesthetics* (pp. 102-130). New York: Plenum.

Piaget, J. (1969), *Das Erwachen der Intelligenz beim Kinde.* Stuttgart: Klett.

Reason, J. (1990). *Human error.* New York: Cambridge University Press.

Roberts, T. L., & Moran, T. P. (1983). The evaluation of text editors: Methodology and empirical results. *Communications of the ACM, 26,* 265-283.

Sauvageod, F. (1990). *Explorationsfördernde Maßnahmen im Computertraining.* Unpublished thesis (Diplomarbeit), University of Osnabrück.

Schölmerich, A. (1990). Der Erwerb neuer Information im Verlauf des Explorationsprozesses: Eine sequentielle Analyse von Handlungsketten. Unpublished Dissertation, University of Osnabrück.

Skell, W. (1980). Erfahrungen mit Selbstinstruktionstraining beim Erwerb kognitiver Regulationsgrundlagen. In W. Volpert (Hrsg.), *Beiträge zur Psychologischen Handlungstheorie* (S. 50-70). Bern: Huber.

Voss, H. G., & Keller, H. (1983). *Curiosity and exploration: Theories and results.* New York: Academic Press.

Voss, H. G., & Keller, H. (1986). Curiosity and exploration. A program of investigation. *German Journal of Psychology, 10*(4), 327-337.

Wohlwill, J. F. (1984). Relationships between exploration and play. In T. D. Yawkey & A. D. Pellegrini (Eds.), *Child's play: Developmental and applied* (pp. 143-170). Hillsdale: Erlbaum.

Wehner, T. & Stadler, M. (1989). Fehler und Fehlhandlungen. in S. Greif, H. Holling & N. Nicholson (Hrsg.), *Arbeits- und Organisationspsychologie. Internationales Handbuch in Schlüsselbegriffen* (S. 219-222). München: Psychologie Verlags Union.

Zapf, D. , Lang, T., & Wittmann, A. (1991). Untersuchungen zum Prozeß der Fehlerbewältigung bei einem Textverarbeitungsprogramm. In E. Ackermann & E. Ulich (Hrsg.), *Software-Ergonomie'91* (S. 332-341). Stuttgart: Teubner.

Zimolong, B. (1990). Fehler und Zuverlässigkeit. In C. Graf Hoyos & B. Zimolong (Hrsg.), *Enzyklopädie der Psychologie: Themenbereich D, Serie III. Wirtschafts-, Organisations- und Arbeitspsychologie: Band 2. Ingenieurpsychologie* (S. 313-345). Göttingen. Hogrefe.

CHAPTER **V.16**

Urban Development for Children Reexploring a New Research Area*

Dietmar Görlitz, Richard Schröder

Apropos

I think the ordinary is more valuable for the observer. If the interesting is served to him with the explanations that go with it so that he comprehends quickly and simply—that goes down fast. The commonplace, however, first has to be plowed up. (Kiefer, 1990, p. 27; translation by the authors).

The field of work presented in this chapter and the resulting research plans for studying "urban development for children" grew out of the pleasure of dealing with the commonplace and watching children "dawdling or hurrying," for, after all, "they don't stroll" (Hessel, 1932). For all the spontaneity of youngsters who are growing up, this research project also sprang from the hope that one can plan, nay, shape the circumstantial context.

To support children in their curiosity and exploration, we shall report on *our* curiosity as researchers, our curiosity about large entities. There are utopias conveying how ideal cities are to be designed. (As Kruft [1989, pp. 10-11] points out, such visions have not been around for long.) They are based on concepts of a state or society that are oriented at most to the citizen who has already come of age, not to

---

* The project described in the present chapter was planned and executed in cooperation with two colleagues at the Technical University of Berlin, Heinz Wagner, professor of architecture, and Martin Daub, professor of urban and regional planning. The authors are grateful to them and especially to the city fathers of Herten as well as our translator and editor, David Antal.

children. What is appropriate for children seems to be something other than the involved meaning of what serves their development (Weinert, 1979). This topic brings up the relative simplicity of the culturally and temporally rooted idea of childhood in which experiences, trust, desires, and other components have always mixed.

The reader will not find any of the deliberate or necessary differentiations between curiosity, exploration, and play, that have been the focus in the past (Görlitz & Wohlwill, 1987). In this chapter, we, the authors, chart the terrain on which children and what they do will enable us to make careful distinctions.

Psychologist's Curiosity about the City and its Children

Special Features and Innovations

Urban research from the perspective of developmental psychology focuses on entities, not primarily on the user. Nevertheless, it is neither about the city as part of the legacy of human works nor about developmental psychology as a field of interest in the totality of the human being's life. In the present project, lifespan is confined to work involving children. They are the partners of psychologists, architects, and city planners, and they allow these specialists to share in their everyday realm of experience in the city. Of what use can psychology be in the process? What can psychologists want to know?

The psychologists of today cannot rely on the axioms of classical psychology in Wundt's day, namely, that more complex entities of the psychomental realm may ultimately be composed of elements (analogous in our project to distinctly separate situations; see Saegert & Winkel, 1990). After reflecting on our positions *and* methods as researchers, we must instead be courageous enough to explore the larger, more inclusive spaces in which humans live. In terms of position, that means acting on the psychologist's epistemological interest in the experiences and behavior (including interactions) of human beings in the course of their lives. It also means becoming a partner in perspectives on reality that the researcher does not share in the main but tries to learn about with passionate curiosity, the kind of curiosity that spawns theories about the structures and organization of the urban world as a *child* lives it.

On the Tradition of the Subject

There is a rich tradition behind the topic of observable activities on the streets and squares of a city, in and around its residential and business buildings. It includes not only reputable predecessors and partners but also loafers and paid loiterers, idlers whom the big city has stylized into a literary type, police informers, and newspaper reporters. Of course, urban research from the perspective of developmental psychology does not follow everything from their spectrum of interests. Moreover, our project focuses on the immediate physical partners in everyday activity (buildings, streets, and squares) and the environs.

Part of what the project researchers have had in mind was to pursue *child*-centered, developmental urban research in the tradition of old Berlin pedagogy (Görlitz, 1993; support for this perspective is found in Schröder, 1987). The idea has been to continue, in a modern way, the sort of work embodied in Schwabe and Bartholomäi's initial survey (1870) of what first graders in Berlin knew about their urban environment. Another example of this tradition is Tews's work on the metropolis as an educator (1911), on what the city unintentionally teaches children as they grow up. The second thrust of the current project has been to pick up on Muchow's analyses of life space (1935; see also Wohlwill, 1985). By contrast, Rolf Lindner's (1990) resourceful work entitled *Entdeckung der Stadtkultur* (Discovering urban culture), which is rich in ideas on the subject, teaches how much such curiosity (among psychologists, too) is due to the Chicago school of sociology under Robert Ezra Park in the 1920s. Park used to instruct his students to "explore the city on foot, talk with the people, and record their [own] observations in detail" (Lindner, 1990, p. 116; all translations of Lindner are by the authors). Such "ethnography of the street," the intent of which was to convert the unfiltered "acquaintance with" into systematically ordered "knowledge about," glorified the big city and its "fragmentation and segmentation into a mosaic of little worlds" (Lindner, 1990, p. 105), conceived of it "as a sociological laboratory for the study of human behavior" (Lindner, 1990, p. 89), and stuck to adult subjects who were able and willing to speak.

An Articulated Interest

The present project, however, is concerned with *children's* views of life in the city. We, the researchers, cannot meet that more complex aspiration of using the city as "a sociological laboratory," as Lindner (1990, p. 173) characterizes the line of the Chicago school of sociology under Park. We are charting "tracks for knowledge" into a broader entity by becoming partners with a city's children aged between three and fourteen years, by letting them take us "in hand." Put in simplified and formal terms, our curiosity as urban researchers has led us to ask who does what, where, when, with whom, and with what. And what does this "what" do to that person? Which chances for changing things are seen or desired by that person, and how are they acted upon? And we "ask" children by observing what they do (as well as what they know and how they represent that knowledge).

Working with children definitely means working with persons living within specific legal parameters. Children, especially those in the city, live and act in preexisting, predetermined, assigned, and material conditions, not primarily in circumstances that they shape by and for themselves. That is one of the basic differences between them and adults. The person responsible for children is always someone who is not a child (even though the actual work of care-giving is performed in most cultures by older siblings; see Valsiner, 1989, p. 230). That is not usually the case for adults. Except for the homeless, the hobo, and the tramp, who, like children, are not part of society's conventional patterns, adults generally do not have anyone to look after them. Jaan Valsiner (see his theoretical work, 1987, pp. 76-86) has developed decisive lines of argumentation on how a culture — through the arrangement of its material aspects and the activity of the mediating, thereby limiting, older educator — moulds its younger members, who nevertheless continue to be active in co-constructing cultural meaning. How does the city "work in" the children and adolescents living in it, and which potential to shape their own experiential realm does it permit them? Children's preferred form of activity is playing, a thesis that was once central in Martha Muchow's work (see Muchow, 1935).

There is another aspect as well. The focus of the project is not on the processes leading to a quasi natural fit and assimilation. Urban projects pursue *cultural* work, they involve learning how culture is passed on. After all, children do not go about encountering things in their city like astronauts landing on the moon — although we, as researchers of their behavior, need to keep the distance of strangers in order to take the gray, ordinary triviality of life's everyday episodes and make them into something that can be studied in their spatial context. Children do not experience their city that way. They encounter stratified things of foreign origin, they meet in cultural realms of material things, and they see and experience their use. It is, therefore, legitimate in its own way for researchers to question children, to see things and areas as if for the first time, trying to understand them by the way they are used by children.

Seeking a town suitable for such studies, we came across Herten. What makes this community an appropriate place for the project on children's activities and needs?

A City Discovers its Children. Herten as a Model

With a population of 69,000, the German town of Herten is located on the northern fringe of the Ruhr District in an area marking the transition from the country's industrial heartland to the rural countryside known as the *Münsterland*. Together with nine other towns, Herten lies in Recklinghausen, the most densely populated county in the Federal Republic of Germany. Two collieries, at which approximately 10,000 miners haul 30,000 tons of bituminous coal a day, make Herten one of the largest mining towns in western Europe.

Kinderfreunde "Children's Friends"

History

In 1978 Herten's town council unanimously voted to conduct a project oriented to the needs and welfare of Herten's children. Named *Aktion kinderfreundliche Stadt*, this undertaking has been jointly sponsored by the German Automobile Club (ADAC), the association known as *Mehr Platz für Kinder*, and the city of Herten with the objective of documenting and steadily improving all aspects of the city as a setting in terms of the quality of life it offers children. The thrust has not been to develop sensational models but rather to enhance the urban setting of children through a variety of complementary measures and activities. Bearing the needs, wishes, and welfare of children in mind is not regarded as an isolated goal, but as part of the effort to make Herten an attractive, comfortable, and desirable place to live for all its citizens.

Such child-friendly urban development requires one to deal with all the problems of urban living from the special perspective of children and adolescents. Some task areas of major importance for children lie outside the traditional sphere of work with young people. They include housing, the design of residential areas, transport planning, and the partnership between children and adults. On the long way to becoming a child-friendly town, Herten has created several bodies to help shape commensurate local policies (see Fig. 1).

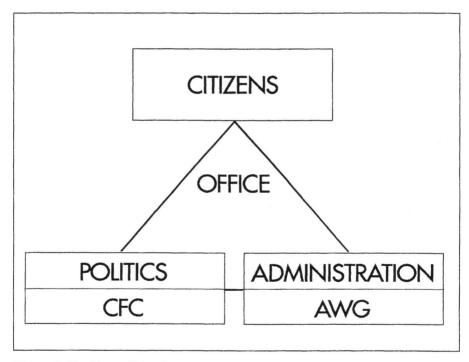

Figure 1. The Herten Triangle

Formal Organization

The "Children's Friends" Office

The "Children's Friends" Office, an integral part of the municipal administration, is the pivotal point of contact and coordination for the various activities undertaken by the citizenry, council, and administration. Its specific responsibilities are:
1. to act as a responsive partner for children, adolescents, and other citizens; associations; and initiatives;

2. to promote and handle sponsorship projects and grass-roots initiatives and coordinate citizen involvement in related projects;
3. to design public relations campaigns for greater consideration of children's needs and wishes in Herten; and
4. to act as a lobby for children in municipal planning projects.

"Children's Friends": The Administrative Working Group. The Administrative Working Group (AWG) consists of employees from a wide variety of departments in the municipal administration (e.g., Youth Affairs Office, Planning Office, and the Office of Cultural Affairs) who elaborate for their superiors recommendations for considering children's needs and wishes.

The "Children's Friends" Committee. Operating at the parliamentary level, the Children's Friends Committee (CFC) pursues a child-centered policy encompassing all spheres of municipal life. This body deals with areas that fall outside the traditional framework of youth affairs. For example, a major focus of the committee's work is to consider the role that children in large families play in areas planned for development and the situation of such children in old quarters that need modernization.

Citizens. The citizens of the town have organized many initiatives for children and parents. Often, measures designed to improve the setting in which children live can be developed and carried out jointly with a large pool of concerned citizens.

Activities of "Children's Friends"

In the past twelve years, several hundred activities of various scopes have been conducted. A number of them have sought to help children discover their town. After all, adults are not the only ones to whom urban settings offer a range of interesting squares, buildings, and social contacts. These spaces must be designed in such a way that children, too, can safely probe the world in which they live. A few examples will illustrate what a town can do to pique the curiosity and encourage the exploration of its youngest inhabitants.

City Map for Children

As they grow older, children and young people are likely to want a working knowledge of the place in which they live and to which they have developed a specific relation in the course of their development. As the years pass, the increasing mobility that busses and bicycles give children eventually enables them to reach interesting places anywhere in the town. A town should try to acquaint children with its layout, including the facilities important for them.

As part of *Aktion kinderfreundliche Stadt* eleven years ago, a project group consisting of teachers from Herten and a team from the design department at the Technical College of Aachen, techniques were developed to present the city in a way geared to children. A special map of the town shows all the things of interest to young people. Pedestrian zones and special facilities for young citizens were highlighted and given their own pictograms (see Fig. 2).

The color scheme, the highlighting of individual houses and areas in which road traffic is relatively safe, the clarity of the symbols used (e.g., a forest represented by small trees), and the symbolization of facilities especially interesting for children (playground, swimming pools, and public bathrooms) make this city map particularly legible for children. All information on the map was also printed in the languages of the town's foreign citizens (primarily Turks and Greeks) as well as in English. The city map has also been used in the school classroom for lessons on "our town." Now ten years old, the first edition of this children's map of the town may need to be updated. The new edition should build on the experience with its forerunner.

A special problem of the first Herten city map for children has been the fact that it is a handdrawn original. Updating it to reflect the constantly changing urban reality of Herten has always meant totally recreating the map. The idea of producing a children's map as a variation of the existing "adult map" is currently being considered. Prepared master copies (and alterations of them) could be used. It would then be necessary only to draw up corresponding masters for the information intended specifically for children. In a new map of Herten, "appealing symbols" are to be added to the existing iconographic ones to call attention not only to special facilities for children but also to fields and open spaces for flying kites or to ponds where children can sail their boats.

The present map depicts the administrative area of the town as an island. The new map is to contain information on adjacents areas, too, for children often seek them out as well.

Figure 2a. Children's Map of Herten (excerpt)

316 CHAPTER V

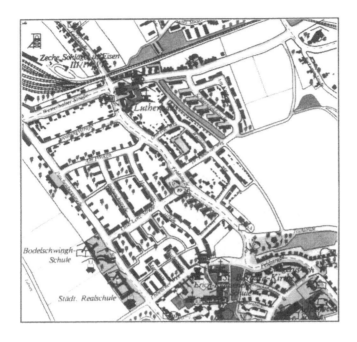

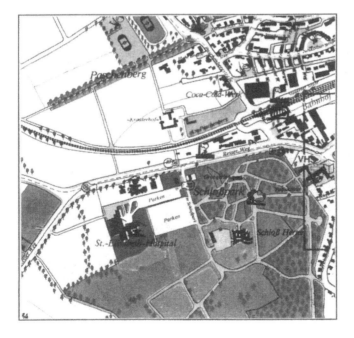

Figure 2b. Children's Map of Herten (excerpt)

Map of Routes for Bicycle Excursions. Experience has shown that children in Herten use bicycles less as a means of transport than as toys. Riding bicycles is one of the favorite leisure activities of children. Just as with the children's city map, the map for bicycle excursions that has been developed by Children's Friends is intended to give children the opportunity to explore their town. In addition to the customary routes for bicycle excursions, the three maps currently available show other interesting routes for bicycling, hiking, and walking all over town. The individual bicycle routes are also described to enable one to picture the route before setting out. Like the city map for children, the bicycle excursion map is provided by Herten free of charge and is very popular not only among children but among families and senior citizens as well.

City Game: "Let's Tour Herten." "Let's Tour Herten" is a board game designed to familiarize people with the town. The two, three, or four players move their tokens along routes marked on a city map of Herten. Each player has a separate starting position and a predetermined destination.

Figure 3. "Let's Tour Herten" Game

At various locations on the board during the game, the players collect cards containing a wide variety of information about those specific items of the town. As in real life, the players can be compelled to stop at traffic lights, road closures, and railroad signals. Individual facilities having to do with dwellings, work, leisure, health, sports, meeting places, transport, and administration are introduced and described in detail. The thirty-six information cards can be used separately from the city map game to play a matching game named "Exploring Herten." As with the children's city map, "Let's Tour Herten" has been used in the classroom for lessons on "our town."

Child-friendly Living

Basics of Planning and Herten's Philosophy

Practice articulates values and affects decisions even before any research begins. The following text concerns those prescriptive sets. Decision-makers in our society share the opinion that children need room to play both indoors and outdoors if their development is to be healthy. In a country with a cool climate, like Germany, and especially in urban areas with little open space, a child's dwelling plays a particularly important role.

Many families with several children must presently live in dwellings whose size, cut, and other architectural aspects fail to meet the needs of the inhabitants. Moreover, it is frequently the case that the dwelling's immediate environs do not provide children with adequate space for playing and socializing. The design of the dwelling and the environs is one of the factors that enhances or detracts from the exploratory behavior of children. Small children in particular depend on the immediate environs for space to play and explore. The following illustration of a residential area that meets children's needs in these respects is based on a catalogue of criteria developed by the Children's Friends Committee in 1986 and used in all planning stages of the construction project known as "Child-friendly Living" (*Kinderfreundliches Wohnen*, 1989-90).

The Settlement Area. A residential area also oriented to the needs, wishes, and welfare of children should not be too large for children and should have its own identifying features. It was the conviction of the committee that such a design could help the inhabitants develop a feeling of belongingness and identification so that a settlement of this kind could come to be recognized as home.

A lively settlement must avoid being only a "bedroom community." A certain multifunctionality is ideal, that is, a dynamic blend of various purposes. Such multifunctionality includes planned areas of interaction that promote contact and communication. The demographic structure should be optimally balanced in order to offer children, too, the broadest possible scope of social experience.

Opportunities for play. In principle, a dwelling's immediate environs should be accessible for games. Instead of fewer large playing areas, there should be many possibilities for play distributed all over the grounds. They need not be playgrounds only; the spaces before the front door and open grass areas and squares in the settlement can serve as play area and is frequently more useful than an elaborate playground. Fallow areas that have retained as much of their natural state as possible are often the most interesting ones for play (trees for climbing and shrubbery for playing hide-and-seek, for instance). Playing areas near the house help avoid traffic accidents involving small children.

Roads and highways. Separation between traffic and immediate residential zones is intended to ensure the possibility of safe play at many spots. Where the various kinds of traffic cannot be separated, it is necessary to take precautionary measures that underscore the residential character of the settlement and instill especially cautious behavior (reducing the volume and speed of traffic and introducing play streets). The streets adjacent to the settlement area are to be considered in such thinking if they must be crossed in order to reach attractive recreational facilities located within or beyond the settlement area.

Child-friendly Architecture. The "child-friendly architecture" described in the following sections is based on the project entitled "Child-friendly Living," which was conducted in Herten in 1989-90. There is no pretension that the ideas presented here constitute the child-friendly architecture. They were developed through long discussions and built as a realistic solution. In this sense, the present result can be regarded as a kind of model worth imitating—one that can be imitated. The arrangement and structure of the buildings and the design of the immediate living environs are oriented to a program developed by the town of Herten as part of its efforts to achieve a "child-friendly city" together with the dfh-Siedlungsbau (Worms) and the Moerser Verein "Wohnen mit Kindern." Many of this association's ideas were integrated into the construction plans and carried out.

The Family-oriented Floor Plan. The opportunities for each member of a family to develop and the interaction between the family members can be promoted by the given living arrangement. They can also be hampered, however. Architecture oriented to the needs, wishes, and welfare of children must always show the same regard for the family, too. After all, the well-being of the family as a whole has an especially significant impact on the children. It is necessary not only to see the needs of the children at their various age levels but also to consider what is important to parents.

There are two basic stipulations in this respect:
1. *An adequate number and size of individual and common rooms.* Dwellings are family-oriented to the extent that they enable the individual family members to choose between being alone and being together, depending on their own needs.
2. *Flexibility in the utilization of living space.* It should be possible to use dwellings flexibly in order to accommodate the individual needs of each family as well as the desires for use as they change in the course of the life cycle.

"Child-friendly Living": The Project

The Immediate Settlement Area. The project entitled "Child-friendly Living" encompasses 13 two-storey private family homes standing on slightly more than one acre. Because only families with several children were accepted for the project, more than thirty children come to play in this settlement. The presence of several children of the same age can be seen as a premise for frequent outdoor play. Social contact within the settlement is encouraged by the similar structures of the family living there, and it becomes easier to watch over each other's children. The very social structure of the settlement is thus a step toward child-friendly and family-oriented living (see Figure 4).

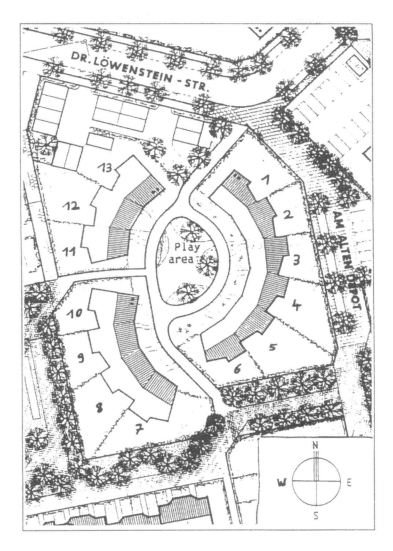

Figure 4. The Child-friendly Settlement. A Contribution to Child-friendly Urban Development

Unlike the development in the surrounding area, this housing tract is not built as a series of right angles. Instead, the three sets of buildings are grouped around an oval area intended as an enclosed place for children in particular to meet, play, and interact. The houses, which offer living space ranging from about 1,150 square feet (108 m^2) to about 1,350 square feet (127 m^2) are juxtaposed in such a way that individually shaped zones can develop in the private backyards.

In contrast to customary floor plans, both sizes of house provide for a large family room connected to the upper floor by an open stairway that assumes a key role not only as a connector between the two levels but as an element of communication. The ground floor has the parlor for receiving guests and withdrawing privately. The parlor and the family room open directly on the backyard so as to encourage outdoor play. The area of the parlor has been limited so that the family room could be larger.

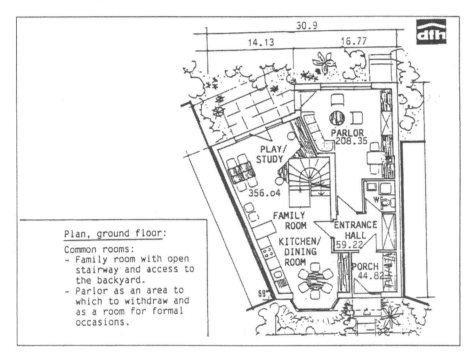

Figure 5. The Child-friendly Settlement. Layout of Ground Floor

The parents' room on the upper floor is soundproofed to allow one to retire for quite solitude even when the house is full of activity. The children's rooms are of sufficient size. They serve as bedroom, individual play area, and space for being alone. They can be enlarged if galleries are added, a possibility that can increase not only the area and stimulation for play, but also the attraction of the rooms in general. On the whole, putting the private rooms on one floor and the common rooms on the other helps to meet the variety of needs for quiet felt by all family members. It increases the protection from noise.

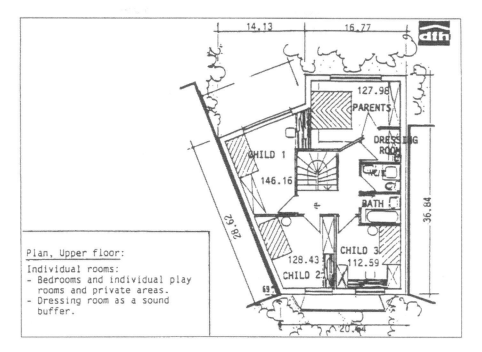

Figure 6. The Child-friendly Settlement. Layout of Upper Floor

It should be possible to interchange the parents' and the children's areas, depending on the family's ideas and phase in the family cycle. The installation of a second bathroom on the ground floor enables one to separate the parlor from one's own living area. Be it the teenage son, the grown-up daughter now in training, or the grandmother living alone, the possibility of separation can give family members in the house their "own territory" in which to live while still being included in the rest of the family.

The Settlement Area as a Whole. Long before the construction project "Child-friendly Living" had come to be, a project group had formed within the town administration to push for a child-friendly orientation in the development plans for the entire 17-acre settlement. Except for one street that was laid according to the conventional principle of separation, all roads within the settlement area were planned either as residential streets with restricted traffic or as walkways. They thereby serve the residents as zones in which to meet and interact. Care was also taken to link the construction area to the town's network of north-south and east-west bicycle paths.

In the immediate vicinity of the settlement project, the town built a public playground nearly half an acre large to serve the recreational needs of various age groups (a sandbox area and wooden toys for small children; a fenced play corner equipped with weight-lifting equipment for larger children; a large, open grass field for running games; and a grass area with a backstop). The community house (*Bürgerhaus*) built on the grounds was provided with a children's room, day-care facilities, a branch of the town library and of the town music school, and a disco for teenagers.

All these measures are intended to help make this settlement a generally child-oriented residential quarter. Follow-up social science studies are planned to evaluate the results of the model after the beginning of cooperation with the Technical University of Berlin (TUB).

A University and a Town as Partners - Child-centered Planning as a Cooperative Task

On the Cooperation between the Town of Herten and the Technical University of Berlin

In response to a small interdisciplinary group consisting of psychologists, architects, and city planners (Görlitz, Wagner, & Daub, 1989), the Technical University of Berlin has decided to enter into a two-year research partnership with Herten, this medium-sized town in the Ruhr District. The purpose is to draw on interdisciplinary experience in opening a new professional perspective to psychologists *and* to assist the residents of the town, particularly the children, in managing their day-to-day lives, beginning first of all by acquainting them with the possibilities for shaping the spatiophysical conditions they experience.

Delimited to analyses of the living *environs* of urban settlement areas, the logic behind the stages of the Herten design is simple, but the execution is complicated in part by changes in season. After a planning phase that had to be abbreviated for extraneous reasons, the project began in January 1990, when TUB opened an office for field research in Herten.

In the field of developmental psychology, urban research concerned primarily with analyses of living environs from the perspective of children residing there and spending a good deal of time outdoors is research in the act of "doing the splits," even though current funding of the project precludes the longitudinal observation of children for one thematic phase. This kind of "gymnastics" does not suggest the Herculean act of weight-lifting as the focus of research. For the city is very much an active subject in itself, and that on a scale far greater than our techniques of observational analysis can grasp. Our challenge is that unusual, external actors set the pace and make for unusual complications in the planning and decision-making for this project as well as its execution. These actors are the seasons of the year, school vacations, precocious weather, and the forbearance of the city fathers, who must keep changing interests in mind for two-year periods as much as for any other. So it is that urban research, even the sort that follows children in their use of open space, comes to be "research doing the splits." (To use the same metaphor, though, even the splits leaves one's hands free.) Urban research and the topic treated here, "urban design also suited to children," need and live from the supporting philosophy of the community and its political decision-makers (as in proposition V by Bronfenbrenner, n.d., p. 9). That philosophy helps to withstand such tension.

The Four Stages of the Herten Design

The straightforward, two-year design of the Herten project calls for four stages of research after a preliminary phase. In the preliminary phase, which Weinert (1990) would see as part of the tradition of research on the city as space for living (*Stadt als Lebensraum*), all Herten children from 3 to 14 years of age (with children younger than 7 years being assisted by their parents) were requested to respond to a questionnaire that gave them the opportunity to identify special areas in the city in which the balance between the zones (i.e., the interiors of their homes, the vicinity of their homes, and their immediate part of the city) seemed to be especially good or especially poor. In the second part of the questionnaire, the children are requested to rate the importance they attach to these same things inside and outside their homes. On the basis of such overall characteristics of the way in which the children subjectively experienced these realms, the project planners are selecting and localizing individual districts in which a reasonable amount of detailed observation could be conducted. It is in these areas that the various stages of the project are being carried out.

In the first stage, the outdoor activities that children engaged in within the vicinity of their homes were recorded in the selected settlements to see how they used that space. The subject in this part was, as Weinert (1990) put it, the city as a setting for action and experience (*Handlungs- und Erlebnisraum*). The idea behind the profile of the way the children used the free space in the vicinity of their homes is to compile the information through many different methods in order to permit a comparison of the veridicality and comparability of children's views about the reality with which they are familiar. The analyses of use are based on participant observation from walks through the area at various times of day. The analyses also draw on data from questionnaires that now specified locations of certain activities. These questionnaires are to be distributed to all households of the selected settlements. (In practice, this part of the method had to be postponed and will be replaced by time-budget analyses consisting of diary entries and by rankings of impressions based on semantic differentials.) Lastly, the analyses in the first stage of the project will use the results of exploratory walks with the children themselves as well as information from personal interviews with the children and their parents in their homes — insofar as the researchers gain personal access through previous contact to the subjects.

The results gleaned from this variously compiled information on the uses to which children put the regions in the vicinity of their homes are to be compared for methodology. They are also to be checked for whether these regions of urban settlements have integrating or segregating areas preferred or avoided by children for their outdoor activity on the basis of age, sex, or ethnic background. The researchers also wish to know how the functional specifics and man-made and natural characteristics of these regions can be described. Lastly, this part of the survey must examine whether such characteristics also apply to other types of settlements beyond the physical and social context in which they have been observed.

The second stage is concentrated on the topic of environmental knowledge, mental representation of the environment among children from the settlements studied during the first phase of the project (Liben & Downs, 1989, present the state of the art). Furthermore, the focus shifts from the shared reality of living "there" to the level of three-dimensional architectural models. Children from the first stage now let the researcher become familiar with the experiential realm (*Erlebnisraum*) of their living environs (as Weinert conceives of it). Acting on general instructions only, the children interact with siblings or friends if possible in planning and constructing (or this arrangement is controlled in various ways with individual experiments), using malleable materials or more or less realistic models of houses (scale, 1:100), objects of nature, and utensils made available to them. In an otherwise free setting, they are asked to build "what it looks like, the place where they live, and the area around it." The models that the children build allow for an abundance of questions that are both interesting in developmental psychology and unclarified in research. The Herten project delves into only one aspect of those questions and makes it the subject of the third stage.

After being confronted with the solutions offered in models prepared by participating adult architects, the children involved in the project had the opportunity to examine their own solutions and to correct them if they chose. The researcher then worked with the children on the model *they* had made of their home's immediate vicinity. The topic of the project's *third* stage is to work with the young partners in this experiment in altering the model of the physical environment so as to improve the opportunities for play, to make the living environs of the respective settlements more responsive to children's needs and wishes. It is a matter of involving children in planning at the level of models. This "reality" of models of concrete reality is still relatively uncharted in developmental psychology. To what extent is the handling of things as models in itself a matter of development? Even less familiar in developmental psychology is knowledge about the degree to which such solutions produced by children working with models are a reliable indication of design specifications for the material environment of the "world for us all." Additionally, the tasks of the third stage are to be worked on solely through model simulations in the Center for Model Simulation at TUB (see Hirche, 1986), giving children the chance to be genuine explorers, albeit in Lilliput.

Design specifications on which consensus has been reached in group discussions with the children will be integrated into the structures to be built by the project planners and architects in the actual settlement regions studied. These activities constitute the fourth stage of the Herten design. The measures taken are then to be evaluated. How much and which reality is there to play with according to the solutions offered by the models that children of different ages have created? Which conflicts of use are to be seen and how are they to be dealt with? What do children do as planners and what does it mean to them? These are the questions that attract us the most in the Herten project. So much for "modelling." The coming months will show what it was like in practice.

Outlook on Researchers Curiosity and Forms of Action Appropriate to Children

The idea for this project developed from visits and lectures in connection with research on child-centered development of the environment, which professors of architecture and psychology had jointly planned. For Herten, it is important that the results of this project be useful for solving practical local problems. For the university, Herten represents the chance to conduct field research. Seminars and opportunities for master's theses devoted to Herten topics are being offered in Berlin. Interns from the participating institutes may work on the project in Herten in a variety of capacities. Just as *Children's Friends* has become a model for other towns, Herten hopes that this unique kind of cooperation between a university and a town can become a model for other local communities.

But the reader will have realized that this chapter has dealt more with the researcher curious about the city than with children exploring it. Can the foregoing text be read in terms of children in an urban environment by putting them in our shoes? Roger Hart (1979, p. 3) reminds us that "[our] greatest period of geographical exploration is that found in each of us — in our childhood." But what children do, including that which they do as inquisitive explorers of the urban environment, enables them to establish other systems of knowledge as products of their action, systems of experiential knowledge other than those built through the epistemological curiosity of the researcher. Thus, the one cannot stand for the other, although they depend on each other like partners, the older even more so than the younger (a relationship expressing a kind of incipient adult patience that Roger Hart illustrated in grand style in *Inavale*, a kind that spurs one to consider one's own position). City administrators are not the only people called upon to include the needs, wishes, and welfare of children in their planning. Representatives of the scientific and academic community are also called upon to theorize with children squarely in mind if they and their concerns are the subject. (Bradley, 1989, has expounded on this point with regard to infants.) But what, then, would be the features of an interdisciplinary theory of exploration rooted in the urban setting and conceived to integrate child(ren)-centered reasoning?

The reader will not find this theory here. If anything, these pages contain more a collection of basic assumptions, such as the premises that the spatiophysical context does make a difference (*Nichtgleichgültigkeit*); that habitual ways of using something — which limit and facilitate orientation — develop by means of spatial exploration; that the position of users, their way of interpreting the environment, and their often child-specific guiding interests are indelible; and that what a person can use as well as its symbolic magnification and social mediation is embedded in a spatial context. With basic assumptions structured as loosely as this, the researcher of curiosity, of exploration and play, is both comprehensible and vulnerable in his "community." He will then have to deal more exclusively with the preconceptions of his time and society, with what constitutes the relative simplicity of a concept like *childhood* and *appropriateness for children* as opposed to the diversity of individual children in the course of their development, and with the question of for whom and in which discretionary situations that concept may guide or legitimate action. If researchers and the policy-maker, educator, or parent share the conviction that it is desirable for children in their development to "make" them into involved, autonomous explorers and experts of their environment or to help them improve in that capacity, then the fixed qualities of the perceivable and treatable environment will be less a topic as far as urban environmental planning is concerned. Attention will be focused more on relational, functional characteristics manifested in the individual interaction between the child and the environment. Such characteristics include availability, accessibility, malleability, changeability, structurability, limitability, manageability, and the capacity to be experienced, structured, and diversified— with the potential for controlled and controllable hazard for the person who is acting in a specific situation.

So much for sketching (see Evans & Cohen, 1987, for a similar intention) a future of child-oriented planning in the urban setting. In a broader sense, Kaminski (1990) calls it "ecopsychological praxeology." Among child-oriented planners it must provide for balance in the sense that "[everything] one teaches the child . . . [the child can] no longer invent or discover itself" (Jean Piaget, according to Bringuier, cited in Kesselring, 1988, p. 65). As Hart (1979, p. 3) has stated, however, "All children have an urge to explore the landscape around them, to learn about it, to give order to it, and to invest it with meaning — both shared and private," values in our culture at least.

Postscript

Books have their histories, and so do projects. Sometimes those histories are not synchronous. For example, much of the planning conveyed in this chapter on work done in Herten has already been acted upon; and the modell for the cooperation between a city and a university has been discussed at length, along with other topics of child-oriented urban planning, at an international forum of a "Herten conference" (see Görlitz et al., 1992). Nonetheless, bearing witness to the spirit of a determined, courageous city and to the changed understanding of children and childhood remains relevant, a message that leaves the preceding pages unaltered.

References

Bradley, B. S. (1989). *Visions of infancy: A critical introduction to child psychology.* Cambridge, UK: PolityPress.
Bronfenbrenner, U. (n.d.). *Cities are for families.* Unpublished manuscript, Ithaca, NY, Cornell University.
Evans, G. W., & Cohen, S. (1987). Environmental stress. In D. Stokols & I. Altman (Eds.), *Handbook of environmental psychology* (Vol. 1, pp. 571-610).
Görlitz, D. (1993). Es begann in Berlin. Entwicklungslinien einer psychologischen Stadtforschung. In H.J. Harloff (Ed.), Psychologie des Wohnungs- und Siedlungsbaus. Psychologie im Dienste von Architektur und *Stadtplanung*. Göttingen: Hogrefe.
Görlitz, D., Harloff, H.J., Valsiner, J., Hinding, B., Mey, G., Ritterfeld, U., & Schröder, R. (1992). The city as a frame of development for children: The Herten Conference. *Children's Environments, 9*, 63-64.
Görlitz, D., Wagner, H., & Daub, M. (1989). Entwicklungspsychologische Stadtforschung und Umweltplanung. Einrichtung einer Außenstelle der Technischen Universität Berlin in Herten (Kreis *Recklinghausen)*. February. Unpublished manuscript, Technical University, Berlin.
Görlitz, D., & Wohlwill, J. F. (Eds.). (1987). *Curiosity, imagination, and play: On the development of spontaneous cognitive and motivational processes.* Hillsdale, NJ: Erlbaum.
Hart, R. (1979). *Children's experience of place.* New York: Irvington.
Hessel, F. (1932). Von der schwierigen Kunst spazieren zu gehen. Republished in F. Hessel, *Ermunterung zum Genuß*. 2nd ed. (pp. 53-61). Berlin: Brinkmann & Bose, 1988.
Hirche, M. (1986). *Architekturdarstellung und ihre Wirkung auf Planungslaien*. Berlin:

Technical University.
Kaminski, G. (1990). *Einige Leitgesichtspunkte für eine ökopsychologische Praxeologie.* Lecture at a colloquium on "Wohnungs- und Siedlungsgestaltung - Eine Aufgabe für Architektur und Psychologie." 5 October, Berlin, Technical University.
Kesselring, T. (1988). *Jean Piaget.* Munich: Beck.
Kiefer, A. (1990, November 16). Das goldene Vlies. Bilderzyklus und Werkstattgespräch [Magazin (supplement), No. 46]. *Süddeutsche Zeitung* (Munich).
Kruft, H.-W. (1989). *Städte in Utopia. Die Idealstadt vom 15. bis zum 18. Jahrhundert zwischen Staatsutopie und Wirklichkeit.* Munich: Beck.
Liben, L. S., & Downs, R. M. (1989). Understanding maps as symbols: The development of map concepts in children. *Advances in Child Development and Behavior, 22,* 145-201.
Linder, R. (1990). *Die Entdeckung der Stadtkultur. Soziologie aus der Erfahrung der Reportage.* Frankfurt am Main: Suhrkamp.
Muchow, M. (1935). *Der Lebensraum des Großstadtkindes.* Hamburg: Martin Riegel. (Reprinted with an introduction by J. Zinnecker, Bensheim: päd.extra Buchverlag, 1980).
Saegert, S., & Winkel, G. H. (1990). Environmental psychology. *Annual Review of Psychology, 41,* 441-477.
Schröder, R. (1987). *Kinderspiel in städtischer Umwelt. Subjektive Sicht und historischer Wandel nach biographischen Interviews.* Unpublished master's thesis, Technical University, Berlin.
Schwabe, H., & Bartholomäi, F. (1870). Ueber Inhalt und Methode einer Berliner Schulstatistik. In: Berlin und seine Entwickelung. *Städtisches Jahrbuch für Volkswirthschaft und Statistik.* Vierter Jahrgang, 1-76.
Tews, J. (1911). *Großstadtpädagogik.* Lectures delivered at the Humboldt Academy in Berlin. Leipzig: Teubner.
Valsiner, J. (1987). *Culture and the development of* children's action: A cultural-historical theory of *developmental psychology.* Chichester: Wiley.
Valsiner, J. (1989). *Human development and culture: The social nature of personality and its study.* Lexington, MA: Heath.
Weinert, F. E. (1979). *Über die mehrfache Bedeutung des Begriffes "entwicklungsangemessen" in der pädagogisch-psychologischen Theorienbildung.* In J. Brandtstädter, G. Reinert, & K. A. Schneewind (Eds.), *Pädagogische Psychologie: Probleme und Perspektiven* (pp. 181-207). Stuttgart: Klett-Cotta.
Weinert, F. E. (1990). *Ökopsychologie des Stadtlebens.*Unpublished manuscript, Max Planck Institute for Psychological Research, Munich.
Wohlwill, J. F. (1985). Martha Muchow and the life space of the urban child. *Human Development, 28*(4), 200-209.

CHAPTER VI

Epilogue

CHAPTER **VI.17**

Applause for Aurora: Sociobiological Considerations on Exploration and Play

Robert Fagen

Human nature involves play, curiosity, exploration, and humor. These features may well be essential, but they were largely ignored, misinterpreted, or misrepresented by a generation of sociobiologists intent on creating their own vision of the human condition. A look at the biology of exploration and play in the seal family Phocidae, and especially in the Northern fur seal, opens new biological perspectives on exploration and curiosity for interdisciplinary discussion. To further develop these themes, I reexamine the fairy tale Sleeping Beauty in biological terms, particularly in the context of its treatment in classical ballet. Whether we analyze fur seals or classical ballet, we reach the same conclusion: play, curiosity and the like are overlooked aspects of human and nonhuman behavior. They point to fundamental limitations of sociobiological approaches, and they offer some deep mysteries of their own. Science rightfully rejects both an uncritical holism and the fictions that follow from attempts to reduce human nature to a gratuitous and facile biology. On this matter, scientists' professional skepticism follows plain common sense. As a result, the twenty-year campaign to move sociobiological theory into social practice by a variety of subtle (and not-so-subtle) means has had little real impact. Of course, rhetoric has ensued, much of it in the form of an outpouring of printed matter that by now should be sufficient to extend around the world several times. Not to mention an associated media explosion whose first broadcast components (c. 1975), travelling away from Earth at the speed of light, have by now reached the Solar System's nearest neighbors in the Galaxy. If these neighbor worlds support intelligent life, let us hope that their inhabitants are amused, and are not in the process of mounting a punitive expedition.
Chekhov, Chaplin, the storied alumni/ae of the Russian circus and

Clown College, and a thousand real and fictional clowns, from classical ballet's Swanilda and Lise to your family Golden Retriever, all had it right: it doesn't pay to take either ourselves or our science too seriously. What is seriously important could well be called curiosity, exploration, play, laughter. Ergo, this volume, and this chapter.

Considerations of play, exploration, and human evolution serve in part to expand and in part to revise a specific view of human nature generally-associated with the term "sociobiology". The term "sociobiology" is far older and broader than the hardened belief system that developed under the name of sociobiology some years after the 1970's saw publication of several books whose titles included the word. Sociobiology seeks evolutionary perspectives on individuals' interactions and relationships, and on the social structures which these relationships build and in which they are embedded. As it rests on specific assumptions about evolution as a whole, and largely ignores play and exploration other than as forms of practice or directed learning, current sociobiology has still not fully come to grips with the original insights of evolutionary social ethology. During this century, field biologists Jane Goodall, John King, and George Schaller, theoreticians John Emlen and Stephen Fretwell, Stuart and Jeanne Altmann, and many others of their generation both in Europe and in North America documented individual distinctiveness, play, curiosity, and innovation in free-ranging, wild animals. They further defined population and evolutionary consequences of these phenomena.

Ecological and evolutionary consequences of behavioral individuality and cultural inheritance in nonhumans are by no means restricted to humans' closest animal relatives, the great apes. Individualistic behavior of nonhuman animals can influence survival and reproduction of other animals in their social group and can even effect radical change in the distribution and abundance of entire species (e.g., Baker, 1978). The well-documented natural history of northern fur seals Callorhinus ursinus illustrates these points especially well. During recent times, northern fur seals began breeding on San Miguel Island, off the coast of southern California about 125 km west of Los Angeles (Peterson, LeBoeuf, & DeLong, 1968). The nearest breeding colonies of this species are thousands of kilometers to the north, in the Bering Sea, on the Pribilof (Alaska, USA) and Commander (Siberia, USSR) Islands. Fur seals from both the USA and the USSR joined the San Miguel colony during its first 3 years. (Wildlife biologists had previously tagged five of these females as pups on the Pribilofs or the Commander Islands.)

After weaning, juvenile fur seals migrate long distances and wander nomadically throughout the North Pacific, hauling-out on strange shores during migration (Baker, 1978, pp. 105-112, 733-736). We can only speculate about factors that sparked the curiosity of a few young fur seals and led them to colonize San Miguel Island, uninhabited by fur seals for many years, rather than returning to Alaskan or Siberian islands. Fur seals appear to enjoy body-surfing (Kostyal, 1988, pp. 25-26), and possibly these pioneering individuals found their perfect wave on the beaches of San Miguel.

Once the first fur seals colonized San Miguel, steady immigration began. The colony grew from less than ten seals in 1965 to over 60 in 1968 (Peterson, LeBoeuf, & DeLong 1968). Although only a few individuals possess the rare characteristics that lead them to colonize a previously uninhabited site, a breeding colony will immediately attract many more individuals from the pool of wandering juveniles and adults. Baker (1978, pp. 111-112) speculates that young fur seals wander widely around the North Pacific, becoming familiar with the ocean and coastlines, visiting remote areas, and checking for breeding colonies that they might possibly join at a later time.

Colonization of previously uninhabited areas may also occur in at least five other members of the seal family Phocidae (Baker 1978, pp. 106-111). Such pioneering ventures may involve movements of thousands of kilometers. It would be fascinating to know enough about the lives and personal histories of these individuals to understand why they alone, of all the members of their species, chose to explore and settle so far from home.

Elsewhere in this volume, Rosemarie Rigol retells the story of the Sleeping Beauty. A classic fairy tale, it represents a case of exploration that had profound negative consequences: "the coming of the sleep for all." In the version of this fairy tale presented in the classic Sleeping Beauty ballet, Princess Aurora's curiosity plays only a proximate role in her fate, and the hundred-year sleep ultimately results from a disastrous faux pas by a minor court bureaucrat who happens to leave a very powerful but not very good fairy named Carabosse off the invitation list for Aurora's christening party. In revenge, Carabosse casts a spell under whose provisions Aurora will prick her finger with a spindle at fifteen and die. Carabosse intended death for Aurora, but timely intervention by Lilac, another fairy invited to the christening, commuted Aurora's sentence to a hundred-year sleep for her and the entire court.

The centennial of the Sleeping Beauty ballet spurred numerous new productions of the work. I was privileged to view the San Francisco Ballet (SFB) and American Ballet Theatre (ABT) productions, and my discussion will be based on these and on the older Kirov production available on home video. Following these productions, a flood of new criticisms and interpretations of the ballet inundated arts columns and pages of dance journals. A full treatment of the topic is beyond the scope of this chapter (and exceeds the limitations of my own dance background), but several comments are apposite.

The good fairies who attend Aurora's christening party bring gifts, variously interpreted in productions of the ballet over the years. One of the gifts is sometimes taken to be play. In the Kirov video, Aurora (I. Kolpakova) plays delightedly with the spindle, hidden in a bouquet, whose touch brings on the hundred-year sleep. In this sense Play's gift and Carabosse's curse seem inevitably linked.

I agree with Rigol that Aurora's story is in fact a story about a character who changes, who develops. This aspect is very clear in each of the three productions I saw, a point made by nearly all of the newspaper and magazine reviews I consulted: we see Aurora as a baby, as an adolescent, as a sleeping princess, and at her wedding. Obviously, the story has many different aspects, but one of these is clearly the development of an individual woman. Let me cite a few details that helped convince me.

When Carabosse confronts Aurora on her fifteenth birthday and mimes "Me, me, me", it seemed (especially in the SFB production with Evelyn Cisneros as Aurora and Jim Sohm as Carabosse) that she was saying to Aurora, "Look in the mirror. What I am is part of you. You will need to come to terms with the darkness in yourself before you can be fully human. This darkness, and the strength that it can give you, is my gift." At fifteen, Aurora is no pushover. Indeed, she has just shown her strength in the famous Rose Adagio section of the ballet, a section which has tested the mettle of ballerinas for over a century. In this section, the juxtaposition of strength and beauty — the steel and the rose, with Tchaikovsky's orchestration leaving no doubt in this observer's mind as to what was intended — seems very clear. But Aurora is only fifteen, Carabosse's steel is supernatural, and Aurora falls.

The spirit of play and curiosity leads Aurora to handle the spindle. Fortunately, unlike her sisters elsewhere, she has the next century to come to terms with her dark side, and the Aurora who emerges from her hundred-year sleep is the adult all of us wish we could be. She is a rose. She is steel. She is tough, blithe, gutsy, fun (and oh, can she dance!). And she is in control, ready to rule her kingdom. Ballerina Amanda McKerrow, an admirable Aurora for ABT, brought out this aspect of the Sleeping Beauty story especially well on opening night (May 3) of her company's 1993 spring season. Aurora dances into the parents' throne room, where parents and entourage wait. "Hi, Mom," she says (dancing). "Hi, Dad. Oh, by the way, I'm getting married today. This is the Prince. I found him in the forest. Bring on the rest of my life." And they all sit down together — to watch fairy tales! No joke. You could look it up. Why? Is this lengthy coda to the ballet just an entertainment and nothing more? Nearly all contemporary productions omit one or more of the fairy tales that appeared in the St. Petersburg original. This is unfortunate. Entertainment may well be the reason for the original length (and economics, or an audience anxious to get out of the city before midnight, the reason for the contemporary truncation), but it makes sense to have the whole suite of fairy-tale characters there at the end for the very reasons that Rigol states: to restate the historical experience with human behavior, to show that children have to be curious, that we can face our own troublesome character traits (our "dark side"), and that curiosity matters. Carabosse's curse was in fact Carabosse's gift: an elegant move choreographed by Lilac, the real genius of the piece. None of the good fairies could give Aurora what Carabosse could: access to her own dark side, to the dark energies that are always in motion in the world, promoting an ideology here, exiling a scientist or a poet there, slipping messages about genetic determinism into weekly newsmagazines, speaking in dulcet tones to the unarmored infancy of the preconscious mind. Aurora, led to the spindle by the gifts of play and curiosity, was also to receive a gift from Lilac, but the nature of Lilac's intended gift is never made explicit in the story. What was Lilac's gift? We are usually told that Lilac acted to save Aurora from death and that this was her gift. But if Lilac actually came to the christening with this gift in mind, why didn't she prevent Carabosse from acting in the first place, unless Carabosse was necessary? To do this question justice is a challenge for future dancers and dance scholars, and for students of play and curiosity. Lilac is not simply a bigger, stronger good fairy. Rather, her power is dialectical. Her gift enables Aurora to incorporate the Carabosse in her own nature and to harness that dark strength, using the gifts of the good fairies (serenity, benevolence, generosity, happiness, and temperament in the ABT production; tenderness, generosity, serenity, playfulness and courage in the SFB version — it makes sense either way). Might we add laughter, humor, and curiosity? Whatever the nature of Lilac's gift, the result is clear: a girl becomes a woman without forsaking the curious child in her. The princess becomes a queen with the strength to be responsible for her relationships and for the effects of her actions on others. Plato notwithstanding, what better prescription for a leader could there be? If Aurora ran for President tomorrow, I'd vote for her.

Sociobiology has done a splendid job of portraying the dark side of humanity, as if it were inevitable. Despite sociobiologists' pro forma utterances that sociobiology is not merely about sex and violence, but also addresses cooperation and altruism, the underlying genetic calculus of selfishness clearly evokes Herbert Spencer and the right-wing theoreticians of the Reagan-Bush years, not John Dewey or Hannah Arendt. After all, in the 1992 presidential campaign who was it that enunciated an explicitly sociobiological platform invoking biologically-based "family values?"

From the perspective of twenty years in sociobiology that began in E. O. Wilson's graduate seminar at the Harvard Biological Laboratories, I see sociobiology, and perhaps many individual sociobiologists, as a kind of Carabosse. In an era when classical naturalists were rudely ejected from science by molecular biology, biophysics, and experimental psychology, they weren't invited to the christening, and oh, did they ever want to get even! And they did, by giving us an unsparing scientific portrait of our dark side, couched in genetic terms and subtly qualified (with allusions to learning, culture, and conditional strategies) so as to distance it from all preceding biological determinisms. With this perspective in mind, and viewing the contributions of generations of dedicated field naturalists as a class of gifts antithetical to Carabosse's, the sociobiologists restate in biological terms a question that is by no means new: what are the uses of the dark energies within us? Enter play, curiosity, and exploration, guests largely excluded from the sociobiological feast: A task for Lilac, if she is willing.

(The musical score and accompanying dance notation for a variant Russian Sleeping Bruin variant of the Sleeping Beauty ballet, very possibly the original version considering that it antedates the Petipa-Tchaikovsky version by at least 50 years, was recently discovered by antiquarians in an old Russian-American Company theatre trunk in Sitka, Alaska. Although the manuscripts are in poor shape after more than a century of burial in acid spruce-hemlock forest soil, they appear to state that Aurora and her parents' court were exiled to Alaska for ten years.

The librettist comments that this would be far worse punishment than simply to sleep for a hundred. However, those few dance scholars who have agreed to examine the material consider the ballet, and indeed this very account, entirely fictional. Despite these unpromising initial evaluations, a small community dance group somewhere in Alaska is proceeding with plans for the first 20th (21st?)-century production of Sleeping Bruin at some indefinite future time.).

References

Baker, R.R. (1978).The evolutionary ecology of animal migration. New York: Holmes & Meier.

Kostyal, K.M. (1988). Animals at play. (National Geographic books for young explorers.) Washington, D.C.: National Geographic Society.

Peterson, R.S., LeBoeuf, B.J., & DeLong, R.L. (1968). Fur seals from the Bering Sea breeding in California. Nature, 219, 899-901.

Rigol, R.M. (This volume.) Fairy tales and curiosity.

Author Index

A

Acker, M., 116
Aggleton, J. P., 56
Ainsworth, M. D. S., 123, 124, 126, 131, 137, 141, 178, 188, 194, 202, 242, 243
Ajzen, I., 266
Altmann, J., 334
Altmann, S., 334
Amaral, D. G., 48, 51
Amastasi, A., 213, 215, 223
Anders, T. F., 125
Andersen, P., 45, 49
Antal, D., 307
Apfel, N., 130
Arend, R., 202
Arend, R. A., 131
Arendt, H., 338
Arnon, R., 282
Atkinson, J. W., 166, 291
Auerswald, M., 151, 244, 264

B

Bagshaw, M. H., 55
Baker, C. T., 222
Baker, R. R., 334, 335
Bakke, H. K., 55
Bakker, E., 57
Baldwin, J. D., 178
Baldwin, J. I., 178
Barnett, S. A., 31, 32, 52
Barthelmy, E., 178
Bartholomäi, F., 309
Bates, J. E., 125
Bateson, G., 123
Bayliss, J., 52
Beall, D., 264
Becker-Carus, C., 35
Beek, P. J., 227
Bell, S. M., 242, 243
Belsky, J., 123, 129, 132, 135, 137, 178, 194, 244
Ben-Zur, H., 277
Berg, C., 216, 219, 222

Berlyne, D. E., 46, 79, 80, 101, 114, 115, 123, 127, 152, 162, 178, 179, 205, 260, 261, 288, 291, 292, 295, 298
Bigelow, A. E., 235
Bingham, G. P., 227
Birch, D., 291
Bischof, N., 124, 144, 178, 201, 292, 293, 295, 297, 298, 299
Blèhar, M. C., 124, 178
Blodgett, H. C., 47
Blanchard, D. C., 58
Blanchard, R. J., 58
Boigs, R., 154, 194, 200, 201, 202, 204, 206, 209, 210, 242, 249, 254
Bornstein, M. H., 216
Bösser, T., 288
Bouman, H., 57
Bower, T. G. R., 34, 129
Bowlby, J., 123, 124, 126, 143, 188, 194, 202, 242
Boyle, G. J., 210, 277
Bradley, B. S., 328
Brazelton, T. B., 129, 134
Bretherton, I., 131
Brodbeck, F. C., 303
Brody, M., 246
Bronfenbrenner, U., 109, 325
Bruner, J. S., 123
Buchholtz, C., 32, 38
Buechter, S., 66
Bühler, K., 88
Butler, R. A., 32

C

Cagle, M., 259
Camp, C. J., 210, 262, 265, 277
Campbell, H., 129
Card, S. K., 288
Carello, C., 227
Carroll, J. M., 287, 288, 289, 290, 291, 292, 296
Caruso, D. A., 223, 283
Case, R., 242

Cattell, R. B., 153, 158
Chapanis, A., 288, 303
Chapieski, M. L., 136
Chapman, R. M., 38
Charlesworth, W. R., 66, 155, 157, 177
Chess, S., 202
Cicchetti, P., 56
Cisneros, E., 336
Clark, C. R., 57
Clark, R. A., 166
Coffman, W., 221
Cohen, L. B., 244
Cohen, S., 329
Coie, J. D., 156, 217
Connolly, K., 236
Coover, G., 55, 56
Corman, P. E., 54
Cowan, P. E., 31, 32, 52
Croiset, G., 57
Csikszentmihalyi, M., 88, 89, 101, 104
Cummins, R. A., 47
Curtis, L. E., 234

D

Dalland, T., 52
Dagleish, M., 236
Darwin, C., 66
Dash, J., 264
Dash, S., 264
Daub, M., 307, 324
Day, H. I., 81, 115
Deane, K. E., 126
Deci, E. L., 88, 89
Delacour, 46, 54
DeLong, R. L., 334, 335
Dember, W. N., 52, 179, 242
Descartes, R., 65, 66
Dewey, J., 88, 338
Dietrich, M. S., 210, 277
Di Franco, D., 129
Dodwell, P. C., 129
Dollinger, S. J., 217
Dostrovsky, J., 51
Douglas, R. J., 52
Downs, R. M., 326
Duncker, K., 303
Dunst, C. J., 125
Dweck, C. S., 223

E

Earl, R. W., 52, 179, 242
Ehlers, D., 301
Eibl-Eibesfeldt, I., 31
Eichorn, D. H., 129
Eilam, D., 47
Emde, R. N., 131, 132, 135
Emlen, J., 334
Endsley, R. C., 178, 195, 223
Ennaceur, A., 46
Evans, G. W., 329
Ewert, J.-P., 34

F

Fawl, C. L., 52
Feirtag, M., 48
Farish, G., 131
Fein, G. G., 123, 130
Feldman, S. S., 135
Fenson, L., 129, 130, 145
Field, J., 129
Field, T., 125
File, S. A., 57
Fink, B., 83, 86, 88, 90, 92, 93, 101, 102, 103, 104, 105, 107, 108, 109, 110, 116, 222
Fischer, M., 81
Fishbein, M., 266
Flohr, B., 155, 244
Föse, B., 101
Forster, P., 101, 106
Fox, A., 31
Frankel, D. G., 238
Freeland, C. A. B., 125
Frese, M., 288, 294, 301, 303
Fretwell, S., 334
Freud, S., 123
Frysinger, R. C., 57
Fukuda, M., 54
Fuller, D., 265

G

Gaiter, J. L., 131, 157
Gallagher, M., 53, 57
Gamradt, J., 155, 177
Gandour, M. J., 125
Garduque, L., 123
Garner, A. P., 178, 223

Gauda, G., 82, 102, 126, 202, 204, 244, 245, 260
Gay, P. E., 54
Gediga, G., 296, 29
Geffen, G. M., 57
Geffen, L. B., 57
Gelman, R., 234
Geschwind, N., 48
Gibson, E. J., 227, 231, 232, 233, 238, 259
Gibson, J. J., 227, 228, 229, 230, 231
Glendenning, K. K., 54
Goddard, G. V., 49
Görlitz, D., 155, 308, 309, 324, 330
Goodall, J., 334
Goode, M. K., 135, 178, 194
Gottfredson, L. S., 102, 109
Gove, F. L., 202
Graham, P. W., 53
Gray, J. A., 50, 51, 179
Greidanus, T. B. V. W., 57
Greif, S., 288, 291, 295, 297, 299, 301
Groos, K., 123
Grossmann, K. E., 243
Grossmann, S. P., 57
Grunwald, M., 287
Gubler, H., 124, 144, 201

H

Haberman, S. J., 190
Hacker, W., 303
Hall, G. S., 123
Halliday, M. S., 46
Halton, A., 129
Hamborg, K.-C., 297
Harlow, H. F., 152
Harmon, R. J., 131, 157
Harris, B. J., 102
Hart, R., 328, 329
Harter, S., 157, 218, 261
Harty, H., 264
Hassenstein, B., 241
Hayes, K. J., 215, 221
Hazen, N. L., 202
Hebb, D. O., 179
Heckhausen, H., 89, 291
Heiland, A., 105
Henderson, B. B., 81, 152, 154, 155, 156, 165, 170, 177, 178, 194, 195, 199, 214, 215, 218, 219, 220, 223, 264, 277, 287
Herrmann, P., 178
Herrmann, T., 58
Hess, V. L., 238
Hessel, F., 307
Hidi, S., 79, 88, 95, 96
Hill, A. J., 51
Hillebrandt, A., 151
Hinde, R. A., 37, 38
Hinton, G. E., 65
Hoff, E. M., 91
Hogarty, P. S., 129
Holland, P. C., 53
Holst, E. v., 32, 34
Hoyos, C., Graf, 303
Hrncir, E., 123
Hubel, D. H., 34
Hughes, M., 129
Hull, C. L., 37
Hungerige, H., 151
Hunt, J.McV., 82, 91, 114, 115, 179, 290
Husarek, B., 155
Husney, R., 235
Hutcherson, M. A., 178, 223
Hutt, C., 82, 129, 151, 152, 153, 164, 170, 190, 260

I

Immelmann, K., 32
Inagaki, K., 217
Inhelder, B., 130
Irmer, C., 303
Isaacs, N., 66
Isaacson, R. L., 55, 58
Iwata, J., 56
Izard, C. E., 65, 66, 102

J

James, W., 152, 179
Janczyk, L., 151
Janikowski, A., 299
Jansen, J., Jr., 45, 49
Jellestad, F. K., 55, 56
Jennings, K. D., 131, 157
Joachimthaler, E. A., 277
Johns, C., 178, 195
Joubert, A., 31

K

Kaada, B. R., 45, 49, 50
Kagan, J., 129, 130, 135, 144, 145, 259, 290
Kalafat, J., 129
Kaminski, G., 329
Kapp, B. S., 57
Karoly, A. J., 52
Kasten, H., 105, 107
Kaufman, A., 219
Kaufman, N., 219
Kearsley, R. B., 129, 130
Keil, C. F., 102
Kellenaers, C. J. J., 142
Keller, H., 82, 101, 102, 116, 126, 154, 194, 199, 200, 201, 202, 203, 204, 206, 207, 209, 210, 222, 227, 241, 242, 244, 245, 254, 255, 259, 260, 264, 265, 271, 288, 291, 295, 302
Keller, J. A., 102, 214
Kellman, P. J., 234
Kerschensteiner, G., 88
Kesner, R. P., 56
Kesselring, T., 329
Kiefer, A., 307
Kieras, D., 288
Kim, E. E., 235
Kim, N. G., 227
King, J., 334
Kinsella-Shaw, J., 227
Klahr, D., 242
Klein, P. S., 265
Koch, M., 210
Köhler, C., 49, 50
Köhler, W., 303
Kolb, B., 58
Kolin, E. A., 164
Kolpakova, I., 336
Krasnegor, N. A., 216
Krapp, A., 79, 83, 89, 90, 91, 95, 96, 102, 103, 104, 105, 107, 108, 116, 222, 242
Krechevsky, I., 38
Kreitler, H., 81, 152, 156, 199, 202, 217, 255, 259, 260, 261, 262, 263, 264, 265, 266, 267, 268, 270, 271, 274, 275, 277, 278, 282
Kreitler, S., 81, 152, 156, 199, 202, 217, 255, 259, 260, 261, 262, 263, 264, 265, 266, 267, 268, 270, 271, 274, 275, 277, 278, 282
Krieger, R., 81, 304
Kruft, W.-H., 307
Krüger, B., 151
Kuder, G. F., 81
Kugler, P. N., 227

L

Lamb, M. E., 125
Lang, T., 303
Lange, D., 167
Lange, K., 151
Langevin, R., 217
Lastovicka, J. L., 277
Lazarus, M., 123
Leaton, R. N., 50
LeBoeuf, B. J., 334, 335
Leckrone, T. G., 92
LeDoux, J. E., 56
Lee, D. N., 231
Leenaars, A. A., 135
Leggett, E. L., 223
Lehwald, G., 81
Lemm, H.-D., 299
Lemond, C., 244, 245
Lemond, L. C., 127, 128, 133, 190
Leone, C. T., 194
Leong, C. Y., 34
Lessac, M. S., 134
Lettvin, J. Y., 34
Levine, S., 55
Levy, N., 38
Lewin, K., 104
Lewis, M., 129
Lezine, I., 130
Liben, L. S., 326
Lindner, R., 309, 310
Lingerfelt, B., 125
Livesey, P. J., 52
Lohmann, D., 299, 300, 301
Lorenz, K., 31, 32, 34, 152, 179, 241, 242, 245, 287
Lounsbury, M. L., 125
Loy, L. L., 194
Lowell, E. L., 166
Lugt-Tappeser, H., 151, 152, 170, 171
Lunk, G., 88
Lounsbury, M. L., 125

M

Maccoby, E. E., 135
Muchow, M., 309, 310
Mack, R. L., 287, 288, 289, 291
Mackowiak, K., 151
Mackworth, N. H., 55
MacLean, P. D. 51
Maddi, S. R., 81
Magoun, M., 45
Main, M., 131, 202
Main, M. B., 124
Malcuit, G., 129
Mangel, I., 301
Markley, R. P., 265
Markowska, A., 55
Martin, C. E., 223
Martin, M. J., 178, 223
Massie, M., 151, 167
Matas, L., 131
Matheny, A. P., 125
Matip, E.-M., 151, 157
Maw, E. W., 156, 213, 217, 262
Maw, W. H., 156, 213, 217, 262
McCann, B., 217
McCall, R. B., 129, 130, 135, 145, 152
McCartney, K., 213, 216
McCleary, R. A., 50, 55
McClelland, D. C., 166
McClelland, J. L., 65
McGrath, M. P., 136
McKerrow, A., 337
Menaker, S., 45
McReynolds, P., 116
Menten, T. G., 244
Messer, S. B., 145
Mesulam, M. M., 48
Meyer, H. J., 199
Meyer, P. M., 54
Meyer, W.-U., 66, 68, 70
Meyrer, D. R., 54
Miller, S. N., 127
Milner, P. M., 50
Minner, D. G., 146
Miranda, D., 82, 102, 126, 244, 245, 260
Mishkin, M., 54, 56
Moch, M., 151, 244, 264
Monecke, U., 299
Monkman, J. A., 44
Montgomery, K. C., 32, 44, 152, 178
Moore, S. G., 81, 152, 154, 155, 165, 170, 199, 215, 219, 220
Moran, T. P., 288
Morgan, G. A., 131, 157
Moruzzi, G., 45
Most, r. K., 129, 135, 137, 178, 194, 244
Muir, D. W., 129
Müller, M., 298
Murphey, R. M., 31
Myhrer, T., 50

N

Nadel, L., 50, 51, 52
Nakamura, K., 54
Nauta, W. J. H., 48
Nelson, V. L., 222
Neuhaus, C., 151, 167
Newell, A., 288
Niepel, M., 68, 70
Nishijo, H., 54, 56
Nishino, H., 56
Nissen, H. W., 39
Nunnally, J. C., 127, 128, 133, 190, 242, 244, 245, 287

O

Oerter, R., 102, 105, 109
O'Keefe, J., 50, 51, 52
Olds, J., 52
Olson, K. R., 210, 262, 265, 277
Ono, T., 54
Overbeeke, C. J., 230

P

Packer, M., 129
Pandya, D. N., 48
Palmer, C. F., 233, 235
Papousek, H., 102
Papousek, M., 102
Park, R. E., 309
Pavlov, I. P., 45
Pearson, P. H., 81
Penney, R. K., 217
Persch, A., 31, 37
Peters, J., 236

Peterson, R. S., 334, 335
Petters, A., 177, 181, 182
Pezalla, C., 299
Piaget, J., 89, 123, 129, 297, 329
Pick, A. D., 238
Piechulla, D., 31
Pielstick, N. L., 218
Pietila, C., 116
Plomin, R., 213
Plutchik, R., 65
Polson, P. G., 288
Pomerleau, A., 129
Poucet, B., 58
Power, T. G., 136
Prenzel, M., 83, 88, 89, 95, 103, 104, 105, 106
Pribram, K. H., 55
Price, L., 164
Prümper, J., 303
Purmann, E., 151, 157

R

Raisman, G., 49, 59
Rapp, P. R., 57
Rathunde, K., 88
Reason, J., 303
Reddish, P. E., 231
Reed, E. S., 227, 228, 230
Renninger, K. A., 87, 88, 89, 92, 95, 96, 106, 242
Reis, D., 56
Rigol, R. M., 335, 336, 337
Roberts, T. L., 288
Rochat, P., 233
Rochberg-Halton, E., 101
Rodgers, J. D.,
Rodrigue, J. R., 265
Roe, A., 102
Rolls, E. T., 51
Romanski, L. M., 56
Rosenblatt, D., 129, 130
Rosvold, H. E., 54
Rubin, K. H., 123, 126, 130
Rudolph, U., 68, 70
Ruff, H. A., 129
Ruge, A., 151
Rumelhart, D. E., 65
Russell, P. A., 46
Ryan, R. M., 88

S

Saegert, S., 308
Saint Paul, U. v., 32, 34
Sandfort, R., 151, 244, 264
Sapper, V., 145
Sauvageod, F., 299
Saxe, R. M., 178, 195
Scarr, S., 213, 215, 216
Schaffer, W., 32
Schaller, G., 334
Scherer, K. R., 66
Schiefele, H., 89, 102, 105
Schiefele, U., 91, 95, 96, 107
Schildbach, B., 243
Schiller, F., 123
Schleidt, W. M., 34
Schmalt, H.-D., 165
Schmidt, R. F., 34
Schneider, K., 151, 152, 153, 155, 165, 171, 178, 181, 190, 199, 244, 264
Schölmerich, A., 82, 101, 102, 126, 206, 222, 244, 245, 249, 260, 288, 291, 295
Schützwohl, A., 68, 72, 70
Schwabe, H., 309
Schwartzbaum J. S., 54, 55
Shaw, R. E., 227
Siegelmann, M., 102
Sinclair, H., 130
Singer, D. G., 259
Singer, J. L., 259
Skar, J., 55
Skell, W., 301
Skinner, B. F., 32
Smets, G. J. F., 230
Smitherman, C. H., 125
Smitsman, A. W., 234, 236
Sohms, J., 336
Smolensky, P., 65
Sokolov, E. N., 44, 50
Solomon, J., 131
Solomon, R. L., 134
Sontag, L. W., 222
Sorce, J. F., 131, 132, 135
Sostek, A. M., 125
Spelke, E. S., 234, 238
Spencer, H., 338
Spencer, W. B., 277

Sroufe, L. A., 131, 202
Stadler, M., 303
Stambak, M., 130
Starkey, P., 234
Stayton, D. J., 124
Sternberg, R. J., 216, 219, 221, 222
Stolba, A., 31
Stollak, G., 178, 195
Stoner, S. B., 277
Stratmann, M. H., 230
Strauss, M. S., 234
Sundberg, H., 45, 46
Sutton-Smith, B., 123, 127
Swanson, L. W., 51

T

Teitelbaum, H., 50
Tembrock, G., 31
Tews, J., 309
Thews, G., 34
Thomas, A., 202
Thompson, R. A., 143
Tinbergen, N., 34, 35, 36
Todt, E., 109
Tomkins, S. S., 65, 70
Tracy, R., 131
Travers, R. M. W., 102, 104, 109, 116
Trieber, F. A., 125
Trudewind, C., 155
Turvey, M. T., 227
Tzuriel, D., 265

U

Unzner, L., 155, 181
Ursin, H., 45, 50, 51, 52, 53, 54, 55, 58
Ursin. U., 45, 52, 53

V

van den Boom, D. C., 125, 137, 143, 145
Valsiner, J., 310
Van der Wilk, R., 89
Van Hoesen, G. W., 48
Vandenberg, B.R., 123, 218, 260
Vanderwolf, C. H., 57
van Leeuwen, C., 236
van Leeuwen, L., 236
van Loosbroek, E., 234
Vauclair, J., 31

Vest, B., 54
Vincente, K. J., 227
Vinogradova, O. S., 50
Voss, H.-G., 82, 94, 101, 116, 152, 199, 227, 241, 242, 255, 259, 260, 288, 291
Vygotsky, L. S., 123, 129

W

Wachs, T. D., 125
Wagner, H., 307, 324
Walker, A. S., 233
Walker, E. L., 52
Wall, S., 124, 178
Wallace, J. G., 242
Walser, R. D., 56
Walsh, R. N., 47
Walther, B., 55
Wapner, S., 108, 109
Ward, A. A., Jr., 45
Warren, W. H., Jr., 235
Waters, E., 124, 126, 178
Wehner, T., 303
Weinert, F. E., 308, 325, 326
Weisler, A., 127, 152
Weiss-Bürger, M., 31
Weizman, F., 135
Welker, W. I., 31, 46
Wenckstern, S., 135, 136, 145
Werka, T., 55, 56
Werner, H., 109
Wester, K., 45, 53
Whishaw, I. Q., 58
White, R. W., 91, 114, 115, 152, 157, 215
Wiedl, K.-H., 81
Wiesel, T. N., 34
Wilcox, B., 234
Wilson, E. O., 338
Wilson, S., 219
Winkel, G. H., 308
Winzenried, G., 56
Wittig, B. A., 131, 242
Wohlwill, J. F., 80, 213, 242, 244, 297, 308, 309
Wood-Gush, D. G. M., 31
Woodruff, A. B., 218
Wozniak, R. H., 88
Wünschmann, A., 31
Wundt, W., 308

Y

Yarrow, L. J., 131, 157

X

Xagoraris, A., 56

Z

Zach, U., 202
Zapf, D., 303
Zeidner, M., 277
Zelazo, P. R., 129, 130
Zigler, E., 152, 156, 199, 218, 255, 260,
    261, 263, 264, 267, 271
Zimbardo, P. G., 32
Zimolong, B., 303
Zoob, I., 164
Zuckerman, M., 81, 164

Subject Index

A

action skill, 233, 235
adaptive value, 2, 4, 6, 179
anxiety scale, 159, 283
arousal system, 10, 45, 124, 126, 133, 144
attachment behavior, 2, 124, 125, 126, 133, 143, 242, 243
attachment classification, 131, 142
attachment theory, 123, 126, 131, 132, 133
attention span, 135, 136, 137, 138, 139

B

behavioral dispositions, 151, 154, 156
behavioral inhibition, 159, 178

C

coercive behaviors, 186, 188, 194
coercive intervention, 190
competence striving, 173
curiosity disposition, 169, 173
curiosity incentives, 11, 154, 166, 178
curiosity motivation, 152, 166, 167, 170, 173, 194
curiosity motive, 152, 154, 156, 157, 165-170, 173
curiosity questionnaire, 157, 165, 170
curiosity scale, 159, 160, 165, 170-173

D

dependency, 144, 253
diversive exploration, 90, 96, 115, 126, 127, 147, 164, 170, 175, 196, 242, 291, 292

E

ecological key situation, 155, 156, 165
effectance motivation, 156, 174
epistemic curiosity, 115, 159, 162, 165, 170, 171, 173

F

felt security, 125, 143
field study, 11, 180, 299
fixation time, 145, 148, 244
four-group design, 134

H

habituation, 45, 46, 49, 53, 59, 60, 125, 149, 244, 247, 256
homeostatic, 9, 58, 144

I

infants, 10, 40, 123, 125, 129-149, 176, 195, 228, 231-235, 238, 239, 242, 243, 255, 290, 329
information aquisition, 9, 242, 245, 247, 251-255
intelligence, 13, 20, 81, 98, 99, 156, 175, 238
interactional quality, 11
intervention, 10, 133, 134, 135, 137, 138, 140-144, 149, 182, 190, 335
irritability, 10, 125, 126, 134, 142, 143, 149

L

level of exploration, 135, 137-139, 279, 282
locomotion, 49, 184, 186, 191, 229, 231, 232
looking time, 233, 244

M

manipulation, 28, 41, 47, 57, 60, 115, 127, 135, 138, 139, 140, 141, 143, 144, 148, 151, 152, 153, 164, 170, 184, 188, 190, 191, 192, 231, 232, 233, 235, 239, 244, 245, 249, 260, 262, 263, 276, 277, 297
manipulative exploration, 12, 148, 241,

249, 250, 252-255
mastery, 147, 156, 174, 175

N

neophobia, 178, 179

O

observation, 1, 67, 90, 91, 92, 107, 112, 129, 135, 137, 138, 139, 140, 142, 149, 154, 155, 156, 165, 170, 171, 180, 181, 184, 186, 187, 188, 190, 191, 192, 194, 246, 248, 260, 261, 263, 277, 287, 289, 292, 309, 325, 326

P

persistence, 87, 88, 92, 99, 103, 104, 136, 147, 153, 176
perceptual exploration, 262, 264, 275, 276, 279, 280, 281
perceptual search, 228, 231
perceptual system, 227, 230, 231, 235, 239
play behavior, 91, 127, 135, 143, 146, 153, 255
play object, 92, 94, 109, 135, 153
preschooler, 11, 151, 153, 156, 157, 166, 167, 170, 177, 194, 195, 268

Q

question asking, 151, 152, 155, 162, 182

S

secure base, 2, 123, 126, 132, 188, 243
sensation seeking, 81, 100, 164, 170, 176
sensitive responsiveness, 10, 131, 132, 133, 135, 137, 138, 143
specific curiosity, 90, 91, 96, 97, 155, 156, 170
subjective uncertainty, 5, 9, 127, 152, 154, 156, 178, 196

T

tactile exploration, 184, 189, 190, 249, 253, 254, 264

touch, 17, 19, 21, 25, 31, 37, 44, 47, 151, 153, 177, 178, 190, 192, 227, 228, 229, 236, 237, 238, 264, 336

V

verbal exploration, 176, 197
visual exploration, 32, 40, 254, 260
visual fixation, 129
visual inspection, 151-153, 244, 245